Nano-Physics & Bio-Electronics: A New Odyssey

Editors:

Prof. T. Chakraborty
Max-Planck-Institut für Physik komplexer Systeme
Nöthnitzer Str. 38
D-01187 Dresden, Germany
E-mail: tapash@mpipks-dresden.mpg.de
and
Institute of Mathematical Sciences
Taramani, Chennai 600 113, India
E-mail: tapash@imsc.ernet.in

Prof. F. Peeters
Department of Physics
University of Antwerp (UIA)
Universiteitsplein 1
2610 Antwerpen, Belgium
E-mail: peeters@uia.ua.ac.be

Prof. U. Sivan
Department of Physics & Solid State Institute
Technion
Haifa 32000, Israel
E-mail: phsivan@techunix.technion.ac.il

Nano-Physics & Bio-Electronics: A New Odyssey

T. Chakraborty
F. Peeters
U. Sivan
Editors

2002
ELSEVIER
Amsterdam-London-New York-Oxford-Paris-Shannon-Tokyo

ELSEVIER SCIENCE B.V.
Sara Burgerhartstraat 25
P.O. Box 211, 1000 AE Amsterdam, The Netherlands

First edition 2002

Library of Congress Cataloging in Publication Data
A catalog record from the Library of Congress has been applied for.

ISBN: 0 444 50993 3

Printed and bound by Antony Rowe Ltd, Eastbourne

Preface

This book is a collection of some of the invited talks presented at the international meeting held at the Max Planck Institut für Physik komplexer Systeme, Dresden, Germany during August 6-30, 2001, on the rapidly developing field of nanoscale science and bio-electronics (http://www.mpipks-dresden.mpg.de/~nanobio/).

Semiconductor physics has experienced unprecedented developments over the second half of the twentieth century. The exponential growth in microelectronic processing power and the size of dynamic memories has been achieved by significant downscaling of the minimum feature size. Smaller feature sizes result in increased functional density, faster speed, and lower costs. In this process one is reaching the limits where quantum effects and fluctuations are beginning to play an important role.

Physicists are already looking beyond this frontier and proposing new approaches that do not rely on downscaling of the existing technology but rather on new devices, e.g. single-electron transistors, that depend on quantum effects like tunneling. The typical feature size in this new technology is inherently in the nanometer scale, in contrast to existing integrated circuits that have feature sizes in the micrometer and submicrometer range. A major reason why the field is so interesting is the realization that it derives from an almost unprecedented combination of scientific challenge and practical utility.

If nanoelectronics succeeds in all of its goals then integrated circuits with 10^{12} devices on a single chip would be possible. In fact, supercomputers could be built on a single chip. At present, one has no clear idea of how to realize such a high density of components on a single chip and therefore it is an important challenge that needs to be explored. On the other hand, biological systems (e.g. DNA [1]) are able to realize such a high density of information storage and parallel processing power. In fact, biology is very different from conventional engineering since it does not aim to translate a well-defined blueprint into a machine. It rather uses a macroscopic number of attempts and selects the successful ones. Molecular electronics might have to follow this line. It is clear that molecular electronics will be self-assembled but self-assembly means inherent errors that we will have to learn how to control or even use. In order to understand the underlying problems and complexity of these systems an interdisciplinary approach will be needed that would involve physicists, chemists and biologists. The aim of the meeting was to bring together such a group of people.

Progress on dimensional downscaling of existing semiconductor systems e.g., quantum dots has also been very significant. Recently, quantum dots have been made reproducible and built with several different materials. Quite remarkably, these quasi-zero-dimensional systems that started out as a unique laboratory for studying fundamental concepts of quantum confinement [2] have entered a new realm of impor-

[1] C. Dekker and M.A. Ratner, Physics World, August 2001, p. 29.

[2] T. Chakraborty, Quantum Dots (Elsevier, 1999)

tant device applications. One other unique nanostructure that holds vast potential for future nano-electronic and nano-mechanical devices, is the carbon nanotube. These are essentially rolled up sheets of carbon hexagons. Depending on how the two-dimensional graphite sheet is rolled up, one can get armchair, zigzag or chiral nanotubes. These are extremely strong materials and also have good thermal conductivity. Depending on their structure, they can be metallic or semiconducting. Carbon nanotubes are interesting models suitable for fundamental studies of one-dimensional systems. At the same time, they are proving to be very important for a wide variety of potential applications. Our book begins with an in-depth review of this system.

We hope that the articles in this book to some extent reflect the achievements of the present times and future directions of research on nanoscopic dimensions. We are grateful to all the participants for their valuable contributions in the "nanobio" meeting and in particular, to those invited speakers who submitted their excellent piece of work for publication in this book. We are thankful to Professor Peter Fulde and the Max Planck Institute, Dresden for their support and excellent cooperation. We thank in particular, Ms Katrin Lantsch for her superb assistance in making the meeting a great success. This book was typeset in Elsevier format by one of us (T. C.). He wishes to thank Ms Ambika Vanchinathan (Chennai) for her expert technical help in transforming the manuscripts written in wide variety of styles to a coherent format. He also thanks the Institute of Mathematical Sciences, Chennai for support and the Elsevier team: Dr Egbert van Wezenbeek, Mrs Linda Versteeg and Dr Donna Wilson for helping with the publication of the book.

Tapash Chakraborty
Francois Peeters
Uri Sivan

Dresden, Germany
Antwerpen, Belgium
Haifa, Israel

December 2001

Contents

Chapter 1

Electronic states and transport in carbon nanotubes

Tsuneya Ando

Institute for Solid State Physics, University of Tokyo
5–1–5 Kashiwanoha, Kashiwa, Chiba 277-8581, Japan,
E-mail: ando@issp.u-tokyo.ac.jp

Abstract

A brief review is given of electronic and transport properties of carbon nanotubes mainly from a theoretical point of view. The topics include a giant Aharonov-Bohm effect on the band gap and a Landau-level formation in magnetic fields, optical absorption spectra, and exciton effects. Transport properties are also discussed including absence of backward scattering except for scatterers with a potential range smaller than the lattice constant, a conductance quantization in the presence of short-range and strong scatterers such as lattice vacancies, and transport across junctions between nanotubes with different diameters. A continuum model for phonons in the long-wavelength limit and the resistivity determined by phonon scattering is reviewed as well.

1. Introduction

Graphite needles called carbon nanotubes (CNs) were discovered recently [1,2] and have been a subject of an extensive study. A CN is a few concentric tubes of two-dimensional (2D) graphite consisting of carbon-atom hexagons arranged in a helical fashion about the axis. The diameter of CNs is usually between 20 and 300 Å and their length can exceed 1 μm. The distance of adjacent sheets or walls is larger than the distance between nearest neighbor atoms in a graphite sheet and therefore electronic properties of CNs are dominated by those of a single layer CN. Single-wall nanotubes are produced in a form of ropes [3,4]. The purpose of this article is to give a brief review of recent theoretical study on electronic and transport properties of carbon nanotubes.

Figure 1 shows a transmission micrograph image of multi-wall nanotubes and Fig. 2 a computer graphic image of a single-wall nanotube. Carbon nanotubes can be either a metal or semiconductor, depending on their diameters and helical arrangement. The condition whether a CN is metallic or semiconducting can be obtained based on the band structure of a 2D graphite sheet and periodic boundary conditions along the circumference direction. This result was first predicted by means of a tight-binding model ignoring the effect of the tube curvature [5–14].

These properties can be well reproduced in a $\boldsymbol{k}\cdot\boldsymbol{p}$ method or an effective-mass approximation [15]. In fact, the effective-mass scheme has been used successfully in the study of wide varieties of electronic properties of CN. Some of such examples are magnetic properties [16] including the Aharonov-Bohm effect on the band gap [15], optical absorption spectra [17,18], exciton effects [19], lattice instabilities in the absence [20] and presence of a magnetic field [21,22], and magnetic properties of ensembles of nanotubes [23].

Transport properties of CNs are interesting because of their unique topological structure. There have been some reports on experimental study of transport in CN bundles [24] and ropes [25,26]. Transport measurements became possible for a single multi-wall nanotube [27–31] and a single single-wall nanotube [32–36]. Single-wall nanotubes usually exhibit large charging effects presumably due to nonideal contacts

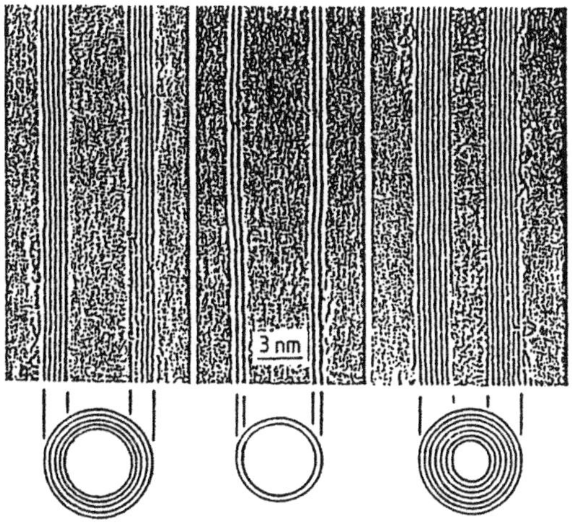

Fig. 1: Some examples of transmission micrograph images of carbon nanotubes [1]. The diameter is 67, 55, and 65 Å from left to right.

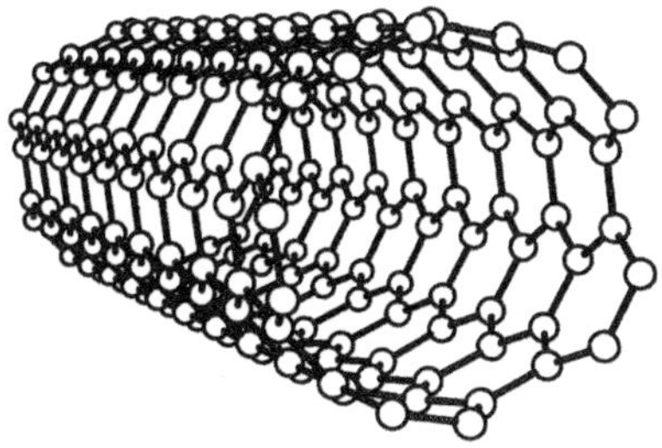

Fig. 2: A computer graphic image of a single-wall armchair nanotube.

[37–41].

In this article we shall mainly discuss electronic states and transport properties of nanotubes obtained theoretically in the $\boldsymbol{k}\cdot\boldsymbol{p}$ method combined with a tight-binding model. It is worth mentioning that several papers giving general reviews of electronic properties of nanotubes were published already [42–47].

In Sect. 2, electronic states are discussed first in a nearest-neighbor tight-binding model. Then, the effective mass equation is introduced and the band structure is discussed with a special emphasis on Aharonov-Bohm effects and formation of Landau levels in magnetic fields. In Sect. 3, optical absorption is discussed in the effective-mass scheme and the nanotube is shown to behave differently in light polarization parallel or perpendicular to the axis. The importance of exciton effects is emphasized and some related experiments are discussed.

In Sect. 4, effects of impurity scattering are discussed and the total absence of

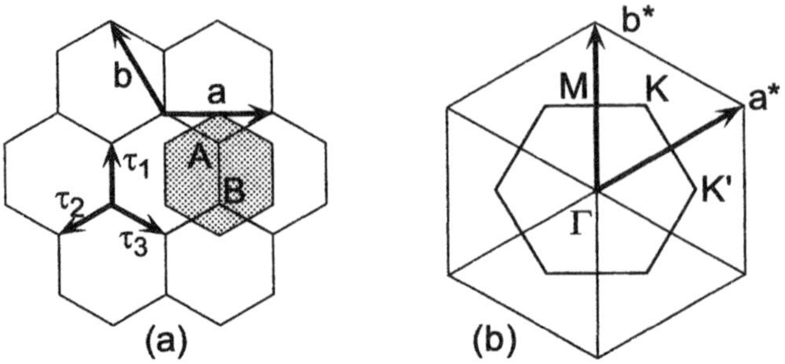

Fig. 3: (a) The lattice structure of a 2D graphite sheet and various quantities. $|\boldsymbol{a}|=|\boldsymbol{b}|=a$. (b) The reciprocal lattice vectors and the first Brillouin zone.

backward scattering is pointed out except for scatterers with a potential range smaller than the lattice constant. Further, the conductance quantization in the presence of lattice vacancies, i.e., strong and short-range scatterers, is also discussed. In Sect. 5, the transport across a junction of nanotubes with different diameters through a pair of topological defects such as five- and seven-member rings is discussed. In Sect. 6, a continuum model for phonons in the long-wavelength limit is introduced and effective Hamiltonian describing electron-phonon interaction is derived. A short summary is given in Sect. 7.

2. Electronic states

2.1 Two-dimensional graphite

The structure of 2D graphite sheet is shown in Fig. 3. We have the primitive translation vectors $\boldsymbol{a}=a(1,0)$ and $\boldsymbol{b}=a(-(1/2),\sqrt{3}/2)$, and the vectors connecting between nearest neighbor carbon atoms $\vec{\tau}_1=a(0,1/\sqrt{3})$, $\vec{\tau}_2=a(-1/2,-1/2\sqrt{3})$, and $\vec{\tau}_3=a(1/2,-1/2\sqrt{3})$. Note that $\boldsymbol{a}\cdot\boldsymbol{b}=-a^2/2$.

The primitive reciprocal lattice vectors $\boldsymbol{a}^*$ and $\boldsymbol{b}^*$ are given by $\boldsymbol{a}^*=(2\pi/a)(1,1/\sqrt{3})$ and $\boldsymbol{b}^*=(2\pi/a)(0,2/\sqrt{3})$. The K and K' points are given as $\boldsymbol{K}=(2\pi/a)(1/3,1/\sqrt{3})$ and $\boldsymbol{K}'=(2\pi/a)(2/3,0)$, respectively. We then have the relations, $\exp(\mathrm{i}\boldsymbol{K}\cdot\vec{\tau}_1)=\omega$, $\exp(\mathrm{i}\boldsymbol{K}\cdot\vec{\tau}_2)=\omega^{-1}$, $\exp(\mathrm{i}\boldsymbol{K}\cdot\vec{\tau}_3)=1$, $\exp(\mathrm{i}\boldsymbol{K}'\cdot\vec{\tau}_1)=1$, $\exp(\mathrm{i}\boldsymbol{K}'\cdot\vec{\tau}_2)=\omega^{-1}$, and $\exp(\mathrm{i}\boldsymbol{K}'\cdot\vec{\tau}_3)=\omega$, with $\omega=\exp(2\pi\mathrm{i}/3)$.

In a tight-binding model, the wave function is written as

$$\psi(\boldsymbol{r})=\sum_{\boldsymbol{R}_A}\psi_A(\boldsymbol{R}_A)\phi(\boldsymbol{r}-\boldsymbol{R}_A)+\sum_{\boldsymbol{R}_B}\psi_B(\boldsymbol{R}_B)\phi(\boldsymbol{r}-\boldsymbol{R}_B), \tag{1}$$

where $\phi(\boldsymbol{r})$ is the wave function of the p_z orbital of a carbon atom located at the origin, $\boldsymbol{R}_A=n_a\boldsymbol{a}+n_b\boldsymbol{b}+\vec{\tau}_1$, and $\boldsymbol{R}_B=n_a\boldsymbol{a}+n_b\boldsymbol{b}$ with integer n_a and n_b. Let $-\gamma_0$ be the transfer integral between nearest-neighbor carbon atoms and choose the energy origin at that of the carbon p_z level. Then, we have

$$\varepsilon\psi_A(\boldsymbol{R}_A)=-\gamma_0\sum_l\psi_B(\boldsymbol{R}_A-\vec{\tau}_l),$$

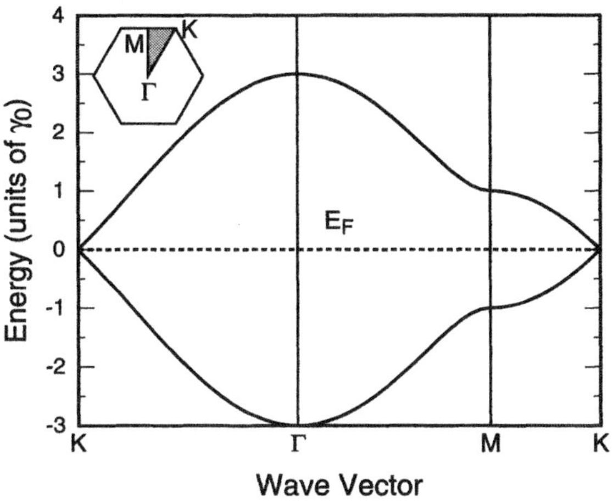

Fig. 4: Calculated band structure of a two-dimensional graphite along K → Γ → M → K shown in the inset.

$$\varepsilon\psi_B(\boldsymbol{R}_B) = -\gamma_0 \sum_l \psi_A(\boldsymbol{R}_B + \vec{\tau}_l), \tag{2}$$

where the energy origin has been chosen at the energy level of the p_z orbital.

Assuming $\psi_A(\boldsymbol{R}_A) \propto f_A(\boldsymbol{k}) \exp(\mathrm{i}\boldsymbol{k}\cdot\boldsymbol{R}_A)$ and $\psi_B(\boldsymbol{R}_B) \propto f_B(\boldsymbol{k}) \exp(\mathrm{i}\boldsymbol{k}\cdot\boldsymbol{R}_B)$, we have

$$\begin{pmatrix} 0 & -\gamma_0 \sum_l \exp(-\mathrm{i}\boldsymbol{k}\cdot\vec{\tau}_l) \\ -\gamma_0 \sum_l \exp(+\mathrm{i}\boldsymbol{k}\cdot\vec{\tau}_l) & 0 \end{pmatrix} \begin{pmatrix} f_A(\boldsymbol{k}) \\ f_B(\boldsymbol{k}) \end{pmatrix} = \varepsilon \begin{pmatrix} f_A(\boldsymbol{k}) \\ f_B(\boldsymbol{k}) \end{pmatrix}. \tag{3}$$

The energy bands are given by

$$\varepsilon_\pm(\boldsymbol{k}) = \pm\gamma_0 \sqrt{1 + 4\cos\left(\frac{ak_x}{2}\right)\cos\left(\frac{\sqrt{3}ak_y}{2}\right) + 4\cos\left(\frac{ak_x}{2}\right)^2}. \tag{4}$$

It is clear that $\varepsilon_\pm(\boldsymbol{K}) = \varepsilon_\pm(\boldsymbol{K}') = 0$. Near the K and K' point, we have $\varepsilon_\pm(\boldsymbol{k}+\boldsymbol{K}) = \varepsilon_\pm(\boldsymbol{k}+\boldsymbol{K}') = \pm\gamma\sqrt{k_x^2+k_y^2}$ with $\gamma = \sqrt{3}a\gamma_0/2$. The band structure is shown in Fig. 4.

In the following, we shall consider the coordinates (x, y) rotated around the origin by η as well as original (x', y') as shown in Fig. 5. For states in the vicinity of the Fermi level $\varepsilon = 0$ of the 2D graphite, we assume that the total wavefunction is written as

$$\begin{aligned} \psi_A(\boldsymbol{R}_A) &= \exp(\mathrm{i}\boldsymbol{K}\cdot\boldsymbol{R}_A) F_A^K(\boldsymbol{R}_A) + \mathrm{e}^{\mathrm{i}\eta} \exp(\mathrm{i}\boldsymbol{K}'\cdot\boldsymbol{R}_A) F_A^{K'}(\boldsymbol{R}_A), \\ \psi_B(\boldsymbol{R}_B) &= -\omega \mathrm{e}^{\mathrm{i}\eta} \exp(\mathrm{i}\boldsymbol{K}\cdot\boldsymbol{R}_B) F_B^K(\boldsymbol{R}_B) + \exp(\mathrm{i}\boldsymbol{K}'\cdot\boldsymbol{R}_B) F_B^{K'}(\boldsymbol{R}_B), \end{aligned} \tag{5}$$

in terms of the slowly-varying envelope functions F_A^K, F_B^K, $F_A^{K'}$, and $F_B^{K'}$.

The above can be written as

$$\psi_A(\boldsymbol{R}_A) = \boldsymbol{a}(\boldsymbol{R}_A)^+\boldsymbol{F}_A(\boldsymbol{R}_A), \quad \psi_B(\boldsymbol{R}_B) = \boldsymbol{b}(\boldsymbol{R}_B)^+\boldsymbol{F}_B(\boldsymbol{R}_B), \tag{6}$$

with

$$\boldsymbol{a}(\boldsymbol{R}_A)^+ = \left(e^{i\boldsymbol{K}\cdot\boldsymbol{R}_A} \quad e^{i\eta}e^{i\boldsymbol{K}'\cdot\boldsymbol{R}_A} \right), \quad \boldsymbol{b}(\boldsymbol{R}_B)^+ = \left(-\omega e^{i\eta}e^{i\boldsymbol{K}\cdot\boldsymbol{R}_B} \quad e^{i\boldsymbol{K}'\cdot\boldsymbol{R}_B} \right), \tag{7}$$

and

$$\boldsymbol{F}_A = \begin{pmatrix} F_A^K \\ F_A^{K'} \end{pmatrix}, \quad \boldsymbol{F}_B = \begin{pmatrix} F_B^K \\ F_B^{K'} \end{pmatrix}. \tag{8}$$

In order to obtain equations for $\boldsymbol{F}$, we first substitute Eq. (6) into Eq. (2). Multiply the first equation by $g(\boldsymbol{r}-\boldsymbol{R}_A)\boldsymbol{a}(\boldsymbol{R}_A)$ and then sum it over $\boldsymbol{R}_A$, where $g(\boldsymbol{r})$ is a smoothing function which varies smoothly in the range $|r|\sim a$ and decays rapidly for $|r|\gg a$. It should satisfy the conditions:

$$\sum_{\boldsymbol{R}_A} g(\boldsymbol{r}-\boldsymbol{R}_A) = \sum_{\boldsymbol{R}_B} g(\boldsymbol{r}-\boldsymbol{R}_B) = 1, \tag{9}$$

and

$$\int d\boldsymbol{r}\, g(\boldsymbol{r}-\boldsymbol{R}_A) = \int d\boldsymbol{r}\, g(\boldsymbol{r}-\boldsymbol{R}_B) = \Omega_0, \tag{10}$$

where Ω_0 is the area of a unit cell given by $\Omega_0=\sqrt{3}a^2/2$. The function $g(\boldsymbol{r}-\boldsymbol{R})$ can be replaced by a delta function when it is multiplied by a smooth function such as the envelopes, i.e., $g(\boldsymbol{r}-\boldsymbol{R})\approx\Omega_0\delta(\boldsymbol{r}-\boldsymbol{R})$. Then, the equation is rewritten as

$$\begin{aligned} &\varepsilon \sum_{\boldsymbol{R}_A} g(\boldsymbol{r}-\boldsymbol{R}_A)\boldsymbol{a}(\boldsymbol{R}_A)\boldsymbol{a}(\boldsymbol{R}_A)^+\boldsymbol{F}_A(\boldsymbol{r}) \\ &= -\gamma_0 \sum_l \sum_{\boldsymbol{R}_A} g(\boldsymbol{r}-\boldsymbol{R}_A)\boldsymbol{a}(\boldsymbol{R}_A)\boldsymbol{b}(\boldsymbol{R}_A-\vec{\tau}_l)^+\Big[\boldsymbol{F}_B(\boldsymbol{r}) - \Big(\vec{\tau}_l\cdot\frac{\partial}{\partial \boldsymbol{r}}\Big)\boldsymbol{F}_B(\boldsymbol{r})+\ldots\Big]. \end{aligned} \tag{11}$$

Noting that

$$\sum_{\boldsymbol{R}_A} g(\boldsymbol{r}-\boldsymbol{R}_A)\boldsymbol{a}(\boldsymbol{R}_A)\boldsymbol{a}(\boldsymbol{R}_A)^+ \approx \begin{pmatrix} 1 & 0 \\ 0 & 1 \end{pmatrix}, \tag{12}$$

we immediately obtain the left hand side of Eq. (11) as $\varepsilon\boldsymbol{F}_A(\boldsymbol{r})$. As for right hand side, we should note first that

$$\sum_{\boldsymbol{R}_A} g(\boldsymbol{r}-\boldsymbol{R}_A)\boldsymbol{a}(\boldsymbol{R}_A)\boldsymbol{b}(\boldsymbol{R}_A-\vec{\tau}_l)^+ \approx \begin{pmatrix} -\omega e^{i\eta}e^{-i\boldsymbol{K}\cdot\vec{\tau}_l} & 0 \\ 0 & e^{-i\eta}e^{-i\boldsymbol{K}'\cdot\vec{\tau}_l} \end{pmatrix}. \tag{13}$$

This immediately leads to the conclusion that the first term in the right hand side of Eq. (11) vanishes identically. To calculate the second term we first note that

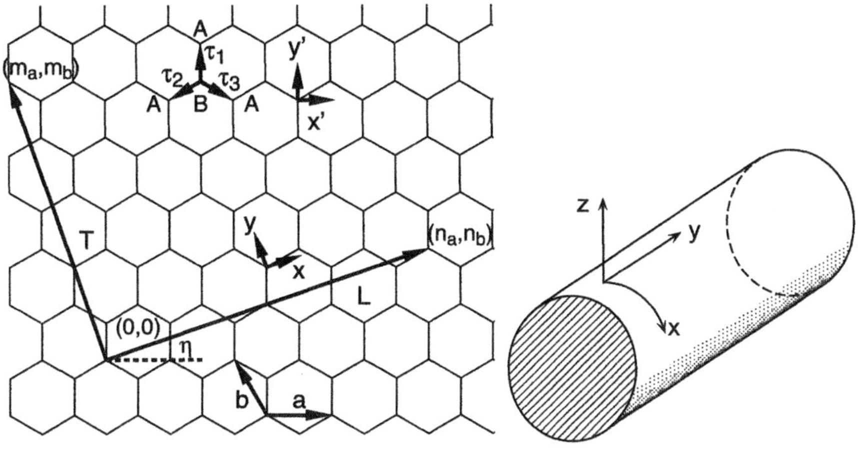

(a) (b)

Fig. 5: (a) The lattice structure of a 2D graphite sheet and various quantities. We shall consider the case that $0 \leq \eta \leq \eta$. The zigzag nanotube corresponds to $\eta = 0$ and the armchair nanotube to $\eta = \pi/6$. (b) The coordinate system on the cylinder surface.

$$
\begin{aligned}
&\sum_l \mathrm{e}^{-\mathrm{i}\boldsymbol{K}\cdot\vec{\tau}_l} \left(\tau_l^x \quad \tau_l^y \right) = \frac{\sqrt{3}}{2} \omega^{-1} a \left(+\mathrm{i} \quad +1 \right), \\
&\sum_l \mathrm{e}^{-\mathrm{i}\boldsymbol{K}'\cdot\vec{\tau}_l} \left(\tau_l^x \quad \tau_l^y \right) = \frac{\sqrt{3}}{2} a \left(-\mathrm{i} \quad +1 \right),
\end{aligned}
\tag{14}
$$

and the final result is

$$
\varepsilon \boldsymbol{F}_A(\boldsymbol{r}) = \gamma \begin{pmatrix} \hat{k}_x - \mathrm{i}\hat{k}_y & 0 \\ 0 & \hat{k}_x + \mathrm{i}\hat{k}_y \end{pmatrix} \boldsymbol{F}_B(\boldsymbol{r}),
\tag{15}
$$

where $\hat{\boldsymbol{k}} = -\mathrm{i}\boldsymbol{\nabla}$.

The equation for $\boldsymbol{F}_B(\boldsymbol{r})$ can be obtained in a similar manner and the full Schrödinger equation is given by

$$
\mathcal{H}_0 \boldsymbol{F}(\boldsymbol{r}) = \varepsilon \boldsymbol{F}(\boldsymbol{r}),
\tag{16}
$$

with

$$
\mathcal{H}_0 = \begin{array}{c} \begin{array}{cccc} KA & KB & K'A & K'B \end{array} \\ \begin{pmatrix} 0 & \gamma(\hat{k}_x - \mathrm{i}\hat{k}_y) & 0 & 0 \\ \gamma(\hat{k}_x + \mathrm{i}\hat{k}_y) & 0 & 0 & 0 \\ 0 & 0 & 0 & \gamma(\hat{k}_x + \mathrm{i}\hat{k}_y) \\ 0 & 0 & \gamma(\hat{k}_x - \mathrm{i}\hat{k}_y) & 0 \end{pmatrix} \end{array},
\tag{17}
$$

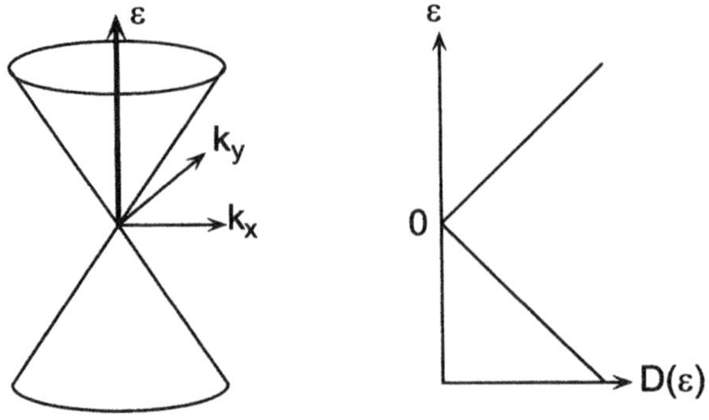

Fig. 6: The energy bands in the vicinity of the K and K' points and the density of states.

where

$$\boldsymbol{F}^K(\boldsymbol{r}) = \begin{pmatrix} F_A^K(\boldsymbol{r}) \\ F_B^K(\boldsymbol{r}) \end{pmatrix}, \quad \boldsymbol{F}^{K'}(\boldsymbol{r}) = \begin{pmatrix} F_A^{K'}(\boldsymbol{r}) \\ F_B^{K'}(\boldsymbol{r}) \end{pmatrix}, \tag{18}$$

and

$$\boldsymbol{F}(\boldsymbol{r}) = \begin{pmatrix} \boldsymbol{F}^K(\boldsymbol{r}) \\ \boldsymbol{F}^{K'}(\boldsymbol{r}) \end{pmatrix}. \tag{19}$$

This is rewritten as

$$\begin{aligned} &\gamma(\hat{\boldsymbol{k}} \cdot \vec{\sigma})\boldsymbol{F}^K(\boldsymbol{r}) = \varepsilon \boldsymbol{F}^K(\boldsymbol{r}), \\ &\gamma(\hat{\boldsymbol{k}}' \cdot \vec{\sigma})\boldsymbol{F}^{K'}(\boldsymbol{r}) = \varepsilon \boldsymbol{F}^{K'}(\boldsymbol{r}), \end{aligned} \tag{20}$$

with $\boldsymbol{\sigma}$ being the Pauli spin matrices, $\hat{k}'_x = \hat{k}_x$, and $\hat{k}'_y = -\hat{k}_y$. These are Dirac equation with vanishing rest mass known as Weyl's equation.

The energy bands are given by $\varepsilon(k) = \pm\gamma|k|$, with $k = \sqrt{k_x^2 + k_y^2}$. The corresponding density of states becomes

$$D(\varepsilon) = \frac{1}{S}\sum_{\boldsymbol{k}} \delta[\varepsilon - \varepsilon(k)] = \frac{|\varepsilon|}{2\pi\gamma^2}. \tag{21}$$

Figure 6 gives a schematic illustration of the conic dispersion and the corresponding density of states. The density of states varies linearly as a function of the energy and vanishes at $\varepsilon = 0$.

The 2D graphite is conventionally called a zero-gap semiconductor because of the vanishing density of states at $\varepsilon = 0$. However, the conductivity calculated with the use of the Boltzmann transport equation is independent of the Fermi energy and the graphite shows a metallic behavior even at $\varepsilon = 0$. This is due to the fact that the scattering probability proportional to the final-state density of states vanishes at $\varepsilon = 0$ and the relaxation time becomes infinite, giving rise to a nonvanishing conductivity even in the absence of states contributing to the transport. A more refined treatment in a self-consistent Born approximation gives the result that the conductivity at $\varepsilon = 0$ is given by a quantum $e^2/\pi\hbar$ and rapidly approaches the Boltzmann result with the deviation from $\varepsilon = 0$ [48].

In a magnetic field B perpendicular to the 2D graphite sheet, we have to replace $\hat{\boldsymbol{k}} = -\mathrm{i}\vec{\nabla}$ by $\hat{\boldsymbol{k}} = -\mathrm{i}\vec{\nabla} + (e/c\hbar)\boldsymbol{A}$ with $\boldsymbol{A}$ being a vector potential ($\boldsymbol{B} = \mathrm{rot}\boldsymbol{A}$). Then, we have $[\hat{k}_x, \hat{k}_y] = -\mathrm{i}l^{-2}$ with $l^2 = c\hbar/eB$. Define $a = (l/\sqrt{2})(\hat{k}_x - \mathrm{i}\hat{k}_y)$ and $a^+ = (l/\sqrt{2})(\hat{k}_x + \mathrm{i}\hat{k}_y)$. Then, we have $[a, a^+] = 1$. In terms of these operators the Hamiltonian for the K point is rewritten as

$$\mathcal{H}_0 = \frac{\sqrt{2}\gamma}{l}\begin{pmatrix} 0 & a \\ a^+ & 0 \end{pmatrix}. \tag{22}$$

We shall define a function $h_n(x, y)$ such that

$$a\, h_0(x, y) = 0, \quad h_n(x, y) = \frac{(a^+)^n}{\sqrt{n!}}\, h_0(x, y). \tag{23}$$

Then, we have

$$a^+ h_n = \sqrt{n+1}\, h_{n+1}, \quad a\, h_{n+1} = \sqrt{n+1}\, h_{n+1}, \quad a^+ a\, h_n = n\, h_n. \tag{24}$$

Therefore, there is a Landau level with vanishing energy ε_0 with the wave function

$$\boldsymbol{F}_0^K = \begin{pmatrix} 0 \\ h_0 \end{pmatrix}, \quad \varepsilon_0 = 0, \tag{25}$$

for the K point. Other Landau levels are at $\varepsilon = \varepsilon_n$ with wave function

$$\boldsymbol{F}_n^K = \frac{1}{\sqrt{2}}\begin{pmatrix} \mathrm{sgn}(n) h_{|n|-1} \\ h_{|n|} \end{pmatrix}, \quad \varepsilon_n = \mathrm{sgn}(n)\frac{\sqrt{2}\gamma}{l}\sqrt{|n|}, \quad (n = \pm 1, \pm 2, \ldots). \tag{26}$$

Similar expressions can be derived for the K' point. The presence of the Landau level at $\varepsilon = 0$ independent of the magnetic-field strength is a remarkable feature of the Weyl equation for a neutrino.

2.2 Nanotubes

Every structure of single tube CNs can be constructed from a monatomic layer of graphite as shown in Fig. 5 (a). Each hexagon is denoted by the chiral vector

$$\boldsymbol{L} = n_a \boldsymbol{a} + n_b \boldsymbol{b} = \Big(n_a - \frac{1}{2} n_b, \frac{\sqrt{3}}{2} n_b\Big). \tag{27}$$

In another convention for the choice of primitive translation vectors, $\boldsymbol{L}$ is characterized by two integers (p, q) with $p = n_a - n_b$ and $q = n_b$ and the corresponding CN is sometimes called a (p, q) nanotube. We shall construct a nanotube in such a way that the hexagon at $\boldsymbol{L}$ is rolled onto the origin. For convenience, we introduce another unit basis vectors $(\boldsymbol{e}_x, \boldsymbol{e}_y)$ as shown in Fig. 5. The direction of $\boldsymbol{e}_x$ or x is along the circumference of CN, *i.e.*, $\boldsymbol{e}_x = \boldsymbol{L}/L$ with

$$L = |\boldsymbol{L}| = a\sqrt{n_a^2 + n_b^2 - n_a n_b}, \tag{28}$$

and $\boldsymbol{e}_y$ or y is along the axis of CN. Further, the origin $x = 0$ is chosen always at a point corresponding to the top side when the sheet is rolled and the point $x = L/2$ at point corresponding to the bottom side as is shown in Fig. 5(b).

A primitive translation vector in the $\boldsymbol{e}_y$ direction is written as

$$\boldsymbol{T} = m_a \boldsymbol{a} + m_b \boldsymbol{b}, \tag{29}$$

with integer m_a and m_b. Now, $\boldsymbol{T}$ is determined by the condition $\boldsymbol{T} \cdot \boldsymbol{L} = 0$, which can be written as

$$m_a(2n_a - n_b) - m_b(n_a - 2n_b) = 0, \tag{30}$$

where use has been made of $\boldsymbol{a} \cdot \boldsymbol{b} = -a^2/2$. This can be solved as

$$p m_a = n_a - 2n_b, \quad p m_b = 2n_a - n_b, \tag{31}$$

where p is the greatest common divisor of $n_a - 2n_b$ and $2n_a - n_b$. The first Brillouin zone of the nanotube is given by the region $-\pi/T \leq k_y < \pi/T$ with

$$T = a\sqrt{m_a^2 + m_b^2 - m_a m_b}. \tag{32}$$

The unit cell is formed by the rectangular region determined by $\boldsymbol{L}$ and $\boldsymbol{T}$.

For nanotubes with sufficiently large diameter, effects of mixing between π bands and σ bands and change in the coupling between π orbitals can safely be neglected. Then, the energy bands of a nanotube are obtained simply by imposing periodic boundary conditions along the circumference direction, i.e., $\psi(\boldsymbol{r} + \boldsymbol{L}) = \psi(\boldsymbol{r})$. This leads to the condition $\exp(i\boldsymbol{k} \cdot \boldsymbol{L}) = 1$, which makes the wave vector along the circumference direction discrete, i.e., $k_x = 2\pi j/L$ with integer j, but the wave vector perpendicular to $\boldsymbol{L}$ arbitrary except that $-\pi/T \leq k < \pi/T$. The number of one-dimensional bands, i.e., j is given by the total number of carbon atoms in a unit cell determined by $\boldsymbol{L}$ and $\boldsymbol{T}$.

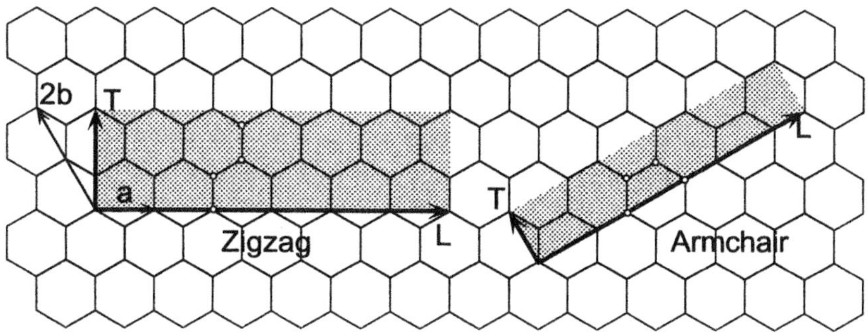

Fig. 7: The chiral vector $\boldsymbol{L}$ and the primitive lattice vector $\boldsymbol{T}$ of a zigzag and armchair nanotube. The number of carbon atoms in a unit cell is given by $4m$ for both nanotubes with $(n_a, n_b) = (m, 0)$ and $(2m, m)$.

The band structure of a nanotube depends critically on whether the K and K' points in the Brillouin zone of the 2D graphite are included in the allowed wave vectors when the 2D graphite is rolled into a nanotube. This can be understood by considering $\exp(\mathrm{i}\boldsymbol{K}\cdot\boldsymbol{L})$ and $\exp(\mathrm{i}\boldsymbol{K}'\cdot\boldsymbol{L})$. We have

$$\begin{aligned}\exp(\mathrm{i}\boldsymbol{K}\cdot\boldsymbol{L}) &= \exp\Big(+\frac{2\pi\mathrm{i}}{3}(n_a+n_b)\Big) = \exp\Big(+\mathrm{i}\frac{2\pi\nu}{3}\Big),\\ \exp(\mathrm{i}\boldsymbol{K}'\cdot\boldsymbol{L}) &= \exp\Big(-\frac{2\pi\mathrm{i}}{3}(n_a+n_b)\Big) = \exp\Big(-\mathrm{i}\frac{2\pi\nu}{3}\Big),\end{aligned} \tag{33}$$

where ν is an integer (0, ±1) determined by

$$n_a + n_b = 3N + \nu, \tag{34}$$

with integer N. This shows that for $\nu = 0$ the nanotube becomes metallic because two bands cross at the wave vector corresponding to K and K' points without a gap. When $\nu = \pm1$, on the other hand, there is a nonzero gap between valence and conduction bands and the nanotube is semiconducting.

For translation $\boldsymbol{r} \to \boldsymbol{r}+\boldsymbol{T}$ the Bloch function at the K and K' points acquires the phase

$$\begin{aligned}\exp(\mathrm{i}\boldsymbol{K}\cdot\boldsymbol{T}) &= \exp\Big(+\mathrm{i}\frac{2\pi\mu}{3}\Big),\\ \exp(\mathrm{i}\boldsymbol{K}'\cdot\boldsymbol{T}) &= \exp\Big(-\mathrm{i}\frac{2\pi\mu}{3}\Big),\end{aligned} \tag{35}$$

where $\mu = 0$ or ±1 is determined by

$$m_a + m_b = 3M + \mu, \tag{36}$$

Type	n_a	n_b	ν	m_a	m_b	T	μ	k_0
Zigzag	m	0	0	1	2	$\sqrt{3}a$	0	0
	m	0	±1	1	2	$\sqrt{3}a$		
Armchair	$2m$	m	0	0	1	a	1	$\pm 2\pi/3a$

Table 1: Parameters for zigzag and armchair nanotubes.

with integer M. When $\nu = 0$, therefore, the K and K' points are mapped onto $k_0 = +2\pi\mu/3T$ and $k'_0 = -2\pi\mu/3T$, respectively, in the one-dimensional Brillouin zone of the nanotube. At these points two one-dimensional bands cross each other without a gap.

A nanotube has a helical structure for general $\boldsymbol{L}$. There are two kinds of nonhelical nanotubes, zigzag with $(n_a, n_b) = (m, 0)$ and armchair with $(n_a, n_b) = (2m, m)$, as shown in Fig. 7. A zigzag nanotube is metallic when m is divided by three and semiconducting otherwise. We have $pm_a = n_a - 2n_b = m$ and $pm_b = 2n_a - n_b = 2m$, which give $m_a = 1$ and $m_b = 2$, and $\mu = 0$ and $T = \sqrt{3}a$. When a zigzag nanotube is metallic, two conduction and valence bands having a linear dispersion cross at the Γ point of the one-dimensional Brillouin zone. On the other hand, an armchair nanotube is always metallic. We have $pm_a = n_a - 2n_b = 0$ and $pm_b = 2n_a - n_b = 3m$, which gives $m_a = 0$ and $m_b = 1$, and $\mu = 1$ and $T = a$. Thus, the conduction and valence bands cross each other always at $k_0 = \pm 2\pi/3a$.

In a tight-binding model, the periodic boundary condition $\psi(\boldsymbol{r}+\boldsymbol{L}) = \psi(\boldsymbol{r})$ is converted into

$$\psi_A(\boldsymbol{R}_A + \boldsymbol{L}) = \psi_A(\boldsymbol{R}_A), \quad \psi_B(\boldsymbol{R}_B + \boldsymbol{L}) = \psi_B(\boldsymbol{R}_B). \tag{37}$$

It is then straightforward to calculate the band structure of nanotubes when a tight-binding model is used. Figures 8 and 9 show some examples for zigzag and armchair nanotubes. Figure 10 shows the band gap of zigzag nanotube as a function of the circumference length.

The periodic boundary conditions Eq. (37) can be converted into those for the envelope functions by substituting Eq. (6) into the above, then by multiplication of them by $g(\boldsymbol{r} - \boldsymbol{R}_A)\boldsymbol{a}(\boldsymbol{R}_A)$ or $g(\boldsymbol{r} - \boldsymbol{R}_B)\boldsymbol{a}(\boldsymbol{R}_B)$, and finally by summation of them over $\boldsymbol{R}_A$ or $\boldsymbol{R}_B$. Explicitly, we have

$$\begin{aligned} \boldsymbol{F}^K(\boldsymbol{r}+\boldsymbol{L}) &= \exp\Big(-\mathrm{i}\boldsymbol{K}\cdot\boldsymbol{L}\Big)\boldsymbol{F}^K(\boldsymbol{r}) = \exp\Big(-\frac{2\pi\mathrm{i}\nu}{3}\Big)\boldsymbol{F}^K(\boldsymbol{r}), \\ \boldsymbol{F}^{K'}(\boldsymbol{r}+\boldsymbol{L}) &= \exp\Big(-\mathrm{i}\boldsymbol{K}'\cdot\boldsymbol{L}\Big)\boldsymbol{F}^K(\boldsymbol{r}) = \exp\Big(+\frac{2\pi\mathrm{i}\nu}{3}\Big)\boldsymbol{F}^{K'}(\boldsymbol{r}). \end{aligned} \tag{38}$$

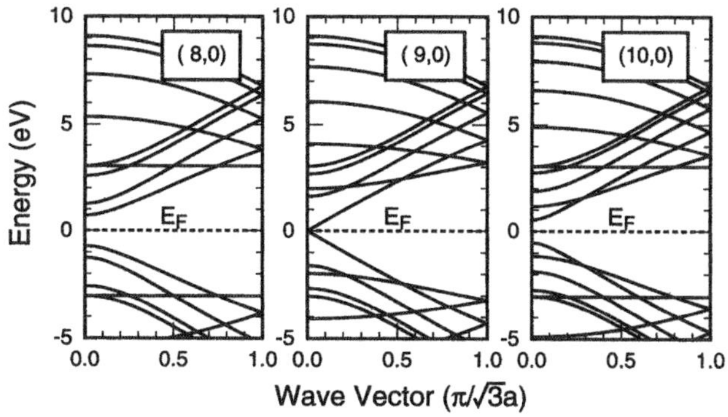

Fig. 8: Some examples of the band structure obtained in a tight-binding model for zigzag nanotubes.

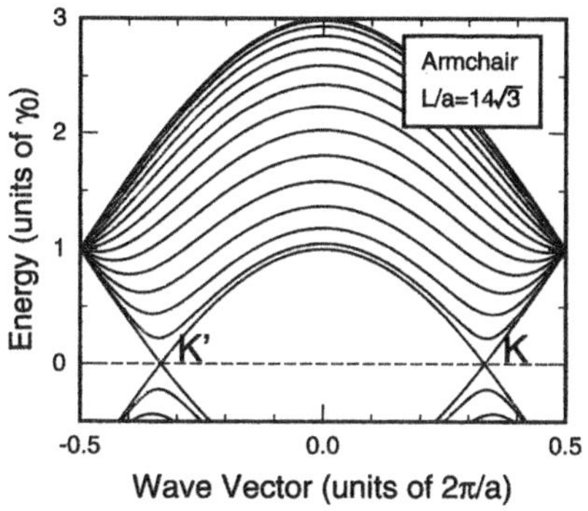

Fig. 9: An example of the band structure of an armchair nanotube.

The wave function on the cylinder surface is given by a plane wave $\boldsymbol{F}^K(\boldsymbol{r}) \propto \exp(\mathrm{i}k_x x+\mathrm{i}k_y y)$. Energy levels in CN for the K point are obtained by putting $k_x = \kappa_\nu(n)$ with

$$\kappa_\nu(n) = \frac{2\pi}{L}\Big(n-\frac{\nu}{3}\Big), \tag{39}$$

and $k_y = k$ in the above $\boldsymbol{k}\cdot\boldsymbol{p}$ equation as [15]

$$\varepsilon_\nu^{(\pm)}(n,k) = s\gamma\sqrt{\kappa_\nu(n)^2+k^2}, \tag{40}$$

where $L = |\boldsymbol{L}|$, n is an integer, and $s = +1$ and -1 represent the conduction and valence bands, respectively. The corresponding wave functions are written as

$$\boldsymbol{F}^K(\boldsymbol{r}) = \frac{1}{\sqrt{2AL}}\begin{pmatrix} sb_\nu(n,k) \\ 1 \end{pmatrix} \exp\left[\mathrm{i}\kappa_\nu(n)x+\mathrm{i}ky\right], \tag{41}$$

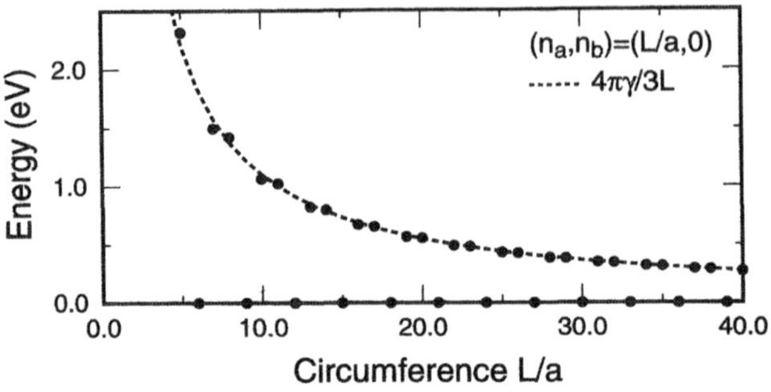

Fig. 10: The band gap of zigzag nanotubes with $(n_a, n_b) = (m, 0)$ as a function of m. The dotted line shows the band gap obtained in a $\boldsymbol{k}\cdot\boldsymbol{p}$ approximation and the black dots those of a tight-binding model.

with

$$b_\nu(n,k) = \frac{\kappa_\nu(n) - ik}{\sqrt{\kappa_\nu(n) + k^2}}, \tag{42}$$

where A is the length of the nanotube. Figure 11 shows the band structure in the vicinity of the K point for $\nu = 0$ and $+1$. For the band $n = 0$ of a metallic nanotube $\nu = 0$, in particular,

$$\boldsymbol{F}^K(\boldsymbol{r}) = \frac{1}{\sqrt{2AL}} \begin{pmatrix} -is(k/|k|) \\ 1 \end{pmatrix} \exp(iky). \tag{43}$$

Because of the one-dimensional energy band, the density of states remains nonzero even for $\varepsilon = 0$ and the system is metallic for $\nu = 0$. In fact, the density of states for the K point is given by

$$D(\varepsilon) = \frac{1}{A}\sum_k \delta(\varepsilon - \gamma|k|) = \frac{2}{2\pi}\int_0^\infty dk\, \delta(\varepsilon - \gamma k) = \frac{1}{\pi\gamma}. \tag{44}$$

This is quite in contrast to the graphite sheet for which the density of states vanishes at $\varepsilon = 0$ even if the band gap vanishes. Each energy band of metallic CNs is two-fold degenerate except those for $n = 0$.

The energy bands and wave functions for the K' point are obtained by replacing ν by $-\nu$ in the above equations. Therefore, the energy bands are completely same at K and K' points and CN becomes metallic for $\nu = 0$ and semiconducting with gap

$$E_g = \frac{4\pi\gamma}{3L}, \tag{45}$$

for $\nu = \pm 1$. Figure 10 compares this gap to that obtained in a tight-binding model.

A magnetic field applied parallel to the axis, i.e., in the presence of a magnetic flux ϕ passing through the cross section as shown in Fig. 12, leads to the change in

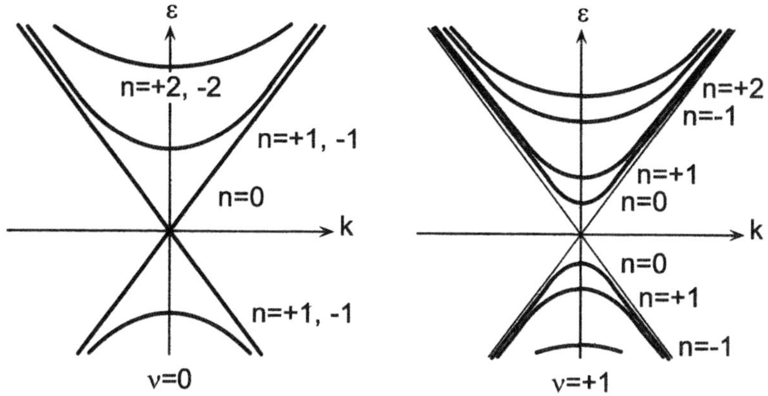

Fig. 11: Energy bands obtained in the effective-mass approximation for $\nu=0$ (left) and $\nu=+1$ (right).

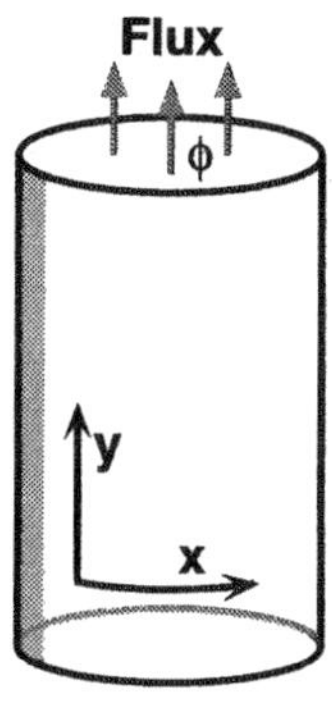

Fig. 12: A cylinder in the presence of magnetic flux ϕ passing through its cross section.

the boundary condition, $\psi(\boldsymbol{r}+\boldsymbol{L})=\psi(\boldsymbol{r})\exp(+2\pi i\varphi)$ with $\varphi=\phi/\phi_0$, where $\phi_0=ch/e$ is the magnetic flux quantum. Consequently, $\kappa_\nu(n)$ is replaced by $\kappa_{\nu\varphi}(n)$ with

$$\kappa_{\nu\varphi}(n)=\frac{2\pi}{L}\Big(n+\varphi-\frac{\nu}{3}\Big). \tag{46}$$

The corresponding result for the K' point is again obtained by the replacement $\nu\to-\nu$. The band gap exhibits an oscillation between 0 and $2\pi\gamma/L$ with period ϕ_0 as shown in Fig. 13. This giant Aharonov-Bohm (AB) effect on the band gap is a unique property of CNs. The AB effect appears also in a tunneling conductance across a finite-length CN [49].

In the presence of a magnetic field B perpendicular to the tube axis as shown in Fig. 14, we can use the gauge

$$\boldsymbol{A}=\Big(0,\frac{LB}{2\pi}\sin\frac{2\pi x}{L}\Big), \tag{47}$$

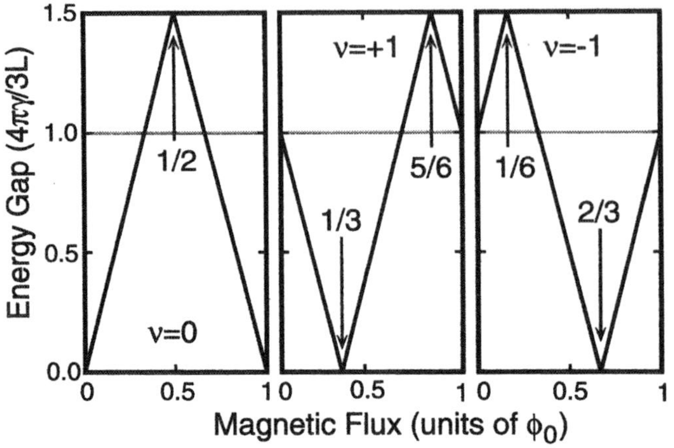

Fig. 13: Aharonov-Bohm effect on the band gap. In the case of a semiconducting nanotube with $\nu = +1$, the gap for $\nu = +1$ corresponds to the K point and that for $\nu = -1$ to the K' point.

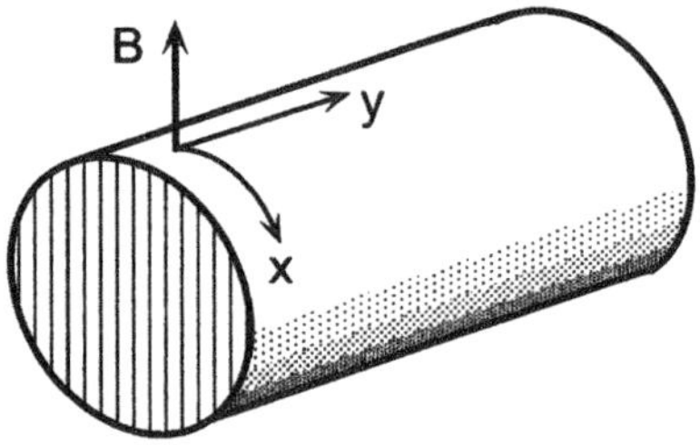

Fig. 14: A nanotube in the presence of a magnetic field B in the direction perpendicular to the axis.

and the effective field for electrons in a CN is given by the component perpendicular to the surface, i.e., $B(x) = B\cos(2\pi x/L)$. The parameter characterizing its strength is given by $\alpha = (L/2\pi l)^2$. In the case of $\alpha \ll 1$, the field can be regarded as a small perturbation, while in the case of $\alpha \gg 1$, Landau levels are formed on the cylinder surface.

Consider a metallic nanotube as an example. The energy levels and wavefunctions are analytically obtained for $\varepsilon = 0$ in this case [50]. First, define a nonunitary matrix

$$P(\boldsymbol{r}) = \begin{pmatrix} \exp[-\alpha(\boldsymbol{r})] & 0 \\ 0 & \exp[+\alpha(\boldsymbol{r})] \end{pmatrix}, \tag{48}$$

with

$$\alpha(\boldsymbol{r}) = \Big(\frac{L}{2\pi l}\Big)^2 \cos\frac{2\pi x}{L}. \tag{49}$$

Then, we have

$$P(\boldsymbol{r})\mathcal{H}P(\boldsymbol{r}) = \mathcal{H}_0, \tag{50}$$

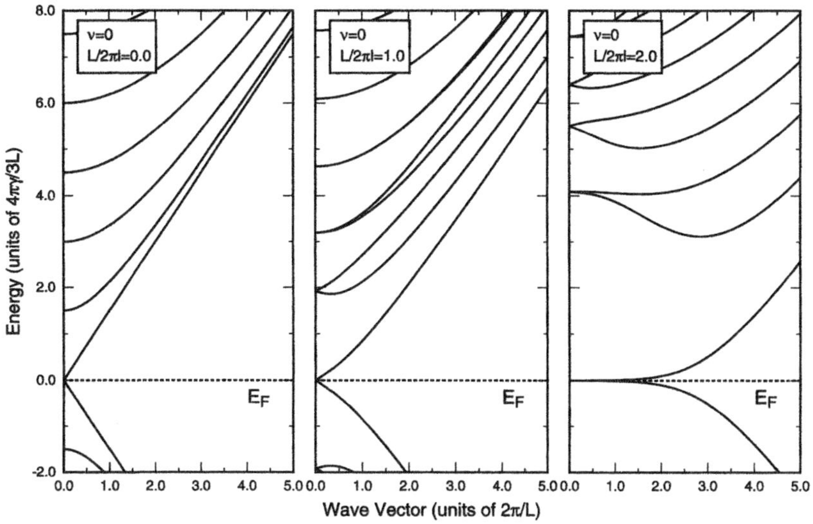

Fig. 15: Some examples of calculated energy bands of a metallic CN in magnetic fields perpendicular to the axis.

for the K point, where $\mathcal{H}_0$ is the Hamiltonian in the absence of a magnetic field. This shows that for an eigen wave-function $\boldsymbol{F}_0^K(\boldsymbol{r})$ of $\mathcal{H}_0$ at $\varepsilon=0$, i.e., $\mathcal{H}_0\boldsymbol{F}_0^K(\boldsymbol{r})=0$, the function $P(\boldsymbol{r})\boldsymbol{F}_0^K(\boldsymbol{r})$ satisfies the corresponding equation $\mathcal{H}P(\boldsymbol{r})\boldsymbol{F}_0^K(\boldsymbol{r})=0$ in nonzero B.

Therefore, the wave functions are given by

$$\boldsymbol{F}_{sk}^K = \frac{1}{\sqrt{2A}}\begin{pmatrix} -\mathrm{i}s(k/|k|)F_-(x) \\ F_+(x) \end{pmatrix}\exp(\mathrm{i}ky), \tag{51}$$

with

$$F_\pm(x) = \frac{1}{\sqrt{LI_0(2\alpha)}}\exp\Big(\pm\alpha\cos\frac{2\pi x}{L}\Big), \tag{52}$$

where $s=+1$ and -1 for the conduction and valence band, respectively, and $I_0(z)$ is the modified Bessel function of the first kind defined as

$$I_0(z) = \int_0^\pi \frac{d\theta}{\pi}\exp(z\cos\theta). \tag{53}$$

In high magnetic fields ($\alpha \gg 1$), $\boldsymbol{F}_-$ is localized around $x=\pm L/2$, i.e., at the bottom side of the cylinder and $\boldsymbol{F}_+$ is localized around the top side $x=0$. The corresponding eigenenergies are given by $\varepsilon_s(k)=s\gamma|k|/I_0(2\alpha)$ which gives the group velocity $v=\gamma/\hbar I_0(2\alpha)$, and the density of states $D(0)=I_0(2\alpha)/\pi\gamma$ at $\varepsilon=0$. We

Circumference (Å)		50	100	200	400	800
Diameter (Å)		16	32	64	127	255
Gap (meV)		541	270	135	68	34
Magnetic	ϕ_0	2080	520	130	32	8
Field (T)	$(L/2\pi l)^2\!=\!1$	1040	260	65	16	4

Table 2: Some examples of actual magnetic-field strength corresponding to the conditions $(L/2\pi l)^2 = 1$ and $\phi/\phi_0 = 1$ as a function of the circumference and the radius. The band gap of a semiconducting nanotube is also shown. ($\gamma = 6.46$ eV·Å).

should note that

$$I_0(2\alpha) \approx \begin{cases} 1+\alpha^2+\cdots & (\alpha \ll 1), \\ e^{2\alpha}/\sqrt{4\pi\alpha} & (\alpha \gg 1). \end{cases} \tag{54}$$

This means that the group velocity for states at $\varepsilon = 0$ decreases and consequently the density of states increases exponentially with the increase of the magnetic field in the high-field regime. The wavefunction for the K' point can be obtained in a similar manner.

Figure 15 gives some examples of energy bands of a metallic CN in perpendicular magnetic fields [15], which clearly shows the formation of flat Landau levels at the Fermi level in high fields. It is worth mentioning that there is no difference in the spectra between metallic and semiconducting CNs and in the presence and absence of an AB flux for $(L/2\pi l)^2 \gg 1$, because the wave function is localized in the circumference direction and the boundary condition becomes irrelevant.

Table 2 shows some examples of actual magnetic-field strength corresponding to the conditions $(L/2\pi l)^2 = 1$ and $\phi/\phi_0 = 1$ as a function of the circumference and the radius. For a typical single-wall armchair nanotube having circumference $L = \sqrt{3}ma$ with $m = 10$, the required magnetic field is too large, but can be easily accessible by using a pulse magnet for typical multi-wall nanotubes with a diameter ~ 50 Å.

For nanotubes with a small circumference, we have to consider higher order $\boldsymbol{k}\cdot\boldsymbol{p}$ terms in the Hamiltonian. A higher order $\boldsymbol{k}\cdot\boldsymbol{p}$ equation was derived in a simple tight-binding model including only a π orbital for each carbon atom as [51]

$$\begin{pmatrix} 0 & \gamma\Big[(\hat{k}_x - i\hat{k}_y) + \frac{a\, e^{3i\eta}}{4\sqrt{3}}(\hat{k}_x + i\hat{k}_y)^2\Big] \\ \gamma\Big[(\hat{k}_x + i\hat{k}_y) + \frac{a\, e^{-3i\eta}}{4\sqrt{3}}(\hat{k}_x - i\hat{k}_y)^2\Big] & 0 \end{pmatrix} \boldsymbol{F}^K = \varepsilon \boldsymbol{F}^K, \tag{55}$$

for the K point, where η is the chiral angle. This gives trigonal warping of the band around the K point in 2D graphite and gives a small correction to the band gap of

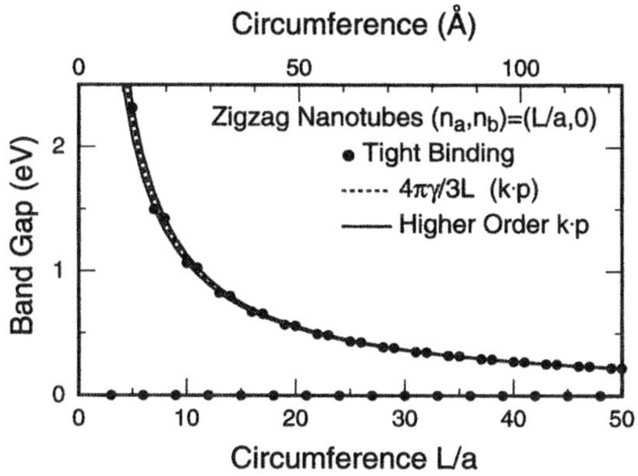

Fig. 16: The band gap of zigzag nanotubes as a function of the circumference L/a. The dots represent tight-binding results and the dotted line the result of the lowest order $\boldsymbol{k}\cdot\boldsymbol{p}$ scheme. When higher-order $\boldsymbol{k}\cdot\boldsymbol{p}$ terms are included, a small deviation present in Fig. 10 is removed almost completely.

CN. In fact, when $\kappa_\nu(n)\neq 0$ and $|k|\ll|\kappa_\nu(n)|$, the bands are written as

$$\varepsilon_s(n,k)=s\gamma\sqrt{\kappa_\nu(n)^2\Big(1+\frac{a\kappa_\nu(n)}{2\sqrt{3}}\cos 3\eta\Big)+\Big(k-\frac{\sqrt{3}a\kappa_\nu(n)^2}{4}\sin 2\eta\Big)^2}, \tag{56}$$

to the lowest order. This shows first that the wave vector k corresponding to extremum points is shifted to the positive direction except in a zigzag nanotube $\eta=0$ $(0\leq\eta\leq\pi/6)$. Further, for semiconducting nanotubes the gap becomes different between $\nu=1$ and -1 and in metallic nanotubes with $\nu=0$ the degeneracy of the bands with positive $+|n|$ and negative $-|n|$ $(n\neq 0)$ is lifted [52].

As shown in Fig. 16, the correction takes account of the deviation from the tight-binding result present in the lowest-order $\boldsymbol{k}\cdot\boldsymbol{p}$ theory almost completely in zigzag nanotubes. In the presence of a magnetic field perpendicular to the axis, the higher order term was shown to cause the appearance of a small band-gap except in armchair nanotubes and a shift of the wave vector corresponding to $\varepsilon=0$ in armchair nanotubes [51].

3. Optical properties

3.1 Dynamical conductivity

We shall consider the optical absorption of CN with an Aharonov-Bohm flux using the linear response theory. We first expand electric field $E_\xi(\theta,\omega)$ and induced current density $j_\xi(\theta,\omega)$ into a Fourier series:

$$E_\xi(\theta,\omega)=\sum_l E_\xi^l(\omega)\exp(il\theta-i\omega t),$$
$$j_\xi(\theta,\omega)=\sum_l j_\xi^l(\omega)\exp(il\theta-i\omega t), \tag{57}$$

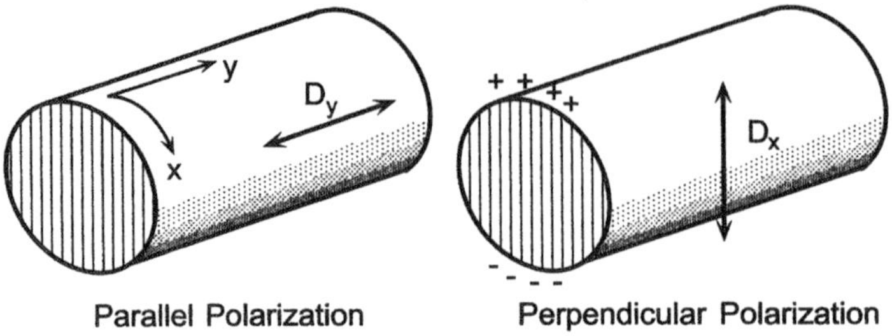

Fig. 17: The schematic picture of the polarized electric field parallel (a) and perpendicular (b) to the tube axis.

where ξ denotes x or y and $\theta = 2\pi x/L$ represents the angle measured from the top side of the nanotube. It is quite straightforward to show that the induced current has the same Fourier component as that of the electric field as follows:

$$j^l_\xi(\omega) = \sigma^l_{\xi\xi}(\omega) E^l_\xi(\omega), \tag{58}$$

where $\sigma^l_{\xi\xi}(\omega)$ is the dynamical conductivity. The dynamical conductivity is calculated using the Kubo formula as

$$\sigma^l_{\xi\xi}(\omega) = \frac{1}{\mathrm{i}\omega}[K^l_{\xi\xi}(\omega) - K^l_{\xi\xi}(0)], \tag{59}$$

with

$$K^l_{\xi\xi}(\omega) = -\frac{2}{AL}\sum_{nk}\sum_{v,w=\pm} \frac{f[\varepsilon^{(v)}_{\nu\varphi}(n,k)] - f[\varepsilon^{(w)}_{\nu\varphi}(n+l,k)]}{\varepsilon^{(v)}_{\nu\varphi}(n,k) - \varepsilon^{(w)}_{\nu\varphi}(n+l,k) + \hbar\omega + \mathrm{i}\delta}$$
$$\times |(n,k,v|\hat{j}^l_\xi|n+l,k,w)|^2 g_0[(\varepsilon^{(v)}_{\nu\varphi}(n,k))] g_0[(\varepsilon^{(w)}_{\nu\varphi}(n+l,k))], \tag{60}$$

where $f(\varepsilon)$ is the Fermi distribution function, the factor 2 comes from the spin degeneracy, and $g_0(\varepsilon)$ is a cutoff function. The cutoff function has been introduced to get the contribution of the electronic states for which the $\boldsymbol{k}\cdot\boldsymbol{p}$ approximation is valid.

The current-density operator $\hat{j}^l_\xi$ at the K point is given by

$$\hat{j}^l_\xi = \frac{-e}{\mathrm{i}\hbar}[\hat{\xi}, \gamma\vec{\sigma}\cdot\hat{\boldsymbol{k}}]\mathrm{e}^{-\mathrm{i}l\theta} = -\frac{e\gamma}{\hbar}\sigma_\xi \mathrm{e}^{-\mathrm{i}l\theta}. \tag{61}$$

At the K' point, operator $\hat{j}^l_x$ is the same as that at the K' point but j^l_y has the opposite sign of that at the K point. The factor $|(n,k,v|j^l_\xi|n+l,k,w)|^2$, however, provides the same value for both K and K' points. Substituting Eq. (60) into Eq. (59) we get the conductivity at the zero temperature

$$\sigma^l_{\xi\xi}(\omega) = \frac{4\hbar}{\mathrm{i}AL}\sum_{nk}\sum_{v,w=\pm} \frac{f[\varepsilon^{(w)}_{\nu\varphi}(n+l,k)]\{1 - f[\varepsilon^{(v)}_n(k)]\}2\hbar\omega}{[\varepsilon^{(v)}_{\nu\varphi}(n,k) - \varepsilon^{(w)}_{\nu\varphi}(n+l,k)]^2 - (\hbar\omega)^2 - \mathrm{i}\hbar^2\omega/\tau}$$

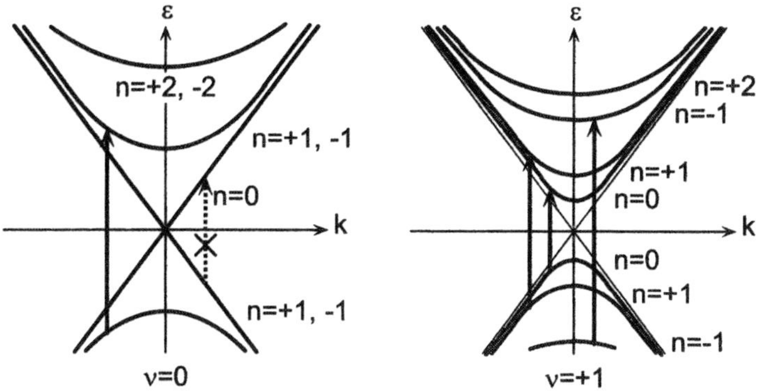

Fig. 18: The band structures of a metallic and semiconducting CN. The allowed optical transitions for the parallel polarization are denoted by arrows.

$$\times \frac{|(n,k,v|\hat{j}_\xi^l|n+l,k,w)|^2}{\varepsilon_{\nu\varphi}^{(v)}(n,k)-\varepsilon_{\nu\varphi}^{(w)}(n+l,k)} g_0[\varepsilon_{\nu\varphi}^{(v)}(n,k)]g_0[\varepsilon_{\nu\varphi}^{(w)}(n+l,k)], \tag{62}$$

where a phenomenological relaxation time τ has been introduced.

3.2 Parallel polarization

When the polarization of external electric field $\mathbf{D}$ is parallel to the tube axis as is shown in Fig. 17, the Fourier components of a total field are written as

$$E_y^l = D_y \delta_{l,0}.$$

Thus the absorption in a unit area is given by

$$P_y(\omega) = \frac{1}{2}\frac{1}{2\pi}\sum_l \int_0^{2\pi} \mathrm{d}\theta \, \mathrm{Re}\,[j_y^l(\omega)E_y^{l\,*}] = \frac{1}{2}\mathrm{Re}\,[\sigma_{yy}^{l=0}(\omega)]D_y^2. \tag{63}$$

For $l = 0$, transitions occur between bands with the same band index n as is seen from Eq. (62). Since all the conduction bands are specified by different n's, there is no transition within conduction bands and within valence bands. At a band edge $k=0$, in particular, the wave function is an eigen function of σ_x and therefore transitions between valence and conduction bands having the same index n are all allowed. An exception occurs for bands with $n=0$ in a metallic nanotube. In this case the wave function is an eigenfunction of σ_y because $\kappa_{\nu\varphi}(n) = 0$ and therefore there is no matrix element between the conduction and valence bands with $n=0$.

In the limit $\tau \to \infty$ absorption spectra except for intraband Drude terms is proportional to

$$\mathrm{Re}\,[\sigma_{yy}^{l=0}(\omega)] = \frac{e^2}{h}\sum_n \Big[\frac{2\gamma\kappa_{\nu\varphi}(n)}{\hbar\omega}\Big]^2$$

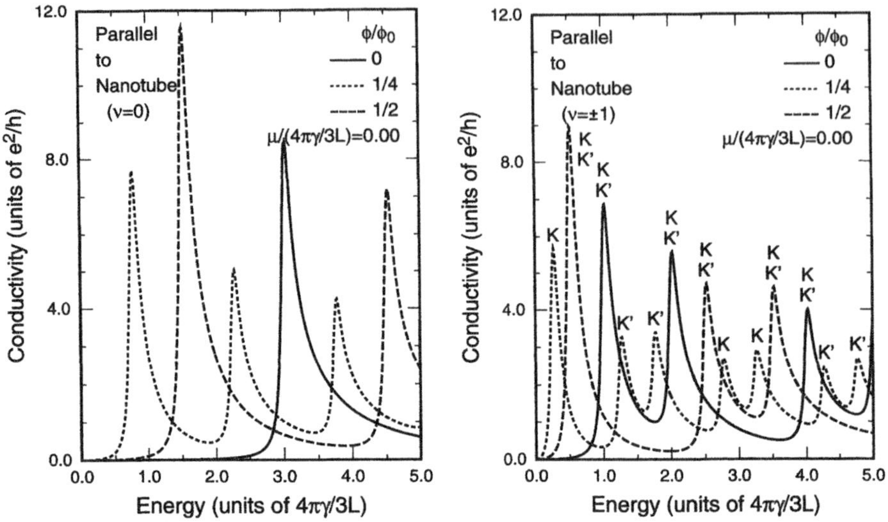

Fig. 19: Calculated optical absorption spectra for the parallel polarization in a metallic (left) and semiconducting (right) CN.

$$\times \frac{2\gamma(2\pi/L)}{\sqrt{(\hbar\omega)^2-[2\gamma\kappa_{\nu\varphi}(n)]^2}}\theta[|\hbar\omega|-2\gamma|\kappa_{\nu\varphi}(n)|]g_0(|\hbar\omega|/2)^2, \tag{64}$$

where $\theta(x)$ is the step function defined by $\theta(x)=1$ for $x\geq 0$ and $\theta(x)=0$ for $x<0$. Figure 19 shows the calculated results of Re $\sigma_{yy}^{l=0}(\omega)$ of a metallic CN for $\varphi=0$, 1/4, and 1/2.

3.3 Perpendicular polarization

When an external electric field is polarized in the direction perpendicular to the CN axis, the electric field has components $l=\pm 1$ and therefore transitions between states with $\Delta n=\pm 1$ become allowed in general. Figure 20 shows allowed transitions. When an external electric field is polarized in the direction perpendicular to the CN axis, effects of an electric field induced by the polarization of nanotubes should be considered. This depolarization effect is quite significant for absorption spectra.

Suppose an external electric field $D_x^l \exp(\mathrm{i}l\theta - \mathrm{i}\omega t)$ is applied in the direction normal to the tube axis and let j_x^l be the induced current. With the use of the equation of continuity

$$\frac{\partial}{\partial t}\rho^l e^{\mathrm{i}l\theta-\mathrm{i}\omega t}+\frac{2\pi}{L}\frac{\partial}{\partial\theta}j_x^l e^{\mathrm{i}l\theta-\mathrm{i}\omega t}=0, \tag{65}$$

the corresponding induced charge density localized on the cylinder surface is written as

$$\rho^l = \frac{2\pi}{L}\frac{l}{\omega} j_x^l. \tag{66}$$

The potential formed by line charge with the density ρ at distance r is given by

$$\phi(r) = -\frac{2\rho}{\kappa}\ln r, \tag{67}$$

where

$$r = \frac{L}{\pi}\sin\left|\frac{\theta-\theta'}{2}\right|. \tag{68}$$

The static dielectric constant κ describes effects of polarization of σ bands and π bands except those lying in the vicinity of the Fermi level. Then it is found that the induced charge leads to potential

$$\phi(\theta) = -2\frac{L}{2\pi}\int_0^{2\pi} \mathbf{d}\theta' \frac{\rho^l}{\kappa}\exp(\mathrm{i}l\theta')\ln\left|\frac{L}{\pi}\sin\frac{\theta-\theta'}{2}\right| = \frac{L}{\kappa|l|}\rho^l \mathrm{e}^{\mathrm{i}l\theta}, \tag{69}$$

and the Fourier component of the potential is written as

$$\phi^l = \frac{L}{\kappa|l|}\rho^l. \tag{70}$$

The potential gives rise to electric field $-(2\pi/L)(\partial\phi^l/\partial\theta)$ and therefore the total electric field is obtained as

$$E_x^l = D_x^l - \frac{2\pi}{L}\frac{\partial\phi^l}{\partial\theta} = D_x^l - \mathrm{i}|l|\frac{4\pi^2}{\kappa L\omega} j_x^l. \tag{71}$$

With the use of $j_x^l = \sigma_{xx}^l E_x^l$ we get

$$j_x^l = \tilde{\sigma}_{xx}^l D_x^l, \tag{72}$$

with

$$\sigma_{xx}^l\left(1 + \mathrm{i}|l|\frac{4\pi^2}{\kappa L\omega}\sigma_{xx}^l\right)^{-1}. \tag{73}$$

For the external field being $\mathbf{D} = (D_x\sin\theta, 0)$, the Fourier components of the external field and the induced current are written as

$$\begin{aligned} D_x^l &= \frac{D_x}{2\mathrm{i}}\delta_{l,1} - \frac{D_x}{2\mathrm{i}}\delta_{l,-1}, \\ j_x^l &= \frac{D_x}{2\mathrm{i}}\sigma_{xx}^{l=1}(\omega)\delta_{l,1} - \frac{D_x}{2\mathrm{i}}\sigma_{xx}^{l=-1}(\omega)\delta_{l,-1}. \end{aligned} \tag{74}$$

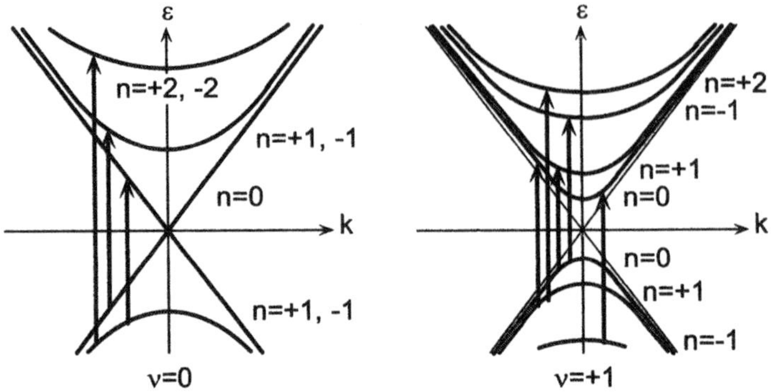

Fig. 20: The band structures of a metallic and semiconducting CN. The allowed optical transitions for the perpendicular polarization are denoted by arrows.

Thus the absorption is given by

$$P_x(\omega)=\frac{1}{2}\frac{1}{2\pi}\sum_l\int_0^{2\pi}\mathrm{d}\theta\mathrm{Re}\,[j_x^l(\omega)E_x^{l\,*}]=\frac{1}{4}\mathrm{Re}\,[\tilde{\sigma}_{xx}(\omega)]D_x^2, \tag{75}$$

with

$$\tilde{\sigma}_{xx}=\frac{1}{2}(\tilde{\sigma}_{xx}^{l=1}+\tilde{\sigma}_{xx}^{l=-1}). \tag{76}$$

In Fig. 21, Re $\tilde{\sigma}_{xx}$ (indicated by 'Self-Consistent') and Re σ_{xx} (indicated by 'Perturbation') are shown for a metallic CN with magnetic flux $\varphi=0$, 1/4, and 1/2. The peaks around $2\pi\gamma/L$ correspond to the allowed transitions at $k=0$ discussed above. In magnetic flux $\varphi=0$, the peak of σ_{xx} is suppressed in comparison with the others. This is because of the absence of the divergence in the joint density of states at the band edge. It is quite interesting that these peaks disappear almost completely if the depolarization effect is taken into account.

To understand the strong suppression of absorption peaks for perpendicular polarization, we consider a simple model in which the real part of conductivity is proportional to a joint density of states of one-dimensional materials and the oscillator strength is constant. The model conductivity is written as

$$\mathrm{Re}\ \sigma_{xx}(\omega)=\frac{ne^2}{m}\frac{\theta(\omega-\omega_1)\theta(\omega_2-\omega)}{\sqrt{(\omega-\omega_1)(\omega_2-\omega)}}\qquad(0<\omega_1<\omega_2,\ \omega>0), \tag{77}$$

where n is the electron density in an unit area and m is the mass of the electron. This conductivity satisfies the intensity sum rule given by

$$\frac{1}{\pi}\int_0^{\infty}\mathrm{d}\omega\mathrm{Re}\ \sigma_{xx}(\omega)=\frac{ne^2}{m}. \tag{78}$$

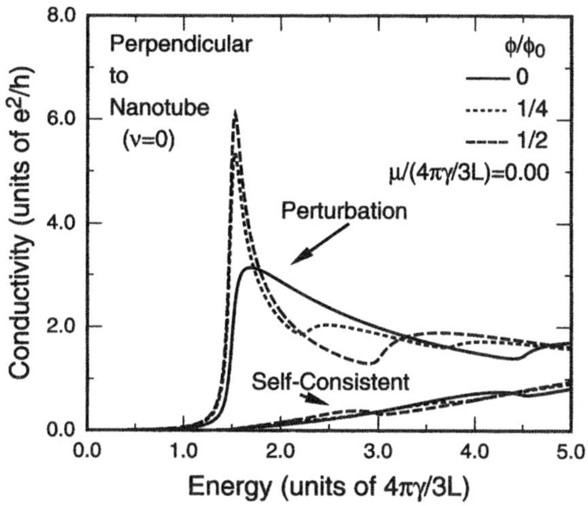

Fig. 21: Calculated real part of σ_{xx} (indicated by 'Perturbation') and $\tilde{\sigma}_{xx}$ (indicated by 'Self-Consistent') of undoped metallic CNs for φ=0, 1/4, and 1/2.

Using the Kramers-Krönig relation, we get the imaginary part of the conductivity as

$$\mathrm{Im}\sigma_{xx}(\omega) = -\frac{\mathrm{P}}{\pi}\int_{-\infty}^{\infty} \mathrm{d}\omega' \frac{\mathrm{Re}\ \sigma_{xx}(\omega')}{\omega'-\omega} = \begin{cases} 0 & \omega_1 \leq \omega \leq \omega_2 \\ -\dfrac{ne^2}{m}\dfrac{1}{\sqrt{(\omega-\omega_1)(\omega-\omega_2)}} & 0 \leq \omega < \omega_1 \\ \dfrac{ne^2}{m}\dfrac{1}{\sqrt{(\omega-\omega_1)(\omega-\omega_2)}} & \omega > \omega_2. \end{cases} \tag{79}$$

As is seen from Eq. (75) absorption peaks exist at the frequency where Re $[\tilde{\sigma}_{xx}(\omega)]$ diverges. From Eq. (72) it is found that an absorption peak occurs at a frequency ω_p higher than the absorption edge ω_2. For $\omega_\mathrm{p} \gg \omega_2$, we have

$$\omega_\mathrm{p}^2 = \frac{\pi}{L}\frac{4\pi ne^2}{m}, \tag{80}$$

which is nothing but a plasma frequency corresponding to the three-dimensional electron density $\pi n/L$. This is the reason for the strong suppression of absorption peaks for perpendicular polarized field.

3.4 Exciton

It is well known that the exciton binding energy becomes infinite in the limit of an ideal one-dimensional electron-hole system [53,54]. This means that the exciton effect can be quite important and modify the absorption spectra drastically. Exciton energy levels and corresponding optical spectra have been calculated in the conventional screened Hartree-Fock approximation within a **k·p** scheme [19].

In the **k·p** scheme, all physical quantities become universal if the length is scaled

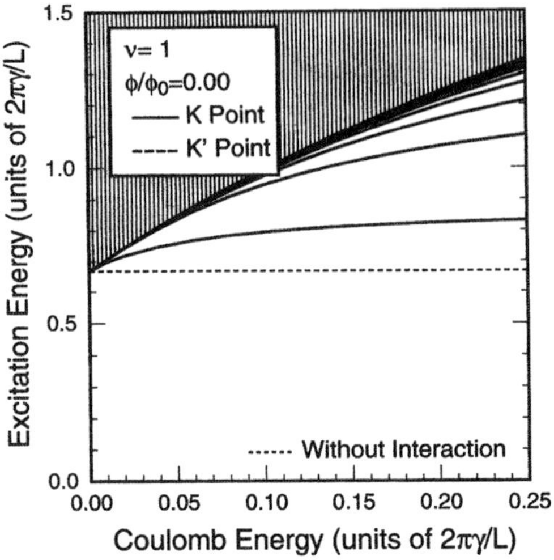

Fig. 22: Interband excitation spectra calculated in a screened Hartree-Fock approximation.

by L and the energy by $2\pi\gamma/L$. The strength of the Coulomb interaction is specified by $(e^2/\kappa L)/(2\pi\gamma/L)$, which turns out to be independent of the circumference length L. This parameter is estimated as follows for $\gamma = 6.46$ eV·Å, which corresponds to $\gamma = \sqrt{3}a|\gamma_0|/2$ with $\gamma_0 = -3.03$ eV and $a = 2.46$ Å:

$$\frac{e^2}{\kappa L}\frac{L}{2\pi\gamma} = \frac{1}{2\pi\kappa}\frac{e^2}{2a_B}\frac{2a_B}{\gamma} = 0.3545 \times \frac{1}{\kappa}. \tag{81}$$

The static dielectric constant κ describing contributions from states except those lying in the vicinity of the Fermi level is not known and therefore will be treated as a parameter in the following.

Figure 22 shows some examples of calculated exciton energy levels for a semiconducting CN ($\nu = 1$) versus the strength of the Coulomb interaction. With the increase in the interaction, the number of exciton bound states increases and their energy levels are shifted to the higher energy side in spite of the fact that their binding energy increases. The reason is in the considerable enhancement of the band gap due to the Coulomb interaction. It is interesting to notice that the energy of the lowest excitonic state varies very little as a function of the strength of the Coulomb interaction.

Figure 23 shows calculated absorption spectra in a semiconducting CN in the absence of a magnetic flux for several values of the interaction parameter $(e^2/\kappa L)/(2\pi\gamma/L)$. The energy levels of excitons are denoted by vertical straight lines. The considerable optical intensity is transferred to the lowest exciton bound states. For a sufficiently larger strength of the Coulomb interaction, transitions to exciton excited states become appreciable (the transition to the first excited states are forbidden due to parity).

Further, in addition to excitons associated with the highest valence and the lowest

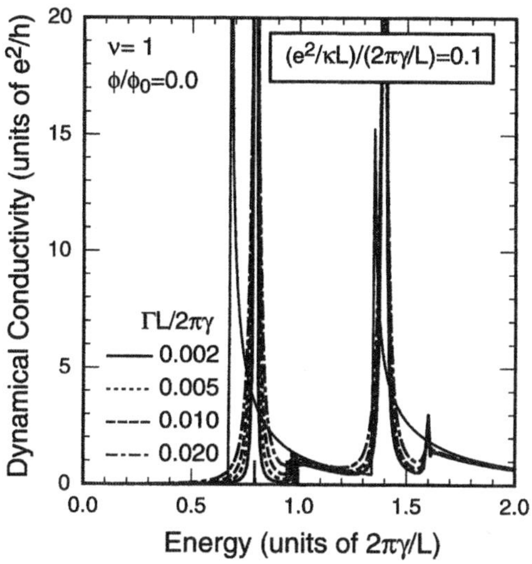

Fig. 23: Examples of interband optical absorption spectra in the presence of electron-electron interaction.

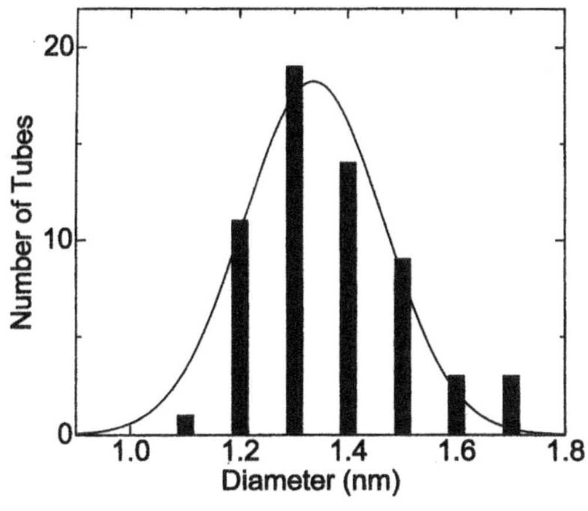

Fig. 24: The distribution of diameter of nanotubes used for the measurement of optical absorption spectra.

conduction bands, exciton effects are important for transitions to excited bands. In fact, the exciton binding energy and the intensity transfer is larger for the transition to the first excited conduction band than to the lowest conduction band. This arises presumably because the effective mass along the axis direction is larger for the excited conduction band. This peculiar feature is the origin of the large enhancement of the oscillator strength of the one-dimensional exciton.

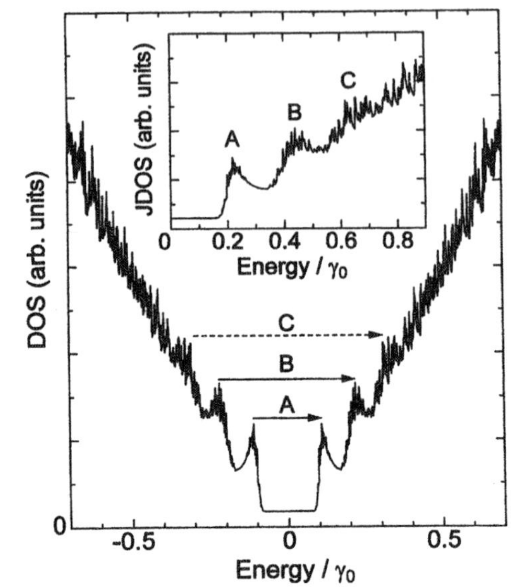

Fig. 25: The averaged density of states for the ensemble of nanotubes with the diameter distribution given by Fig. 24.

3.5 Experiments

The one-dimensional electronic structure of nanotubes was directly observed by scanning tunneling microscopy (STM), scanning tunneling spectroscopy [55], and resonant Raman scattering [56,57]. However, little effort has been made to investigate experimentally optical properties until very recently. Optical absorption spectra of thin film samples of single-wall nanotubes were observed and analyzed by assuming a distribution of their chirality and diameter [58,59]. Careful comparison of the observed spectrum with calculated in a simple tight-binding model suggested the importance of excitonic effects [59].

Figure 24 shows the observed histogram giving the diameter distribution of single-wall nanotubes. The mean diameter and standard deviation are 1.34 nm and 0.13 nm, respectively. Assuming random distribution of chirality of nanotubes and using the diameter distribution, we can calculate the density of states in a tight-binding model with a single parameter γ_0 and show the results in Fig. 25. The inset shows the joint density of states corresponding to the band-to-band optical absorption.

In Fig. 26 the observed optical absorption spectrum is compared with the joint density of states in which the position of peak B is fitted to the second absorption peak with $\gamma_0 = 2.75 \pm 0.05$ eV. Comparing the observed spectrum with the calculated one in the fundamental absorption region, the observed absorption band at 0.68 eV is higher by 0.08 eV than the calculated energy of the band-to-band transition. A peak energy to the calculated band gap energy is ~ 1.13. This roughly corresponds to $(e^2/\kappa L)/(2\pi\gamma/L) \sim 0.05$ in Fig. 22. This result strongly suggests that the exci-

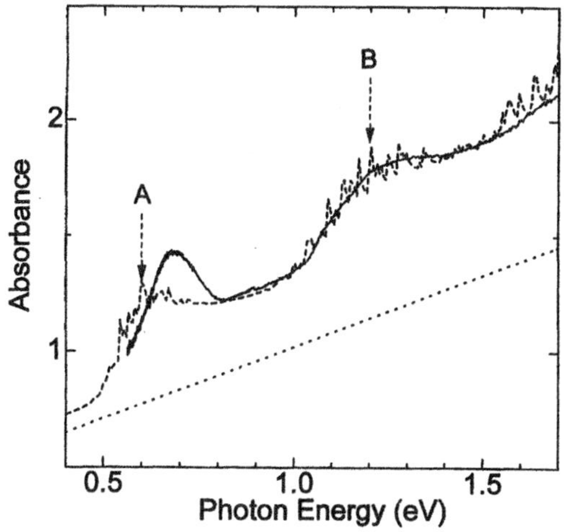

Fig. 26: Observed absorption spectra (solid line) and the joint density of states for the ensemble (dashed line). The dotted line represents a background absorption.

ton effect plays an important role in the optical transition near the fundamental absorption edge in semiconducting nanotubes.

4. Transport properties

4.1 Effective Hamiltonian

In the presence of impurity potential, the equation of motion (2) is replaced by

$$[\varepsilon - u_A(\mathbf{R}_A)]\psi_A(\mathbf{R}_A) = -\gamma_0 \sum_l \psi_B(\mathbf{R}_A - \vec{\tau}_l),$$
$$[\varepsilon - u_B(\mathbf{R}_B)]\psi_B(\mathbf{R}_B) = -\gamma_0 \sum_l \psi_A(\mathbf{R}_B + \vec{\tau}_l), \tag{82}$$

where $u_A(\mathbf{R}_A)$ and $u_B(\mathbf{R}_B)$ represent local site energy. When being multiplied by $g(\mathbf{r}-\mathbf{R}_A)\mathbf{a}(\mathbf{R}_A)$ and summed over $\mathbf{R}_A$, the term containing this impurity potential $u_A(\mathbf{R}_A)$ becomes

$$\sum_{\mathbf{R}_A} g(\mathbf{r}-\mathbf{R}_A)\ u_A(\mathbf{R}_A)\mathbf{a}(\mathbf{R}_A)\mathbf{a}(\mathbf{R}_A)^+\mathbf{F}_A(\mathbf{r}) = \begin{pmatrix} u_A(\mathbf{r}) & \mathrm{e}^{\mathrm{i}\eta}u'_A(\mathbf{r}) \\ \mathrm{e}^{-\mathrm{i}\eta}u'_A(\mathbf{r})^* & u_A(\mathbf{r}) \end{pmatrix} \mathbf{F}_A(\mathbf{r}), \tag{83}$$

with

$$u_A(\mathbf{r}) = \sum_{\mathbf{R}_A} g(\mathbf{r}-\mathbf{R}_A)u_A(\mathbf{R}_A),$$
$$u'_A(\mathbf{r}) = \sum_{\mathbf{R}_A} g(\mathbf{r}-\mathbf{R}_A)\mathrm{e}^{\mathrm{i}(\mathbf{K}'-\mathbf{K})\cdot\mathbf{R}_A}u_A(\mathbf{R}_A). \tag{84}$$

Similarly, the term containing this impurity potential $u_B(\mathbf{R}_B)$ becomes

$$\sum_{\mathbf{R}_B} g(\mathbf{r}-\mathbf{R}_B)\, u_B(\mathbf{R}_B)\mathbf{b}(\mathbf{R}_B)\mathbf{b}(\mathbf{R}_B)^{+}\mathbf{F}_B(\mathbf{r}) = \begin{pmatrix} u_B(\mathbf{r}) & -\omega e^{-i\eta}u'_B(\mathbf{r}) \\ -\omega^{-1}e^{i\eta}u'_B(\mathbf{r})^* & u_B(\mathbf{r}) \end{pmatrix} \mathbf{F}_B(\mathbf{r}), \tag{85}$$

with

$$u_B(\mathbf{r}) = \sum_{\mathbf{R}_B} g(\mathbf{r}-\mathbf{R}_B)u_B(\mathbf{R}_B),$$
$$u'_B(\mathbf{r}) = \sum_{\mathbf{R}_B} g(\mathbf{r}-\mathbf{R}_B)e^{i(\mathbf{K}'-\mathbf{K})\cdot\mathbf{R}_B}u_B(\mathbf{R}_B). \tag{86}$$

Therefore, the 4×4 effective potential of an impurity is written as [60]

$$V = \begin{pmatrix} u_A(\mathbf{r}) & 0 & e^{i\eta}u'_A(\mathbf{r}) & 0 \\ 0 & u_B(\mathbf{r}) & 0 & -\omega^{-1}e^{-i\eta}u'_B(\mathbf{r}) \\ e^{-i\eta}u'_A(\mathbf{r})^* & 0 & u_A(\mathbf{r}) & 0 \\ 0 & -\omega e^{i\eta}u'_B(\mathbf{r})^* & 0 & u_B(\mathbf{r}) \end{pmatrix}. \tag{87}$$

When the potential range is much shorter than the circumference L and the potential at each site is sufficiently small, we have

$$u_A(\mathbf{r}) = u_A\delta(\mathbf{r}-\mathbf{r}_A), \qquad u'_A(\mathbf{r}) = u'_A\delta(\mathbf{r}-\mathbf{r}_A),$$
$$u_B(\mathbf{r}) = u_B\delta(\mathbf{r}-\mathbf{r}_B), \qquad u'_B(\mathbf{r}) = u'_B\delta(\mathbf{r}-\mathbf{r}_B), \tag{88}$$

with

$$u_A = \frac{\sqrt{3}a^2}{2}\sum_{\mathbf{R}_A} u_A(\mathbf{R}_A), \quad u'_A = \frac{\sqrt{3}a^2}{2}\sum_{\mathbf{R}_A} e^{i(\mathbf{K}'-\mathbf{K})\cdot\mathbf{R}_A}u_A(\mathbf{R}_A),$$
$$u_B = \frac{\sqrt{3}a^2}{2}\sum_{\mathbf{R}_B} u_B(\mathbf{R}_B), \quad u'_B = \frac{\sqrt{3}a^2}{2}\sum_{\mathbf{R}_B} e^{i(\mathbf{K}'-\mathbf{K})\cdot\mathbf{R}_B}u_B(\mathbf{R}_B), \tag{89}$$

where $\mathbf{r}_A$ and $\mathbf{r}_B$ are the center-of-mass position of the effective impurity potential and $\sqrt{3}a^2/2$ is the area of a unit cell. The integrated intensities u_A, etc. given by Eq. (89) have been obtained by the $\mathbf{r}$ integral of $u_A(\mathbf{r})$, etc. given by Eqs. (84)–(86). This short-range potential becomes invalid when the site potential is as strong as the effective band width of 2D graphite as will be shown later.

4.2 Absence of backward scattering

In the vicinity of $\varepsilon = 0$, we have two right-going channels $K+$ and $K'+$, and two left-going channels $K-$ and $K'-$. The matrix elements are calculated as [60]

$$V_{K\pm K+} = V_{K'\pm K'+} = \frac{1}{2}(\pm u_A + u_B),$$
$$V_{K\pm K'+} = V^*_{K'\pm K+} = \frac{1}{2}(\mp e^{i\eta} u'_A - \omega^{-1} e^{-i\eta} u'_B). \tag{90}$$

When the impurity potential has a range larger than the lattice constant, we have $u_A = u_B$ and both u'_A and u'_B become much smaller and can be neglected because of the phase factor $e^{i(\mathbf{K}'-\mathbf{K})\cdot\mathbf{R}_A}$ and $e^{i(\mathbf{K}'-\mathbf{K})\cdot\mathbf{R}_B}$. This means that intervalley scattering between K and K' points can be neglected for such impurities as usually assumed in the conventional $\boldsymbol{k}\cdot\boldsymbol{p}$ approximation. Further, the above shows that the backward scattering probability within each valley vanishes in the lowest Born approximation.

Figure 27 gives an example of calculated effective potential u_A, u_B, and u'_B as a function of d/a for a Gaussian potential $(u/\pi d^2)\exp(-r^2/d^2)$ located at a B site and having the integrated intensity u. Because of the symmetry corresponding to a 120° rotation around a lattice point, we have $u'_A = 0$ independent of d/a. When the range is sufficiently small, u_B and u'_B stay close to $2u$ because the potential is localized only at the impurity B site. With the increase of d the potential becomes nonzero at neighboring A sites and u_A starts to increase and at the same time both u_B and u'_B decrease. The diagonal elements u_A and u_B rapidly approach u and the off-diagonal element u'_B vanishes.

Figure 28 shows calculated averaged scattering amplitude, given by $AL\sqrt{\langle|V_{K\pm K+}|^2\rangle}$ and $AL\sqrt{\langle|V_{K'\pm K+}|^2\rangle}$ where $\langle|V_{K\pm K+}|^2\rangle$ and $\langle|V_{K'\pm K+}|^2\rangle$ are the squared matrix elements averaged over impurity position, as a function of d in the absence of a magnetic field. The backward scattering probability decreases rapidly with d and becomes i exponentially small for $d/a \gg 1$. The same is true of the intervalley scattering although the dependence is slightly weaker because of the slower decrease of u'_B shown in Fig. 27.

This absence of the backward scattering for long-range scatterers disappears in the presence of magnetic fields. In the presence of a magnetic field perpendicular to the axis, the matrix elements for an impurity located at $\mathbf{r} = \mathbf{r}_0$ are calculated as

$$V_{K\pm K+} = \frac{1}{2A}[\pm u_A F_-(x_0)^2 + u_B F_+(x_0)^2],$$
$$V_{K'\pm K'+} = \frac{1}{2A}[\pm u_A F_+(x_0)^2 + u_B F_-(x_0)^2],$$
$$V_{K'\pm K+} = \frac{1}{2A}[\mp e^{-i\eta} u'^*_A - \omega e^{i\eta} u'^*_B] F_+(x_0) F_-(x_0), \tag{91}$$
$$V_{K\pm K'+} = \frac{1}{2A}[\mp e^{i\eta} u'_A - \omega^{-1} e^{-i\eta} u'_B] F_+(x_0) F_-(x_0),$$

where the wave functions in a magnetic field are defined in Eq. (51). In high magnetic fields, the intervalley scattering is reduced considerably because of the reduction in the overlap of the wavefunction, but the intravalley backward scattering remains nonzero because $|F_+(x_0)| \neq |F_-(x_0)|$.

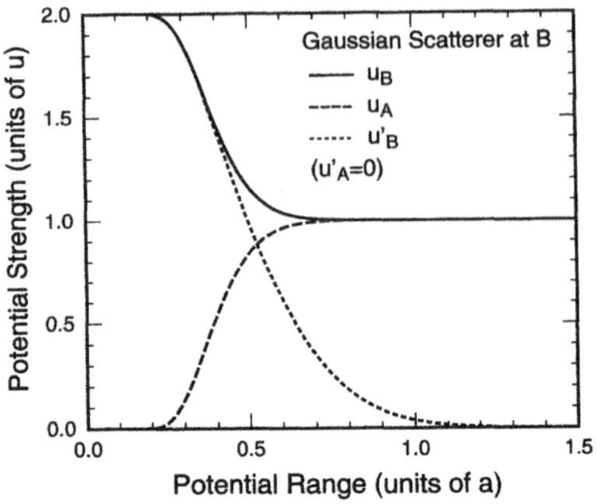

Fig. 27: Calculated effective strength of the potential for a model Gaussian impurity at a B site. After Ref. [60].

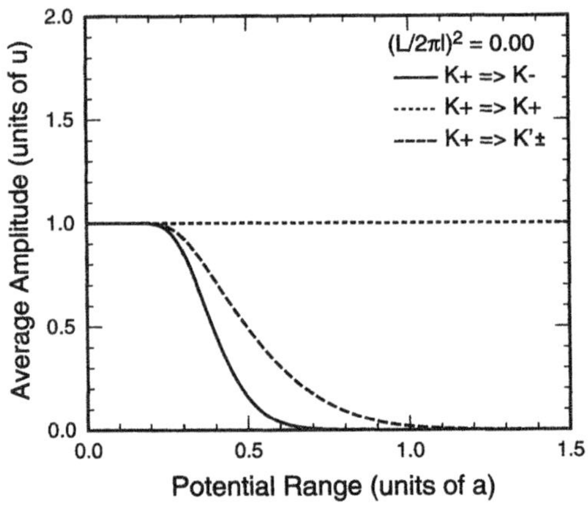

Fig. 28: Calculated effective scattering matrix elements versus the potential range at $\varepsilon=0$ in the absence of a magnetic field. After Ref. [60].

It is straightforward to calculate a scattering matrix for an impurity given by Eq. (88) and a conductance of a finite-length nanotube containing many impurities, combining S matrices [50,60]. Figure 29 shows some examples of calculated conductance at $\varepsilon=0$ in the case that the impurity potential has a range larger than the lattice constant, i.e., $u_A=u_B=u$. The conductance in the absence of a magnetic field is always quantized into $2e^2/\pi\hbar$ because of the complete absence of backward scattering. With the increase of the magnetic field the conductance is reduced drastically and the amount of the reduction becomes larger with the increase of the length.

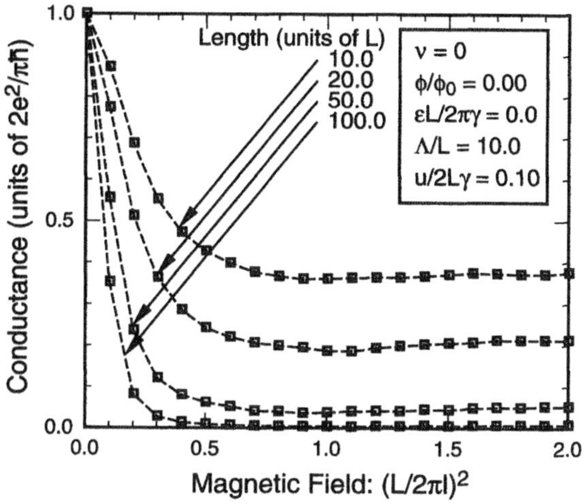

Fig. 29: Calculated conductance of finite-length nanotubes at $\varepsilon=0$ as a function of the effective strength of a magnetic field $(L/2\pi l)^2$ in the case that the effective mean free path Λ is much larger than the circumference L. The conductance is always given by the value in the absence of impurities at $H=0$. After Ref. [60].

4.3 Berry's phase

It has been proved that the Born series for back-scattering vanish identically [60]. This can be ascribed to a spinor-type property of the wave function under a rotation in the wave vector space [61]. In fact, an electron in the nanotube can be regarded as a neutrino, which has a helicity, i.e., its spin is always quantized into the direction of its wave vector, and therefore each scattering corresponds to a spin rotation. When the potential range is sufficiently large, i.e., $u_A(\mathbf{r})=u_B(\mathbf{r})$ and $u'_A(\mathbf{r})=u'_B(\mathbf{r})=0$, a matrix element for scattering is separable into a product of that of the impurity potential and a spin rotation.

A back-scattering corresponds to a spin rotation by $+(2n+1)\pi$ with n being an appropriate integer and its time reversal process corresponds to a spin rotation by $-(2n+1)\pi$. The spinor wave function after a rotation by $-(2n+1)\pi$ has a signature opposite to that after a rotation by $+(2n+1)\pi$ because of a well-known property of a spin-rotation operator. On the other hand, the matrix element of the time reversal process has an identical spatial part. As a result, the sum of the matrix elements of a back-scattering process and its time reversal process vanishes due to the complete cancellation, leading to the absence of backward scattering.

In the effective-mass approximation, the eigen wave functions and energies of this Hamiltonian are written as

$$\mathbf{F}_{s\mathbf{k}}(\mathbf{r})=\frac{1}{\sqrt{LA}}\exp(i\mathbf{k}\cdot\mathbf{r})\,\mathbf{F}_{s\mathbf{k}},\qquad \varepsilon_s(\mathbf{k})=s\gamma|\mathbf{k}|. \tag{92}$$

In metallic CNs the wave vector in the k_x direction is quantized into $k_x=\kappa(n)$ with

$\kappa(n)=2\pi n/L$ where n is an integer. In general, we can write eigenvector $\mathbf{F}_{s\mathbf{k}}$ as

$$\mathbf{F}_{s\mathbf{k}} = \exp[\mathrm{i}\phi_s(\mathbf{k})]\, R^{-1}[\theta(\mathbf{k})]\, |s), \tag{93}$$

where $\phi_s(\mathbf{k})$ is an arbitrary phase factor, $\theta(\mathbf{k})$ is the angle between wave vector $\mathbf{k}$ and the k_y axis, i.e., $k_x+\mathrm{i}k_y=\mathrm{i}|\mathbf{k}|\mathrm{e}^{\mathrm{i}\theta(\mathbf{k})}$ and i $k_x-\mathrm{i}k_y=-\mathrm{i}|\mathbf{k}|\mathrm{e}^{-\mathrm{i}\theta(\mathbf{k})}$, $R(\theta)$ is a spin-rotation operator, given by

$$R(\theta) = \exp\Big(\mathrm{i}\frac{\theta}{2}\sigma_z\Big) = \begin{pmatrix} \exp(+\mathrm{i}\theta/2) & 0 \\ 0 & \exp(-\mathrm{i}\theta/2) \end{pmatrix}, \tag{94}$$

with σ_z being a Pauli matrix, and $|s)$ is the eigenvector for the state with $\mathbf{k}$ in the positive k_y direction, given by

$$|s) = \frac{1}{\sqrt{2}}\begin{pmatrix} -\mathrm{i}s \\ 1 \end{pmatrix}. \tag{95}$$

Obviously, we have

$$R(\theta_1)R(\theta_2) = R(\theta_1+\theta_2), \quad R(-\theta) = R^{-1}(\theta). \tag{96}$$

Further, because $R(\theta)$ describes the rotation of a spin, it has the property

$$R(\theta\pm 2\pi) = -R(\theta), \tag{97}$$

which gives $R(-\pi) = -R(+\pi)$. In order to define states and corresponding wave functions uniquely, we shall assume in the following that

$$-\pi < \theta(\mathbf{k}) \leq +\pi. \tag{98}$$

By choosing the phase $\phi_s(\mathbf{k})$ in an appropriate manner, the wave function can be chosen as either continuous or discontinuous across the point corresponding to $\theta=+\pi$ and $-\pi$ in the $\mathbf{k}$ space. The results are certainly independent of such choices.

In the following we shall consider a back scattering process $\mathbf{k}\to-\mathbf{k}$ due to an arbitrary external potential having a range larger than or comparable to the lattice constant in a 2D graphite sheet, where $\mathbf{k}=(0,k)$. Only difference arising in nanotubes is discretization of the wave vector in the k_x direction as mentioned above. We shall confine ourselves to states in the vicinity of the K point, but the extension to states near a K' point is straightforward.

Introduce a T matrix defined by

$$T = V + V\frac{1}{\varepsilon-\mathcal{H}_0}V + V\frac{1}{\varepsilon-\mathcal{H}_0}V\frac{1}{\varepsilon-\mathcal{H}_0}V + \cdots, \tag{99}$$

where V is the impurity potential given by a diagonal matrix, i.e.,

$$V = \begin{pmatrix} V(\mathbf{r}) & 0 \\ 0 & V(\mathbf{r}) \end{pmatrix}, \tag{100}$$

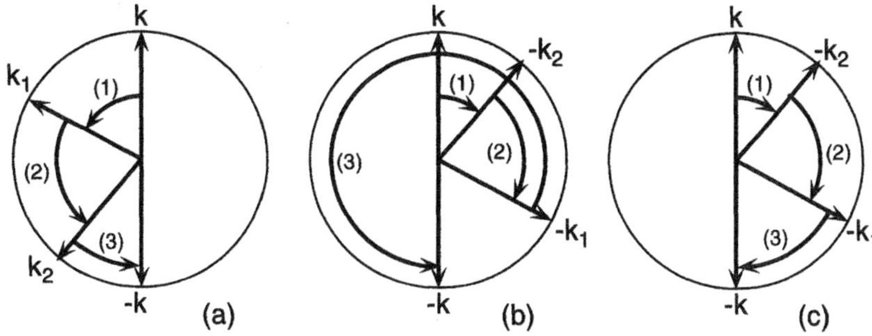

Fig. 30: Schematic illustration of the spin rotation corresponding to the scattering process $\mathbf{k}\to\mathbf{k}_1\to\mathbf{k}_2\to-\mathbf{k}$ (a) and its time-reversal process $\mathbf{k}\to-\mathbf{k}_2\to-\mathbf{k}_1\to-\mathbf{k}$. We have chosen $\theta(\mathbf{k})=0$ and $\theta(-\mathbf{k})=+\pi$. In process (c), the final state is chosen as that obtained by $\theta(-\mathbf{k})\to\theta(-\mathbf{k})-2\pi=-\pi$.

ε is the energy, and $\mathcal{H}_0=\gamma\vec{\sigma}\cdot\mathbf{k}$ is the Hamiltonian in the absence of the potential. The $(p+1)$th order term of the T matrix is written as

$$(s,-\mathbf{k}|T^{(p+1)}|s,+\mathbf{k})=\frac{1}{LA}\sum_{s_1\mathbf{k}_1}\frac{1}{LA}\sum_{s_p\mathbf{k}_p}\frac{V(-\mathbf{k}-\mathbf{k}_p)\cdots V(\mathbf{k}_1-\mathbf{k})}{[\varepsilon-\varepsilon_{s_p}(\mathbf{k}_p)]\cdots[\varepsilon-\varepsilon_{s_1}(\mathbf{k}_1)]}$$
$$\times \mathrm{e}^{-i\phi_s(-\mathbf{k})}(s|R[\theta(-\mathbf{k})]R^{-1}[\theta(\mathbf{k}_p)]|s_p)\cdots(s_1|R[\theta(\mathbf{k}_1)]R^{-1}[\theta(\mathbf{k})]|s)\mathrm{e}^{i\phi_s(\mathbf{k})}, \tag{101}$$

where $V(\mathbf{k}_i-\mathbf{k}_j)$ is a Fourier transform of the impurity potential and phase factors $\exp[i\phi_{s_j}(\mathbf{k}_j)]$ have been cancelled out for all the intermediate states $j=1,\ldots,p$. We have $\theta(\mathbf{k})=0$ and $\theta(-\mathbf{k})=+\pi$.

Define

$$S^{(p+1)}=(s|R[\theta(-\mathbf{k})]R^{-1}[\theta(\mathbf{k}_p)]|s_p)\cdots(s_1|R[\theta(\mathbf{k}_1)]R^{-1}[\theta(\mathbf{k})]|s). \tag{102}$$

This quantity describes matrix elements of a rotation in the spin space and can be illustrated by a diagram as shown in Fig. 30 (a). For each term in Eq. (102), there is a term obtained through the replacement

$$(s_1,\mathbf{k}_1)\to(s_p,-\mathbf{k}_p),\quad (s_2,\mathbf{k}_2)\to(s_{p-1},-\mathbf{k}_{p-1}),\quad \text{etc.}, \tag{103}$$

corresponding to the electron motion of a time-reversal path (see Fig. 31). Both matrix elements of the impurity potential and energy denominators remain unchanged by this replacement in Eq. (102) except that S should be replaced by S' given by

$$S^{(p+1)\prime}=(s|R[\theta(-\mathbf{k})]R^{-1}[\theta(-\mathbf{k}_1)]|s_1)\cdots(s_p|R[\theta(-\mathbf{k}_p)]R^{-1}[\theta(\mathbf{k})]|s). \tag{104}$$

This process is given by a diagram as shown in Fig. 30 (b).

Instead of the above we consider the quantity $S^{(p+1)\prime\prime}$ given by

$$S^{(p+1)\prime\prime}=(s|R[\theta(-\mathbf{k})-2\pi]R^{-1}[\theta(-\mathbf{k}_1)]|s_1)\cdots(s_p|R[\theta(-\mathbf{k}_p)]R^{-1}[\theta(\mathbf{k})]|s), \tag{105}$$

which has been obtained from Eq. (104) through the replacement $\theta(-\mathbf{k})\to\theta(-\mathbf{k})-2\pi$ and can be represented by a diagram shown in Fig. 30 (c). Using Eq. (97), we have $S^{(p+1)''}=-S^{(p+1)'}$. According to the present definition of the angle given by Eq. (98), we have

$$\theta(-\mathbf{k}_j)=\begin{cases}\theta(\mathbf{k}_j)-\pi & \text{if } \theta(\mathbf{k}_j)>0,\\ \theta(\mathbf{k}_j)+\pi & \text{if } \theta(\mathbf{k}_j)\le 0.\end{cases} \tag{106}$$

However, we can always put $\theta(-\mathbf{k}_j)=\theta(\mathbf{k}_j)-\pi$ for $j=1,\ldots,p$ in the expression of $S^{(p+1)''}$ because of Eq. (97) since $\theta(-\mathbf{k}_j)$ always appears in a pair. By making the replacement $\theta(\mathbf{k})\to\theta(-\mathbf{k})-\pi$ and $\theta(-\mathbf{k})\to\theta(\mathbf{k})+\pi$, we can immediately obtain

$$S^{(p+1)''}=-S^{(p+1)'}=S^{(p+1)*}. \tag{107}$$

To see the above explicitly, we have

$$\begin{aligned}&(s|R[\theta(-\mathbf{k})-2\pi]R^{-1}[\theta(-\mathbf{k}_1)]|s_1)=(s_1|R[\theta(\mathbf{k}_1)]R^{-1}[\theta(\mathbf{k})]|s)^*,\\ &\cdots\\ &(s_p|R[\theta(-\mathbf{k}_p)]R^{-1}[\theta(\mathbf{k})]|s)=(s|R[\theta(-\mathbf{k})]R^{-1}[\theta(\mathbf{k}_p)]|s_p)^*.\end{aligned} \tag{108}$$

Further, we have

$$(s|R(\theta)R^{-1}(\theta')|s')=\begin{cases}\cos[(\theta-\theta')/2] & (s=s'),\\ -\mathrm{i}\sin[(\theta-\theta')/2] & (s\neq s'),\end{cases} \tag{109}$$

which is either real or pure imaginary. Because the number of matrix elements for different bands is always even in the expression of $S^{(p+1)}$, we see immediately that $S^{(p+1)}$ is given by a real number. Therefore, we have

$$S^{(p+1)''}=-S^{(p+1)'}=S^{(p+1)}. \tag{110}$$

Although states of 2D graphite in the vicinity of the Fermi level are formally described by Weyl's equation, two components of the wave function do not describe states of the actual electron spin but correspond only to the amplitude at two carbon sites in a unit cell. Therefore, the signature change due to the rotation of the wave vector around the origin should rather be regarded as a Berry's phase [62,63].

Consider the case that the Hamiltonian contains a parameter s. Let the parameter s change from $s(0)$ to $s(T)$ as a function of time t from $t=0$ to $t=T$ and assume that $\mathcal{H}[s(T)]=\mathcal{H}[s(0)]$. Note that $s(T)$ is not necessarily same as $s(0)$. When there is no degeneracy in states, the state at $t=T$ is completely same as that at $t=0$ when the process is sufficiently slow and adiabatic. Therefore, the wave function at $t=T$ is equal to that at $t=0$ except for the presence of a phase factor $\exp(-\mathrm{i}\varphi)$. This extra phase called Berry's phase is given by

$$\varphi=-\mathrm{i}\int_0^T dt\,\Big\langle\psi[s(t)]\Big|\frac{d\psi[s(t)]}{dt}\Big\rangle. \tag{111}$$

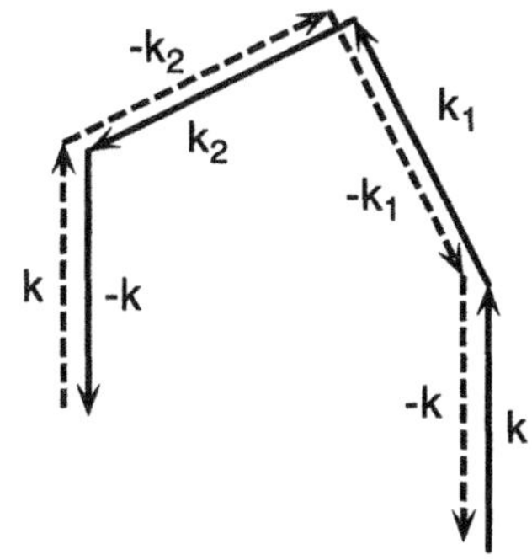

Fig. 31: Schematic illustration of a back scattering process (solid arrows) and corresponding time-reversal process (dashed arrows).

Consider a wave function given by

$$\psi_s(\mathbf{k}) = \frac{1}{\sqrt{2}} \begin{pmatrix} -\mathrm{i}s \exp[-\mathrm{i}\theta(\mathbf{k})] \\ 1 \end{pmatrix}. \tag{112}$$

This is the "spin" part of an eigenfunction of the $\boldsymbol{k}{\cdot}\boldsymbol{p}$ equation, obtained by choosing $\phi_s(\mathbf{k}) = -\theta(\mathbf{k})/2$ in such a way that the wave function becomes continuous as a function of $\theta(\mathbf{k})$ [48]. When the wave vector $\mathbf{k}$ is rotated in the anticlockwise direction adiabatically as a function of time t around the origin for a time interval $0<t<T$, the wavefunction is changed into $\psi_s(\mathbf{k})\exp(-\mathrm{i}\gamma)$, where φ is Berry's phase given by

$$\varphi = -\mathrm{i} \int_0^T \mathbf{d}t \left\langle \psi_s[\mathbf{k}(t)] \middle| \frac{\mathbf{d}}{\mathbf{d}t} \psi_s[\mathbf{k}(t)] \right\rangle = -\pi. \tag{113}$$

This shows that the rotation in the $\mathbf{k}$ space by 2π leads to the change in the phase by $+\pi$, i.e., a signature change. Note that $R^{-1}[\theta(\mathbf{k})]|s)$ is obtained from Eq. (112) by continuously varying the direction of $\mathbf{k}$ including Berry's phase.

A similar phase change of the back scattering process is the origin of the so-called anti-localization effect in systems with strong spin-orbit scattering [64]. In the absence of spin-orbit interactions, the scattering amplitude of each process and that of corresponding time-reversal process are equal and therefore their constructive interference leads to an enhancement in the amplitude of back scattering processes. This gives rise to the so-called quantum correction to the conductivity in the weak localization theory [65]. In the presence of a strong spin-orbit scattering, however, scattering from an impurity causes a spin rotation and resulting phase change of the wave function leads to a destructive interference. As a result the quantum correction has a sign opposite to that in the absence of a spin-orbit scattering [66]. This anti-localization effect was observed experimentally [67,68].

4.4 Experiments

The magnetoresistance of bundles of multi-wall nanotubes was measured and a negative magnetoresistance was observed in sufficiently low magnetic fields and at low temperatures [24]. With the increase of the magnetic field the resistance starts to exhibit a prominent positive magnetoresistance. This positive magnetoresistance is in qualitative agreement with the theoretical prediction of the reappearance of backscattering in magnetic fields.

Measurements of the resistance of a single multi-wall nanotube were reported [27–31] and irregular oscillation analogous to universal conductance fluctuations was observed [27]. A resistance oscillation more regular like that of an Aharonov-Bohm type was observed in a magnetic field parallel to the CN axis [69]. Recently observed resistance oscillation in a parallel field was shown to be consistent with the Aharonov-Bohm oscillation of the band gap due to a magnetic flux passing through the CN cross section [70].

Because of the presence of large contact resistance between a nanotube and metallic electrode, the measurement of the conductance of a nanotube itself is quite difficult and has not been so successful. However, from measurements of single electron tunneling due to a Coulomb blockade and charging effect, important information can be obtained on the effective mean free path and the amount of backward scattering in nanotubes [32–36]. Discrete quantized energy levels were measured for a nanotube with ~ 3 μm length, for example, showing that the electron wave is coherent and extended over the whole length.

It was shown further that the Coulomb oscillation in semiconducting nanotubes is quite irregular and can be explained only if nanotubes are divided into many separate spatial regions in contrast to that in metallic nanotubes [71]. This behavior is consistent with the presence of considerable amount of backward scattering leading to a strong localization of the electrons wave function in semiconducting tubes. In metallic nanotubes, the wave function is extended throughout the whole region of a nanotube because of the absence of backward scattering. The voltage distribution in a metallic nanotube with current was measured by electrostatic force microscopy, which showed clearly that there is essentially no voltage drop in the nanotube [72].

A conductance quantization was observed in multi-wall nanotubes [73]. This quantization is likely to be related to the absence of backward scattering shown here, but much more works are necessary including effects of magnetic fields and problems related to contacts with metallic electrode before complete understanding of the experimental result. At room temperature, where the experiment was performed, phonon scattering is likely to play an important role [25,74–76].

4.5 Lattice vacancies – Strong and short-range scatterers

A point defect leads to peculiar electronic states. STM images of a graphene sheet, simulated in the presence of a local point defect, showed that a $\sqrt{3}\times\sqrt{3}$ interference pattern is formed in the wave function near the impurity [77]. This is intuitively

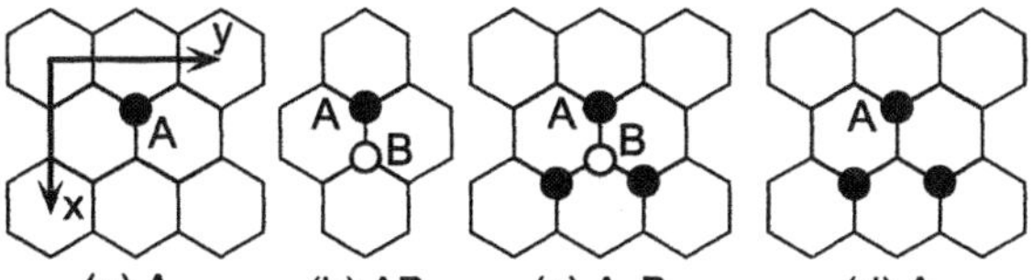

Fig. 32: Schematic illustration of vacancies in an armchair nanotube. The closed and open circles denote A and B lattice points, respectively. (a) A, (b) AB, (c) A_3B, and (d) A_3.

understood by a mixing of wave functions at K and K' points.

Some experiments suggest the existence of defective nanotubes of carpet-roll or papiermâché forms [78,79]. These systems have many disconnections of the π electron network governing transport of CNs and therefore are expected to exhibit properties different from those in perfect CNs. In a graphite sheet with a finite width, for examples, localized edge states are formed at $\varepsilon = 0$, when the boundary is in a certain specific direction [80–82].

Effects of scattering by a vacancy in armchair nanotubes have been studied within a tight-binding model [83–85]. Figure 32 shows three typical vacancies: (a) vacancy I, (b) vacancy IV, and (c) vacancy II. In the vacancy I a single carbon site (site A) is removed, in the vacancy IV one A site and three B sites are removed, and in the vacancy II a pair of A and B sites are removed. The vacancy (d) is equivalent to (c). It has been shown that the conductance at $\varepsilon=0$ in the absence of a magnetic field is quantized into zero, one, or two times of the conductance quantum $e^2/\pi\hbar$ for vacancy IV, I, and II, respectively [84,85].

Figure 33 shows the calculated conductance as a function of the Fermi energy between $\varepsilon = 0$ and $\varepsilon = \varepsilon(1)$, where $\varepsilon(1)$ is the energy of the bottom of the first excited band [$\varepsilon(1) = 2\pi\gamma/L$ in the $\boldsymbol{k}\cdot\boldsymbol{p}$ scheme]. For the vacancy I, the conductance at $\varepsilon = 0$ is the half of that in a defect free system. Both intra- and inter-valley components have an equal amplitude for both transmission and reflection processes, i.e., $|t_{\mu\nu}|^2 = |r_{\mu\nu}|^2 = 1/4$. The conductance increases as a function of ϵ for $0 < \epsilon < \epsilon(1)$ and reaches $2e^2/\pi\hbar$ at $\epsilon = \epsilon(1)$, where a perfect transmission occurs, i.e., $t_{KK} = t_{K'K'} = 1$. Except at $\epsilon = 0$ and $\epsilon = \epsilon(1)$, the conductance increases with the increase of the circumference.

In CNs with the vacancy IV, the conductance at $\varepsilon = 0$ vanishes as shown in Fig. 33 (b) and a perfect reflection occurs within the same valley, i.e., $|r_{KK}| = |r_{K'K'}| = 1$. The conductance increases with increasing ϵ and reaches $2e^2/\pi\hbar$ at $\varepsilon = \varepsilon(1)$. Except at $\varepsilon = 0$ and $\varepsilon = \varepsilon(1)$, the conductance increases with increasing L.

In CNs with the vacancy II, the conductance around $\epsilon = 0$ is slightly smaller than $2e^2/\pi\hbar$ and gradually increases and approaches $2e^2/\pi\hbar$ with the increase of the radius as shown in Fig. 33 (c). The deviation from the perfect conductance $2e^2/\pi\hbar$ decreases with the increase of L in proportion to $(a/L)^2$. The conductance exhibits a dip at the energy slightly below $\epsilon = \epsilon(1)$, but reaches $2e^2/\pi\hbar$ at $\varepsilon = \varepsilon(1)$. The back scattering within each valley r_{KK} and $r_{K'K'}$ and the transmission between different valleys $t_{KK'}$ and $t_{K'K}$ are absent because of a mirror symmetry about a plane containing the axis.

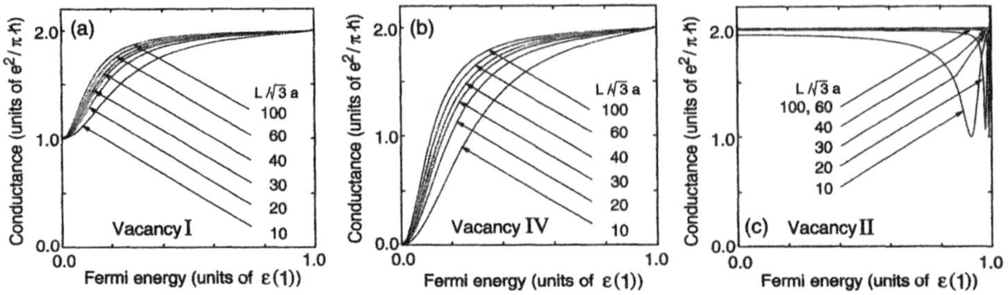

Fig. 33: Calculated conductance in units of $e^2/\pi\hbar$ as a function of the Fermi energy for CNs with vacancy I (a), IV (b), and II (c), where the energy is scaled by $\epsilon(1)$, which corresponds to the bottom of the second conduction bands. After Ref. [84].

Numerical calculations were performed for about 1.5×10^5 vacancies and demonstrated that such quantization is quite general [85]. Let N_A and N_B be the number of removed atoms at A and B sublattice points, respectively, and $\Delta N_{AB}=N_A-N_B$. Then, the numerical results show that for vacancies much smaller than the circumference, the conductance vanishes for $|\Delta N_{AB}|\geq 2$ and quantized into one and two times of $e^2/\pi\hbar$ for $|\Delta N_{AB}|=1$ and 0, respectively. Figure 34 shows calculated histogram of the conductance.

Effects of impurities with a strong and short-range potential can be studied also in a $\boldsymbol{k}\cdot\boldsymbol{p}$ scheme [86,87]. For an impurity localized at a carbon A site $\mathbf{r}_j$ and having the integrated intensity u, we have [60]

$$V(\mathbf{r}) = V_j\delta(\mathbf{r}-\mathbf{r}_j), \tag{114}$$

with

$$V_j = u\begin{pmatrix} 1 & 0 & e^{i\phi_j^A} & 0 \\ 0 & 0 & 0 & 0 \\ e^{-i\phi_j^A} & 0 & 1 & 0 \\ 0 & 0 & 0 & 0 \end{pmatrix}, \tag{115}$$

and $\phi_j^A=(\mathbf{K}'-\mathbf{K})\cdot\mathbf{r}_j+\eta$. For an impurity at a carbon B site we have

$$V_j = u\begin{pmatrix} 0 & 0 & 0 & 0 \\ 0 & 1 & 0 & e^{i\phi_j^B} \\ 0 & 0 & 0 & 0 \\ 0 & e^{-i\phi_j^B} & 0 & 1 \end{pmatrix}, \tag{116}$$

with $\phi_j^B=(\mathbf{K}'-\mathbf{K})\cdot\mathbf{r}_j-\eta+(\pi/3)$.

The scattering matrix can be written formally as

$$S = S^{(0)} + S^{(1)}, \tag{117}$$

with

$$S^{(0)}_{\alpha\beta} = \delta_{\alpha\beta}, \tag{118}$$

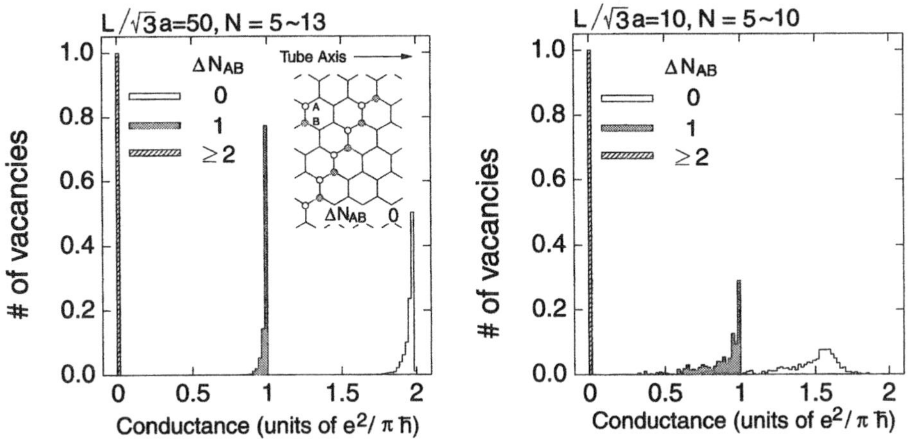

Fig. 34: Calculated distribution of the conductance of CNs with a general vacancy for (a) $L/\sqrt{3}a=50$ and $5\leq N\leq 13$ and for (b) $L/\sqrt{3}a=10$ and $5\leq N\leq 10$. Open, shaded, and hatched portions denote results for $\Delta N_{AB}=0$, 1, and ≥ 2, respectively. The histogram has a width 5.0×10^{-2} in units of $e^2/\pi\hbar$ and is normalized by the total number of vacancies having specified ΔN_{AB}. The inset shows an example of a vacancy which causes a large deviation from the quantized conductance. After Ref. [85].

and

$$S^{(1)}_{\alpha\beta} = -\mathrm{i}\frac{A}{\hbar\sqrt{|v_\alpha v_\beta|}}T_{\alpha\beta}, \tag{119}$$

where v_α and v_β are the velocity of channels α and β. The T matrix satisfies

$$T=V+V\frac{1}{\varepsilon-\mathcal{H}_0+\mathrm{i}0}V+V\frac{1}{\varepsilon-\mathcal{H}_0+\mathrm{i}0}V\frac{1}{\varepsilon-\mathcal{H}_0+\mathrm{i}0}V+\cdots, \tag{120}$$

which is solved as

$$(\alpha|T|\beta) = \sum_{ij}\mathbf{f}_\alpha^+ T_{ij}\mathbf{f}_\beta \exp(-\mathrm{i}\kappa_\alpha x_i - \mathrm{i}k_\alpha y_i)\exp(\mathrm{i}\kappa_\beta x_j + \mathrm{i}k_\beta y_j), \tag{121}$$

where κ_α and κ_β are the wave vectors in the circumference direction corresponding to channel α and β, k_α and k_β are the wave vectors in the axis direction, $\mathbf{f}_\alpha$ and $\mathbf{f}_\beta$ are corresponding eigenvectors defined by

$$\mathbf{F}_\alpha(\mathbf{r}) = \frac{1}{\sqrt{LA}}\mathbf{f}_\alpha \exp(\mathrm{i}\kappa_\alpha x + \mathrm{i}k_\alpha y), \quad \text{etc.}, \tag{122}$$

and

$$T_{ij} = \Big[\Big(1-\frac{1}{AL}VG(\varepsilon+\mathrm{i}0)\Big)^{-1}\frac{1}{AL}V\Big]_{ij}, \tag{123}$$

with $V_{jj'}=V_j\delta_{jj'}$. The Green's function $G_{ij}=G(\mathbf{r}_i-\mathbf{r}_j)$ is written as

$$G(\mathbf{r}) = \frac{-\mathrm{i}A}{2\gamma}\begin{pmatrix} g_0 & g_1 & 0 & 0 \\ \bar{g}_1 & g_0 & 0 & 0 \\ 0 & 0 & g_0 & \bar{g}_1 \\ 0 & 0 & g_1 & g_0 \end{pmatrix}, \tag{124}$$

where

$$\begin{aligned} g_0(x,y) &= \frac{\mathrm{i}\gamma}{\pi}\int dk \sum_n f_c[\kappa(n),k]\frac{\varepsilon\, \mathrm{e}^{\mathrm{i}\kappa(n)x+\mathrm{i}ky}}{(\varepsilon+\mathrm{i}0)^2-\gamma^2[\kappa(n)^2+k^2]}, \\ g_1(x,y) &= \frac{\mathrm{i}\gamma}{\pi}\int dk \sum_n f_c[\kappa(n),k]\frac{\gamma[\kappa(n)-\mathrm{i}k]\, \mathrm{e}^{\mathrm{i}\kappa(n)x+\mathrm{i}ky}}{(\varepsilon+\mathrm{i}0)^2-\gamma^2[\kappa(n)^2+k^2]}, \end{aligned} \tag{125}$$

and $\bar{g}_1(x,y)=g_1(x,-y)$. We have introduced a cutoff function $f_c[\kappa(n),k]$ defined by

$$f_c[\kappa(n),k] = \frac{k_c^2}{k^2+\kappa(n)^2+k_c^2}, \tag{126}$$

in order to exclude contributions from states except in the vicinity of the Fermi level. The cutoff wave vector k_c is determined by the condition that the cutoff wave length $2\pi/k_c$ should be comparable to the lattice constant a, i.e., $2\pi/k_c\approx a$.

At $\varepsilon=0$, in particular, we have

$$g_0(x,y) = 1, \qquad g_1(x,y) = \frac{\cos[\pi(x+\mathrm{i}y)/L]}{\sin[\pi(x+\mathrm{i}y)/L]}. \tag{127}$$

The off-diagonal Green's function g_1 is singular in the vicinity of $\mathbf{r}=0$. Therefore, for impurities localized within a distance of a few times of the lattice constant, the off-diagonal Green's function becomes extremely large. This singular behavior is the origin of the peculiar dependence of the conductance on the difference in the number of vacancies at A and B sublattices.

For a few impurities, the T matrix can be obtained analytically and becomes equivalent to that of lattice vacancies in the limit of strong scatterers, i.e., $|u|/2\gamma L\gg 1$. In particular, the tight-binding results shown in Fig. 33 are reproduced quite well. In the limit of $a/L\to 0$ and a strong scatterer the behavior of the conductance at $\varepsilon=0$ can be studied analytically for arbitrary values of N_A and N_B. The results explain the numerical tight-binding result that $G=0$ for $|\Delta N_{AB}|\geq 2$, $G=e^2/\pi\hbar$ for $|\Delta N_{AB}|=1$, and $G=2e^2/\pi\hbar$ for $\Delta N_{AB}=0$.

The origin of this interesting dependence on N_A and N_B is a reduction of the scattering potential by multiple scattering on a pair of A and B scatterers. In fact, multiple scattering between an A impurity at $\mathbf{r}_i$ and a B impurity at $\mathbf{r}_j$ reduces their effective potential by the factor $|g_1(\mathbf{r}_i-\mathbf{r}_j)|^{-2}\propto (a/L)^2$. By eliminating AB pairs successively, some A or B impurities remain. The conductance is determined essentially by the number of these unpaired impurities.

Such a direct pair-elimination procedure is not rigorous because there are many different ways in the elimination of AB pairs and multiple scattering between unpaired and eliminated impurities cannot be neglected completely because of large off-diagonal Green's functions. However, a correct mathematical procedure can be formulated in which a proper combination of A and B impurities leads to a vanishing scattering potential and the residual potential is determined by another combination of remaining A or B impurities [86].

Effects of a magnetic field were studied for three types of vacancies shown in Fig. 32 [84]. The results show a universal dependence on the field component in the direction of the vacancy position. These results are analytically derived in the effective-mass scheme [87]. There are various other theoretical calculations in tight-binding models of electronic states and transport of tubes containing lattice defects [88] or disorder [89–93].

5. Junctions and topological defects

5.1 Five- and seven-membered rings

A junction which connects CNs with different diameters through a region sandwiched by a pentagon-heptagon pair has been observed in the transmission electron microscope [2]. Figure 35 (a) shows such an example and in Figure 35 (b) bend junctions [94]. Nanotubes can be joined through a structure with finite curvature, which has topological defects at the interfaces adjacent to CNs. If we introduce disclinations such as five-membered ring (pentagon) and seven-membered ring (heptagon), a CN has a finite curvature; pentagon brings positive curvature while heptagon causes negative one. A pair of five- and seven-membered rings make it possible to connect two different types of CNs and thus construct various kinds of CN junctions [95–97].

Some theoretical calculations on CN junctions within a tight-binding model were reported for junctions between metallic and semiconducting nanotubes and those between semiconducting nanotubes [98,97]. In particular tight-binding calculations for junctions consisting of two metallic tubes with different chirality or diameter demonstrated that the conductance exhibits a universal power-law dependence on the ratio of the circumference of two nanotubes [99–101].

A bend-junction was observed experimentally [see Fig. 35 (b)] and the conductance across such a junction between a (6,6) armchair CN and a (10,0) zigzag CN was discussed [94]. Energetics of bend junctions are calculated [102]. Transport measurements of both metal-semiconductor and metal-metal junctions are reported [103]. The bend junction is a special case of the general junction, as shall be explained in the following section.

Junctions can contain many pairs of topological defects. Effects of three pairs present between metallic (6,3) and (9,0) nanotubes were studied [83], which shows that the conductance vanishes for junctions having a three-fold rotational symmetry, but remains nonzero for those without the symmetry. Three-terminal CN systems are also formed with a proper combination of five- and n-membered ($n \geq 8$) rings [104–107].

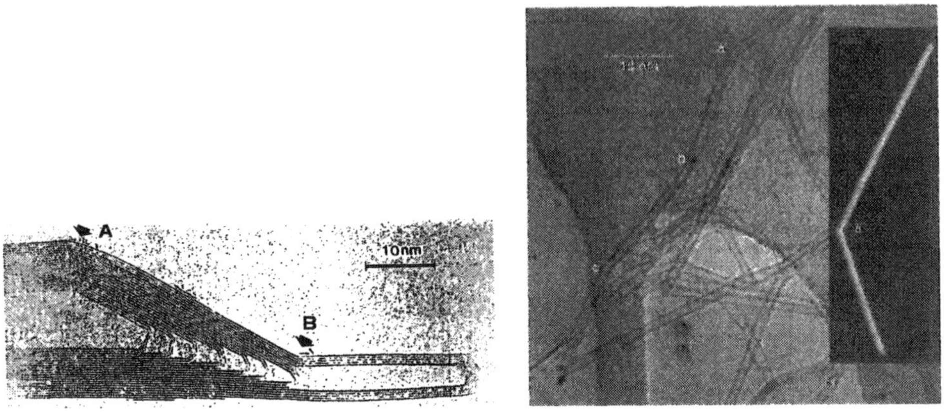

(a) (b)

Fig. 35: Transmission micrograph image of (a) a tip of a carbon nanotube [2] and (b) bend junctions [94].

A Stone-Wales defect or *azupyrene* is a kind of topological defect in graphite network [108]. They are composed of two pairs of five- and seven-membered rings, which can be introduced by rotating a C-C bond inside four adjacent six-membered rings. This type of defects can actually exist quasi-stable, in large fullerenes and CNs [109] and also in a graphite sheet [110]. The conductance of a nantoube contining a Stone-Wales defect was calculated quite recently [111,112].

Besides junctions and Stone-Wales defects, there are many interesting CN networks. For example, a lot of pairs of five- and seven-membered rings aligned periodically along the tube length can form another interesting networks such as toroidals [113–116] and helically-coiled CNs [117–120].

Recently a new type of carbon particles was produced by laser ablation method in bulk quantities. An individual particle is a spherical aggregate of many single-walled tubule-like structures with conical tips with an average cone angle of 20°. The conical tips of individual tubules are protruding out of the surface of the aggregate like horns, and they are called carbon nanohorns [121].

5.2 Boundary conditions

Let us consider a junction having a general structure. Figure 36 shows an example of such junctions [97]. The junction is characterized by two equilateral triangles with a common vertex point and sides parallel to the chiral vectors of two nanotubes. There is a five-member ring (whose position is denoted as $\mathbf{R}_5$) at the boundary of the thicker nanotube and the junction region and a seven-member ring ($\mathbf{R}_7$) at the boundary of the junction region and the thinner nanotube. For any site close to the upper boundary denoted as $(-)$, there exists a corresponding site near the lower

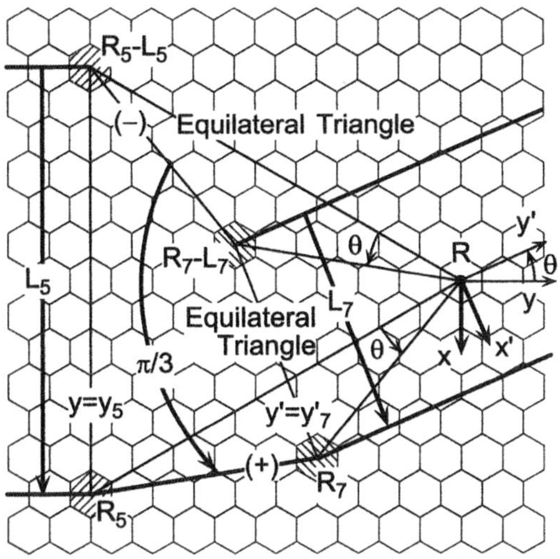

Fig. 36: The structure of a junction consisting of two nanotubes having an axis not parallel to each other (θ is their angle).

boundary denoted as (+) obtained by a rotation around $\mathbf{R}$ by $\pi/3$.

By a proper extrapolation of the wave functions outside of the junction region, we can generalize the boundary conditions as

$$\psi_A(\mathbf{R}'_A) = \psi_B(\mathbf{R}_B), \quad \psi_B(\mathbf{R}'_B) = \psi_A(\mathbf{R}_A), \tag{128}$$

with

$$\mathbf{R}'_A = R(\pi/3)(\mathbf{R}_B - \mathbf{R}) + \mathbf{R}, \quad \mathbf{R}'_B = R(\pi/3)(\mathbf{R}_A - \mathbf{R}) + \mathbf{R}, \tag{129}$$

for all lattice points $\mathbf{R}_A$ and $\mathbf{R}_B$.

In terms of the envelope functions, these conditions can be written explicitly as

$$\begin{aligned}\mathbf{a}(\mathbf{R}'_A)^+\mathbf{F}_A(\mathbf{R}'_A) &= \mathbf{b}(\mathbf{R}_B)^+\mathbf{F}_B(\mathbf{R}_B),\\ \mathbf{b}(\mathbf{R}'_B)^+\mathbf{F}_B(\mathbf{R}'_B) &= \mathbf{a}(\mathbf{R}_A)^+\mathbf{F}_A(\mathbf{R}_A).\end{aligned} \tag{130}$$

In the following we shall choose the origin at $\mathbf{R}$, i.e., $\mathbf{R}=0$. Because $R^{-1}(\pi/3)\mathbf{K}=\mathbf{K}'$, we have

$$\begin{aligned}\exp(\mathrm{i}\mathbf{K}\cdot\mathbf{R}'_A) &= \exp\left(\mathrm{i}[R^{-1}(\pi/3)\mathbf{K}]\cdot\mathbf{R}_B\right) = \exp(\mathrm{i}\mathbf{K}'\cdot\mathbf{R}_B),\\ \exp(\mathrm{i}\mathbf{K}\cdot\mathbf{R}'_B) &= \exp\left(\mathrm{i}[R^{-1}(\pi/3)\mathbf{K}]\cdot\mathbf{R}_A\right) = \exp(\mathrm{i}\mathbf{K}'\cdot\mathbf{R}_A).\end{aligned} \tag{131}$$

Further, we have $R^{-1}(\pi/3)\mathbf{K}'=\mathbf{K}-\mathbf{b}^*$. Noting that $\mathbf{R}_A=n_a\mathbf{a}+n_b\mathbf{b}-\vec{\tau}_3$ and $\mathbf{R}_B=n_a\mathbf{a}+n_b\mathbf{b}+\vec{\tau}_2$ with integers n_a and n_b, we have

$$\exp(\mathrm{i}\mathbf{K}'\cdot\mathbf{R}'_A) = \exp[\mathrm{i}(\mathbf{K}-\mathbf{b}^*)\cdot\mathbf{R}_B] = \omega\exp(\mathrm{i}\mathbf{K}\cdot\mathbf{R}_B),$$

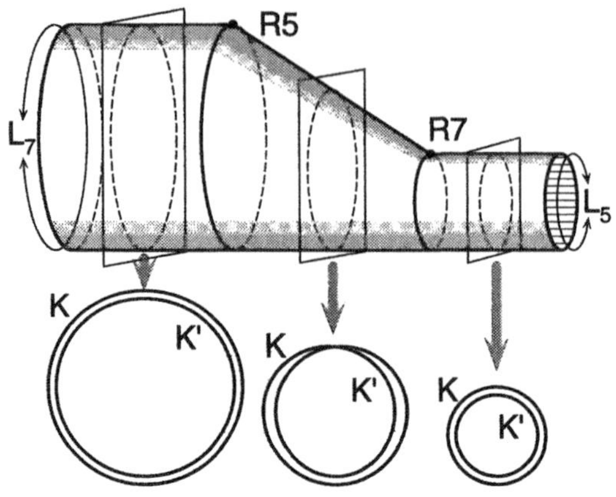

Fig. 37: Schematic illustration of the topological structure of a junction of two carbon nanotubes with different diameter. In the nanotube regions, two cylinders corresponding to spaces associated with the K and K' points are independent of each other and completely decoupled. In the junction region, on the contrary, they are interconnected to each other.

$$\exp(\mathrm{i}\mathbf{K}'\cdot\mathbf{R}'_B)=\exp[\mathrm{i}(\mathbf{K}-\mathbf{b}^*)\cdot\mathbf{R}_A]=\omega^{-1}\exp(\mathrm{i}\mathbf{K}\cdot\mathbf{R}_A), \tag{132}$$

where use has been made of $\exp(-\mathrm{i}\mathbf{b}^*\cdot\vec{\tau}_3)=\omega$ and $\exp(\mathrm{i}\mathbf{b}^*\cdot\vec{\tau}_2)=\omega^{-1}$.

In order to obtain boundary conditions for envelope functions in the junction region, we first multiply $g(\mathbf{r}-\mathbf{R}_B)\mathbf{b}(\mathbf{R}_B)$ from the left on both sides of the first of Eq. (131) and sum them up over $\mathbf{R}_B$ to have

$$\sum_{\mathbf{R}_B} g(\mathbf{r}-\mathbf{R}_B)\ \mathbf{b}(\mathbf{R}_B)\mathbf{a}(\mathbf{R}'_A)^+\mathbf{F}_A[R(\pi/3)\mathbf{r}]$$
$$=\sum_{\mathbf{R}_B} g(\mathbf{r}-\mathbf{R}_B)\mathbf{b}(\mathbf{R}_B)\mathbf{b}(\mathbf{R}_B)^+\mathbf{F}_B(\mathbf{r}). \tag{133}$$

We have

$$\mathbf{b}(\mathbf{R}_B)\mathbf{b}(\mathbf{R}_B)^+=\begin{pmatrix} 1 & -\omega e^{-\mathrm{i}\eta}e^{\mathrm{i}(\mathbf{K}'-\mathbf{K})\cdot\mathbf{R}_B} \\ -\omega e^{\mathrm{i}\eta}e^{-\mathrm{i}(\mathbf{K}'-\mathbf{K})\cdot\mathbf{R}_B} & 1 \end{pmatrix},$$
$$\mathbf{b}(\mathbf{R}_B)\mathbf{a}(\mathbf{R}'_A)^+=\begin{pmatrix} -\omega^{-1}e^{-\mathrm{i}\eta}e^{\mathrm{i}(\mathbf{K}'-\mathbf{K})\cdot\mathbf{R}_B} & -1 \\ 1 & \omega e^{\mathrm{i}\eta}e^{-\mathrm{i}(\mathbf{K}'-\mathbf{K})\cdot\mathbf{R}_B} \end{pmatrix}. \tag{134}$$

Therefore, we have

$$\mathbf{F}_A[R(\pi/3)\mathbf{r}]=\begin{pmatrix} 0 & 1 \\ -1 & 0 \end{pmatrix}\mathbf{F}_B(\mathbf{r}). \tag{135}$$

Similarly, the second equation of Eq. (130) gives

$$\mathbf{F}_B[R(\pi/3)\mathbf{r}]=\begin{pmatrix} 0 & -\omega^{-1} \\ \omega & 0 \end{pmatrix}\mathbf{F}_A(\mathbf{r}). \tag{136}$$

As a result, the boundary conditions for the envelope functions in the junction region are written as [122]

$$\mathbf{F}[R(\pi/3)\mathbf{r}] = T(\pi/3)\,\mathbf{F}(\mathbf{r}), \tag{137}$$

with

$$T(\pi/3) = \begin{pmatrix} 0 & 0 & 0 & 1 \\ 0 & 0 & -\omega^{-1} & 0 \\ 0 & -1 & 0 & 0 \\ \omega & 0 & 0 & 0 \end{pmatrix}, \tag{138}$$

where

$$\mathbf{F}(\mathbf{r}) = \begin{pmatrix} \mathbf{F}^{K}(\mathbf{r}) \\ \mathbf{F}^{K'}(\mathbf{r}) \end{pmatrix}. \tag{139}$$

It is straightforward to show that this should be modified into

$$T(\pi/3) = \begin{pmatrix} 0 & 0 & 0 & e^{i\psi(\mathbf{R})} \\ 0 & 0 & -\omega^{-1}e^{i\psi(\mathbf{R})} & 0 \\ 0 & -e^{-i\psi(\mathbf{R})} & 0 & 0 \\ \omega e^{-i\psi(\mathbf{R})} & 0 & 0 & 0 \end{pmatrix}, \tag{140}$$

with

$$e^{i\psi(\mathbf{R})} = \exp[i(\mathbf{K}'-\mathbf{K})\cdot\mathbf{R}], \tag{141}$$

if $\mathbf{R}$ is not chosen at the origin, i.e., $\mathbf{R} \neq 0$. Physical quantities do not depend on the choice of the origin and consequently on the phase $\psi(\mathbf{R})$.

Figure 37 gives an illustration of the topological structure of the junction [122]. In the nanotube regions, the K and K' points are completely decoupled and therefore belong to different subspaces. In the junction region they are interconnected to each other. In the junction region, the wave function F_A^K turns into $F_B^{K'}$ with an extra phase $e^{-i\psi(\mathbf{R})}$ when being rotated once around the axis. After another rotation, it comes back to F_A^K with an extra phase $\omega^{-1} = \exp(-2\pi i/3)$. On the other hand, F_B^K turns into $F_A^{K'}$ with phase $-\omega e^{-i\psi(\mathbf{R})}$ under a rotation and then into F_B^K with phase $\omega = \exp(2\pi i/3)$ after another rotation.

The above boundary conditions for nanotubes and their junctions have been obtained based on the nearest-neighbor tight-binding model. The essential ingredients of the boundary conditions lie in the fact that the 2D graphite system is invariant under the rotation of $\pi/3$ followed by the exchange between A and B carbon atoms. Therefore, the conditions given by Eq. (135) with Eqs. (138) or (140) are quite general and valid in more realistic models of the band structure.

The present method can be used also to obtain boundary conditions around a five- and seven-member ring schematically illustrated in Fig. 38. First, around the five-member ring, we note that

$$T(5\pi/3)\,T(\pi/3) = T(2\pi) = 1. \tag{142}$$

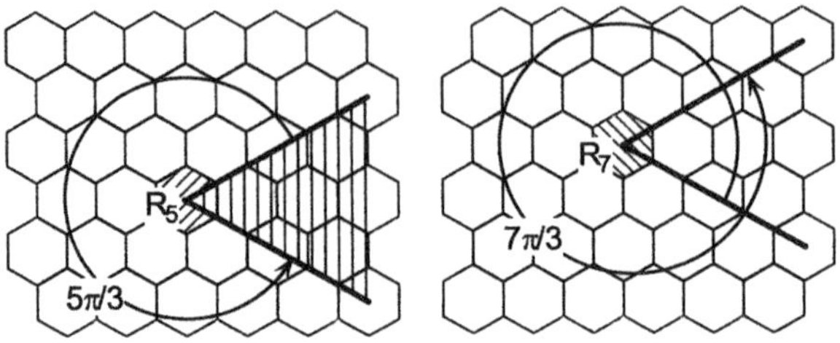

Fig. 38: The structure of a 2D graphite sheet in the vicinity of a five- and seven-member ring.

This immediately gives the conditions

$$\mathbf{F}[R(5\pi/3)\mathbf{r}] = T(5\pi/3)\,\mathbf{F}(\mathbf{r}), \tag{143}$$
$$T(5\pi/3) = T^{-1}(\pi/3) = \begin{pmatrix} 0 & 0 & 0 & \omega^{-1} \\ 0 & 0 & -1 & 0 \\ 0 & -\omega & 0 & 0 \\ 1 & 0 & 0 & 0 \end{pmatrix}.$$

Around the seven member ring, on the other hand, we have

$$\mathbf{F}[R(7\pi/3)\mathbf{r}] = T(7\pi/3)\,\mathbf{F}(\mathbf{r}),$$
$$T(7\pi/3) = T(\pi/3). \tag{144}$$

5.3 Conductance

Next, we consider envelope functions in the junction region for $\varepsilon=0$. First, we should note

$$\hat{k}_+ = \hat{k}_x + \mathrm{i}\hat{k}_y = \frac{2}{\mathrm{i}}\frac{\partial}{\partial z_-}, \quad \hat{k}_- = \hat{k}_x - \mathrm{i}\hat{k}_y = \frac{2}{\mathrm{i}}\frac{\partial}{\partial z_+}, \tag{145}$$

with $z_\pm = x \pm iy$. This means that F_A^K and $F_B^{K'}$ are functions of $z = x + iy$ and F_B^K and $F_A^{K'}$ are functions of $\bar{z} = x - iy$. Therefore, the boundary conditions for F_A^K and $F_B^{K'}$ are given by

$$F_A^K(\sqrt{\omega}z) = \mathrm{e}^{\mathrm{i}\psi(\mathbf{R})} F_B^{K'}(z),$$
$$F_B^{K'}(\sqrt{\omega}z) = \omega \mathrm{e}^{-\mathrm{i}\psi(\mathbf{R})} F_A^K(z). \tag{146}$$

We seek the solution of the form $F_A^K(z) \propto z^{n_A}$ and $F_B^{K'}(z) \propto z^{n_B}$. The substitution of these into the boundary conditions gives $n_A = n_B = 3m+1$ with m being an integer. Similar relations are also obtained for $F_B^K(\bar{z})$ and $F_B^{K'}(\bar{z})$. We have [122]

$$\mathbf{F}_m^{JA}(z) = \frac{1}{\sqrt{L_5}} \begin{pmatrix} 1 \\ 0 \\ 0 \\ (-)^m \mathrm{e}^{-\mathrm{i}\psi'} \end{pmatrix} \left(\frac{+\mathrm{i}z}{L_5}\right)^{3m+1}, \tag{147}$$

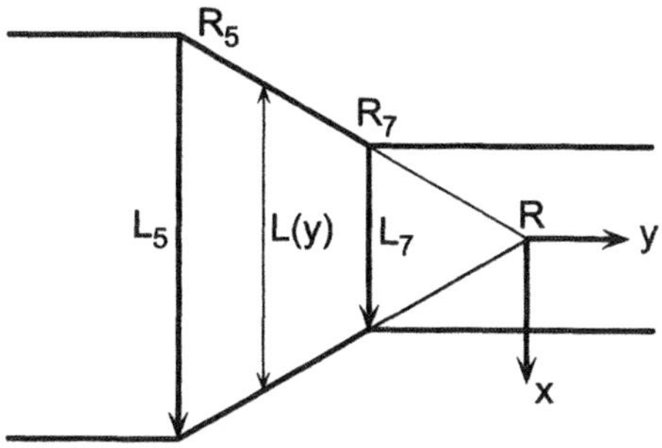

Fig. 39: A schematic illustration of a junction with $\theta=0$ and cordinate systems.

and

$$\mathbf{F}_m^{JB}(\bar{z}) = \frac{1}{\sqrt{L_5}} \begin{pmatrix} 0 \\ 1 \\ (-)^{m+1} e^{-i\psi'} \\ 0 \end{pmatrix} \Big(\frac{-i\bar{z}}{L_5}\Big)^{3m+1}, \tag{148}$$

with $e^{-i\psi'} = \sqrt{\omega} e^{-i\psi(\mathbf{R})}$.

Consider the case $\theta=0$ as illustrated in Fig. 39 for simplicity. The amplitude of the above wave functions decays or grows roughly in proportion to y^{3m+1} with the change of y. In particular, we have

$$\int_{-L(y)/2}^{+L(y)/2} \mathbf{F}_m^{JA*} \cdot \mathbf{F}_m^{JA} dx \propto \Big(\frac{L(y)}{L_5}\Big)^{6m+3}, \tag{149}$$

with $L(y) = -(2/\sqrt{3})y$. This shows that the total squared amplitude integrated over x varies in proportion to $[L(y)/L_5]^{6m+3}$ with the change of $L(y)$. In the case of a sufficiently long junction, i.e., for small L_7/L_5, the wave function is dominated by by that corresponding to $m=0$. This means that the conductance decays in proportion to $(L_7/L_5)^3$, explaining results of numerical calculations in a tight-binding model quite well [99]–[101].

An approximate expression for the transmission T and reflection probabilities R can be obtained by neglecting evanescent modes decaying exponentially into the thick and thin nanotubes. The solution gives

$$T = \frac{4L_5^3 L_7^3}{(L_5^3+L_7^3)^2}, \qquad R = \frac{(L_5^3-L_7^3)^2}{(L_5^3+L_7^3)^2}. \tag{150}$$

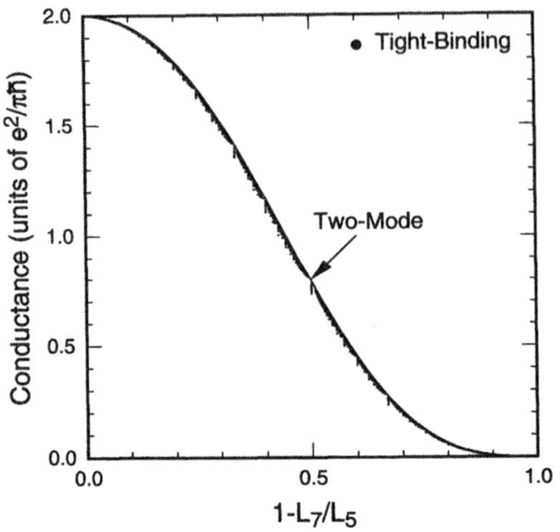

Fig. 40: The conductance obtained in the two-mode approximation and tight-binding results of armchair and zigzag nanotubes versus the effective length of the junction region $(L_5 - L_7)/L_5$. After Ref. ([122]).

We have $T \sim 4(L_7/L_5)^3$ in the long junction ($L_7/L_5 \ll 1$). When they are separated into different components,

$$T_{KK} = T\cos^2\Big(\frac{3}{2}\theta\Big), \qquad T_{KK'} = T\sin^2\Big(\frac{3}{2}\theta\Big), \tag{151}$$

and

$$R_{KK} = 0, \qquad R_{KK'} = R, \tag{152}$$

where the subscript KK means intravalley scattering within K or K' point and KK' stands for intervalley scattering between K and K' points. As for the reflection, no intravalley scattering is allowed. The dependence on the tilt angle θ originates from two effects. One is $\theta/2$ arising from the spinor-like character of the wave function in the rotation θ. Another θ comes from the junction wave function with $m=0$ which decays most slowly along the y axis. Figure 40 shows comparison of the two-mode solution with tight-binding results [99,100] for $\theta=0$.

In actual calculations, we have to limit the total number of eigenmodes in both nanotube and junction regions. In the junction region the wave function for $m \geq 0$ decays and that for $m<0$ becomes larger in the positive y direction. We shall choose cutoff M of the number of eigenmodes in the junction region, i.e., $-M-1 \leq m \leq M$, for a given value of L_7/L_5 in such a way that $(\sqrt{3}L_7/2L_5)^{3M} < \delta$, where δ is a positive quantity much smaller than unity. With the decrease of δ, the number of the modes included in the calculation increases.

Figure 41 shows some examples of calculated transmission and reflection probabilities for $\delta = 10^{-4}$ and 10^{-8}. As for the transmission, contributions from intervalley scattering ($K \to K'$) are plotted together for several values of θ. The dependence

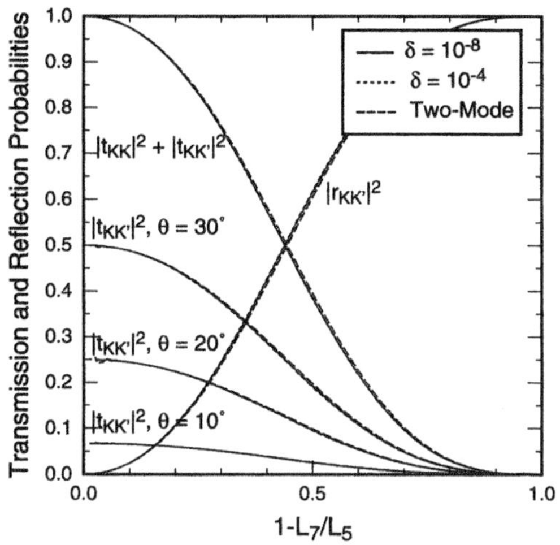

Fig. 41: Calculated transmission and reflection probabilities versus the effective length of the junction region $(L_5–L_7)/L_5$. Contributions of intervalley scattering to the transmission are plotted for $\theta=10°$, $20°$, and $30°$. The results are almost independent of the value of δ. After Ref. ([123]).

on the value of δ is extremely small and is not important at all, showing that the analytic expressions for the transmission and reflection probabilities obtained above are almost exact.

Explicit calculations were performed also for $\varepsilon \neq 0$ [123] and Fig. 42 shows an example. The conductance grows with the energy and has a peak before the first band edge $\varepsilon/\gamma = \pm 2\pi/L_5$. Near the band edge, the conductance decreases abruptly and falls off to zero. This behavior cannot be obtained if we ignore the evanescent modes in the tube region [124]. This implies the formation of a kind of resonant state in the junction region, which would bring forth the total reflection into the thicker tube region. The tight-binding results [124] show a small asymmetry between $\varepsilon > 0$ and $\varepsilon < 0$. The asymmetry arises presumably from the presence of five- and seven-member rings and is expected to be reduced with the increase of the circumference length.

6. Phonons and electron-phonon interaction

6.1 Long wavelength phonons

Acoustic phonons important in the electron scattering are described well by a continuum model [75,76]. The potential-energy functional for displacement $\mathbf{u} = (u_x, u_y, u_z)$ is written as

$$U[\mathbf{u}] = \int \mathbf{d}x\mathbf{d}y \, \frac{1}{2}\Big(B(u_{xx}+u_{yy})^2 + \mu[(u_{xx}-u_{yy})^2+4u_{xy}^2]\Big), \tag{153}$$

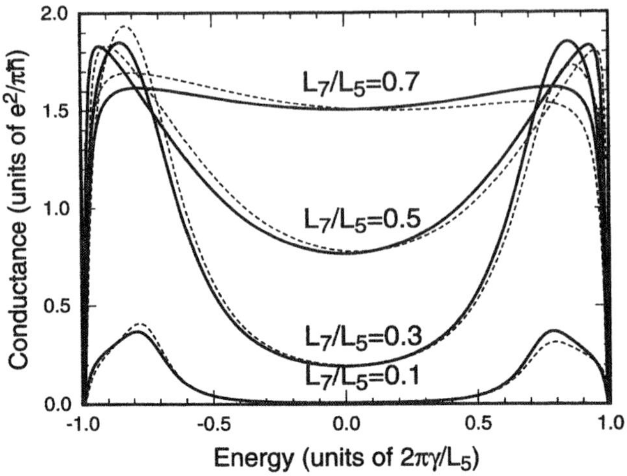

Fig. 42: Calculated conductance versus the energy ε $(-2\pi/L_5<\varepsilon/\gamma<+2\pi/L_5)$ for various values of the junction length. Solid lines represent the results of the $\boldsymbol{k}\cdot\boldsymbol{p}$ method, while dashed lines show tight-binding data for $L_5=50\sqrt{3}a$ (a is the lattice constant). The conductance grows with the energy and has a peak before the first band edge $\varepsilon/\gamma=\pm2\pi/L_5$, followed by an abrupt fall-off. After Ref. ([123]).

with

$$u_{xx}=\frac{\partial u_x}{\partial x}+\frac{u_z}{R},\quad u_{yy}=\frac{\partial u_y}{\partial y},\quad 2u_{xy}=\frac{\partial u_x}{\partial y}+\frac{\partial u_y}{\partial x}, \tag{154}$$

where the term u_z/R is due to the finite radius $R=L/2\pi$ of the nanotube. The parameters $B=\lambda+\mu$ and μ denote the bulk modulus and the shear modulus for a graphite sheet (λ and μ are Lame's constants). The corresponding kinetic energy is written as

$$K[\mathbf{u}]=\int \mathbf{d}x\mathbf{d}y\,\frac{M}{2}[(\dot{u}_x)^2+(\dot{u}_y)^2+(\dot{u}_z)^2], \tag{155}$$

where M is the mass density given by the carbon mass per unit area, $M=9.66\times10^{-7}$ kg/m^2. The corresponding equations of motion are given by

$$\begin{aligned}
M\ddot{u}_x &= (B+\mu)\frac{\partial^2 u_x}{\partial x^2}+\mu\frac{\partial^2 u_x}{\partial y^2}+B\frac{\partial^2 u_y}{\partial x\partial y}+\frac{B+\mu}{R}\frac{\partial u_z}{\partial x},\\
M\ddot{u}_y &= B\frac{\partial^2 u_x}{\partial x\partial y}+(B+\mu)\frac{\partial^2 u_y}{\partial y^2}+\mu\frac{\partial^2 u_y}{\partial x^2}+\frac{B-\mu}{R}\frac{\partial u_z}{\partial y},\\
M\ddot{u}_z &= -\frac{B+\mu}{R}\frac{\partial u_x}{\partial x}-\frac{B-\mu}{R}\frac{\partial u_y}{\partial y}-\frac{B+\mu}{R^2}iu_z.
\end{aligned} \tag{156}$$

The phonon modes are specified by the wave vector along the circumference $\chi(n)=2\pi n/L$ and that along the axis q, $\mathbf{u}(\mathbf{r})=\mathbf{u}_{nq}\exp[\mathrm{i}\chi(n)x+\mathrm{i}qy]$. When $n=0$ and $\chi=0$,

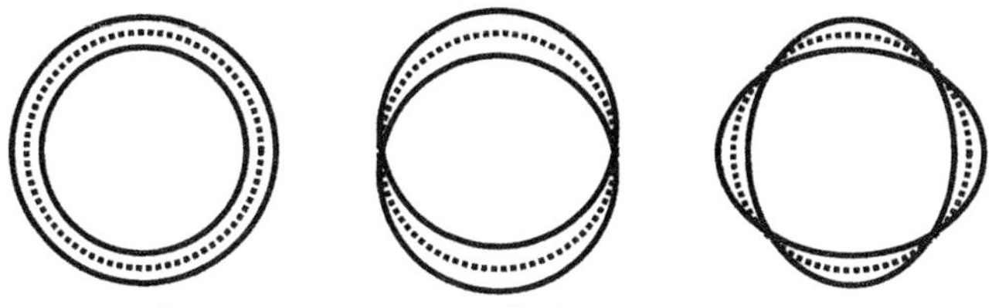

Fig. 43: Some examples of deformation of the cross section of CN with $n=0$, ±1, and ±2.

the eigen equation becomes

$$M\omega^2 \begin{pmatrix} u_x \\ u_y \\ u_z \end{pmatrix} = \begin{pmatrix} \mu q^2 & 0 & 0 \\ 0 & (B+\mu)q^2 & -\mathrm{i}(B-\mu)qR^{-1} \\ 0 & \mathrm{i}(B-\mu)qR^{-1} & (B+\mu)R^{-2} \end{pmatrix} \begin{pmatrix} u_x \\ u_y \\ u_z \end{pmatrix}, \tag{157}$$

which has three eigenmodes called twisting, stretching, and breathing modes. The twisting mode is made of pure circumference-directional deformation and its velocity v_t is equal to that of the TA mode of a graphite sheet,

$$\omega_T(q) = v_T q, \quad v_T = v_T^G = \sqrt{\frac{\mu}{M}}. \tag{158}$$

In the long wavelength limit $q = 0$, the radial deformation generates a breathing mode with a frequency

$$\omega_B = \sqrt{\frac{B+\mu}{M}}\frac{1}{R}, \tag{159}$$

which is inversely proportional to the radius R of the CN. In the case $|qR| \ll 1$, the deformation in the nanotube-axis direction generates stretching modes. When $\omega \ll \omega_B$, we have from the last equation of the above

$$u_z \approx -\mathrm{i}\frac{B-\mu}{B+\mu}qR\,u_y. \tag{160}$$

Upon substitution of this into the second equation, we have

$$\omega_S = v_S q, \quad v_S = \sqrt{\frac{4B\mu}{(B+\mu)M}}. \tag{161}$$

The velocity v_S is usually smaller than that of the LA mode of the graphite $v_L^G = \sqrt{(B+\mu)/M}$. We set $v_L^G = 21.1$ km/s and $v_T^G = 15.0$ km/s, and we obtain $v_s = 21.1$ km/s, $v_t = 15.0$ km/s, and $\hbar\omega_B = 2.04\times10^{-2}$ eV, or 237K for the [10, 10] armchair CN with $R = 6.785$ Å. These values show good agreement with recent results by Saito et al. [125] and can never be reproduced by a zone-folding method.

The above model is too simple when dealing with modes with $n \neq 0$. In order to see this fact explicitly, we shall consider the case with $q = 0$. In this case there is a displacement given by

$$u_x = -\frac{u}{n}\sin\frac{2inx}{L} = -\frac{u}{n}\sin\frac{nx}{R},$$
$$u_z = u\cos\frac{2inx}{L} = u\cos\frac{nx}{R}, \tag{162}$$

with arbitrary u. This displacement gives $u_{xx} = 0$ identically and also $u_{yy} = u_{xy} = 0$, giving rise to the vanishing frequency. For $n = \pm 1$, this vanishing frequency is absolutely necessary because the displacement corresponds to a uniform shift of a nanotube in a direction perpendicular to the axis. For $n > 1$, on the other hand, the displacement corresponds to a deformation of the cross section of the nanotube as shown in Fig. 43. Such deformations should have nonzero frequency in actual graphite because otherwise CN cannot maintain a cylindrical form.

Actually, we have to consider the potential energy due to nonzero curvature of the 2D graphite plane. It is written as

$$U_c[\mathbf{u}] = \frac{1}{2}a^2\Xi\int \mathbf{d}x\mathbf{d}y\Big[\Big(\frac{\partial^2}{\partial x^2} + \frac{1}{R^2} + \frac{\partial^2}{\partial y^2}\Big)u_z\Big]^2, \tag{163}$$

where Ξ is a force constant for curving of the plane. The presence of the term $1/R^2$ guarantees that the deformation with $n = \pm 1$ given in Eq. (163) has a vanishing frequency. This curvature energy is of the order of the fourth power of the wave vector and therefore is much smaller than $U[\mathbf{u}]$ as long as $qR \ll 1$ for $n = 0$ and 1 but becomes appreciable for $n > 1$. Figure 44 shows phonon dispersions calculated in this continuum model.

6.2 Electron-phonon interaction

A long-wavelength acoustic phonon gives rise to an effective potential called the deformation potential

$$V_1 = g_1(u_{xx} + u_{yy}), \tag{164}$$

proportional to a local dilation or dilatation. This term appears as a diagonal term in the matrix Hamiltonian in the effective-mass approximation. Consider a rectangular area $a \times a$. In the presence of a lattice deformation, the area S changes into $S + \delta S(\mathbf{r})$ with $\delta S(\mathbf{r}) = a^2\Delta(\mathbf{r})$, where

$$\Delta(\mathbf{r}) = \frac{\partial u_x}{\partial x} + \frac{\partial u_y}{\partial y}. \tag{165}$$

Therefore, the ion density changes locally by $n_0 \to n_0[1 - \Delta(\mathbf{r})]$. The electron density should change in the same manner due to the charge neutrality condition. Consider a two-dimensional electron gas. The potential energy $\delta\varepsilon(\mathbf{r})$ corresponding to the

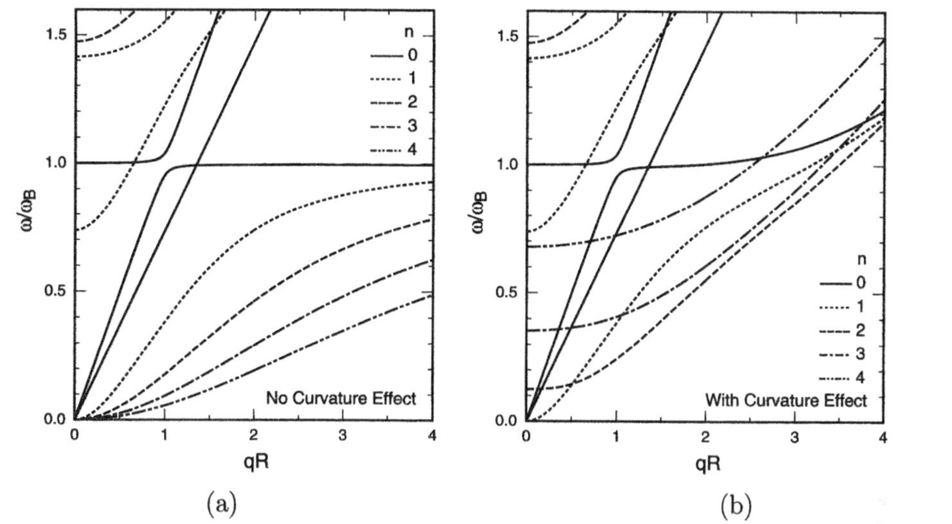

Fig. 44: Frequencies of phonons obtained in the continuum model. (a) Without out-of-plane curvature effect and (b) with out-of-plane curvature effect.

density change should satisfy $\delta\varepsilon(\mathbf{r})D(\varepsilon_F)=n_0\Delta(\mathbf{r})$, where $D(\varepsilon_F)$ is the density of states at the Fermi level $D(\varepsilon)=m/\pi\hbar^2$ independent of energy. Therefore, $n_0 = D(\varepsilon_F)\varepsilon_F$, leading to

$$\delta\varepsilon(\mathbf{r})=\varepsilon_F\Delta(\mathbf{r}). \tag{166}$$

This shows $g_1=\varepsilon_F$. In the two-dimensional graphite, the electron gas model may not be so appropriate but can be used for a very rough estimation of g_1 as the Fermi energy measured from the bottom of the valence bands (σ bands), i.e., 20–30 eV.

A long-wavelength acoustic phonon also causes a change in the distance between nearest neighbor carbon atoms and in the transfer integral. Let $\mathbf{u}_A(\mathbf{R}_A)$ and $\mathbf{u}_B(\mathbf{R}_B)$ be a lattice displacement at A and B site, respectively. Then the transfer integral between an atom at $\mathbf{R}_A$ and $\mathbf{R}_A-\vec{\tau}_l$ changes from $-\gamma_0$ by the amount

$$\begin{aligned}-\frac{\partial\gamma_0(b)}{\partial b}&\Big[|\vec{\tau}_l+\mathbf{u}_A(\mathbf{R}_A)-\mathbf{u}_B(\mathbf{R}_A-\vec{\tau}_l)|-b\Big]\\&=-\frac{\partial\gamma_0(b)}{\partial b}\frac{1}{b}\vec{\tau}_l\cdot[\mathbf{u}_A(\mathbf{R}_A)-\mathbf{u}_B(\mathbf{R}_A-\vec{\tau}_l)],\end{aligned} \tag{167}$$

where $b=|\vec{\tau}_l|=a/\sqrt{3}$. Therefore, the extra term appearing in the right hand side of Eq. (15) is calculated as

$$\begin{aligned}&\sum_l\sum_{\mathbf{R}_A} g(\mathbf{r}-\mathbf{R}_A)\mathbf{a}(\mathbf{R}_A)\mathbf{b}(\mathbf{R}_A-\vec{\tau}_l)^+\\&\qquad\Big(-\frac{\partial\gamma_0(b)}{\partial b}\Big)\frac{1}{b}\vec{\tau}_l\cdot[\mathbf{u}_A(\mathbf{R}_A)-\mathbf{u}_B(\mathbf{R}_A-\vec{\tau}_l)]\mathbf{F}_B(\mathbf{r})\end{aligned}$$

$$= \sum_l \begin{pmatrix} -\omega e^{i\eta} e^{-i\mathbf{K}\cdot\vec{\tau}_l} & 0 \\ 0 & e^{-i\eta} e^{-i\mathbf{K}'\cdot\vec{\tau}_l} \end{pmatrix}$$
$$\Big(-\frac{\partial \gamma_0(b)}{\partial b}\Big)\frac{1}{b}\vec{\tau}_l\cdot[\mathbf{u}_A(\mathbf{r})-\mathbf{u}_B(\mathbf{r}-\vec{\tau}_l)]\mathbf{F}_B(\mathbf{r}). \tag{168}$$

Because $\mathbf{u}_A(\mathbf{r})-\mathbf{u}_B(\mathbf{r}-\vec{\tau}_l)$ involves displacements of different sublattices, it has a contribution of optical modes and $\mathbf{u}_A(\mathbf{r})-\mathbf{u}_B(\mathbf{r}-\vec{\tau}_l) \neq \mathbf{u}(\mathbf{r})-\mathbf{u}(\mathbf{r}-\vec{\tau}_l)$. In the long-wavelength limit, however, we can set

$$\vec{\tau}_l\cdot[\mathbf{u}_A(\mathbf{r})-\mathbf{u}_B(\mathbf{r}-\vec{\tau}_l)] = \alpha\vec{\tau}_l\cdot[\mathbf{u}(\mathbf{r})-\mathbf{u}(\mathbf{r}-\vec{\tau}_l)] = \alpha(\vec{\tau}_l\cdot\nabla)\,\vec{\tau}_l\cdot\mathbf{u}(\mathbf{r}), \tag{169}$$

where α is a constant. This α depends on details of a microsopic model of phonons and becomes $\sim 1/3$ smaller than unity in a valence-force-field model [126]. Now, we shall use the identity

$$\sum_l e^{-i\mathbf{K}\cdot\vec{\tau}_l}\,\big(\,(\tau_l^x)^2 \quad \tau_l^x\tau_l^y (\tau_l^y)^2\,\big) = \frac{\omega^{-1}}{4}a^2\,(\,-1 \quad -i \quad +1\,),$$
$$\sum_l e^{-i\mathbf{K}'\cdot\vec{\tau}_l}\,\big(\,(\tau_l^x)^2 \quad \tau_l^x\tau_l^y \quad (\tau_l^y)^2\,\big) = \frac{1}{4}a^2\,(\,-1 \quad +i \quad +1\,). \tag{170}$$

Then, we have

$$\frac{3\alpha\beta}{4}\gamma_0 \begin{pmatrix} e^{i\eta}\Big(\frac{\partial}{\partial x}+i\frac{\partial}{\partial y}\Big)(u_x+iu_y) & 0 \\ 0 & -e^{-i\eta}\Big(\frac{\partial}{\partial x}-i\frac{\partial}{\partial y}\Big)(u_x-iu_y) \end{pmatrix}, \tag{171}$$

with

$$\beta = -\frac{b}{\gamma_0}\frac{\partial \gamma_0}{\partial b} = -\frac{\mathbf{d}\ln\gamma_0}{\mathbf{d}\ln b}. \tag{172}$$

The above quantities are those in the coordinate system fixed onto the graphite sheet and become in the coordinate system defined in the nanotube

$$u_x \pm iu_y \rightarrow e^{\pm i\eta}(u_x \pm iu_y),$$
$$\frac{\partial}{\partial x} \pm i\frac{\partial}{\partial y} \rightarrow e^{\pm i\eta}\Big(\frac{\partial}{\partial x} \pm i\frac{\partial}{\partial y}\Big). \tag{173}$$

Similar expressions can be obtained for $\mathbf{F}_B$ and the effective Hamiltonian becomes

$$\mathcal{H}'_K = \begin{pmatrix} 0 & V_2 \\ V_2^* & 0 \end{pmatrix}, \quad \mathcal{H}'_{K'} = \begin{pmatrix} 0 & -V_2^* \\ -V_2 & 0 \end{pmatrix}, \tag{174}$$

with

$$V_2 = g_2 e^{3i\eta}(u_{xx} - u_{yy} + 2iu_{xy}), \tag{175}$$

where

$$g_2 = \frac{3\alpha\beta}{4}\gamma_0. \tag{176}$$

Usually, we have $\beta \sim 2$ [126] and $\alpha \sim 1/3$, which give $g_2 \sim \gamma_0/2$ or $g_2 \sim 1.5$ eV. This coupling constant is much smaller than the deformation potential constant $g_1 \sim 30$ eV.

6.3 Resistivity

Apart from the spatial part of the wave function, the (pseudo) spin part is given by

$$|s+\rangle = \frac{1}{\sqrt{2}}\begin{pmatrix} -is \\ 1 \end{pmatrix}, \quad |s-\rangle = \frac{1}{\sqrt{2}}\begin{pmatrix} +is \\ 1 \end{pmatrix}, \tag{177}$$

where $s = +1$ for the conduction band and -1 for the valence band, and + and − for the right- and left-going wave, respectively. Then, we have

$$\left(s - \middle| \begin{pmatrix} V_1 & V_2 \\ V_2^* & V_1 \end{pmatrix} \middle| s + \right) = -\frac{\mathrm{i}}{2}(V_2 + V_2^*) = -\mathrm{iRe}\ V_2. \tag{178}$$

This means that the diagonal deformation-potential term does not contribute to the backward scattering as in the case of impurities and only the real part of the much smaller off-diagonal term contributes to the backward scattering. We have

$$\mathrm{Re}\ V_2 = g_2[\cos 3\eta(u_{xx} - u_{yy}) - 2\sin 3\eta u_{xy}]. \tag{179}$$

In armchair nanotubes with $\eta = \pi/6$, we have Re $V_2 = -2g_2u_{xy}$ and only shear or twist waves contribute to the scattering. In zigzag nanotubes with $\eta = 0$, on the other hand, Re $V_2 = g_2(u_{xx} - u_{yy})$ and only stretching and breathing modes contribute to the scattering.

When a high-temperature approximation is adopted for phonon distribution function, the resistivity for an armchair nanotube is calculated as

$$\rho_A(T) = \frac{\hbar}{e^2}\frac{1}{R}\frac{g_2^2 k_B T}{2\gamma^2 \mu}. \tag{180}$$

At temperatures much higher than the frequency of the breathing mode ω_B, the resistivity of a zigzag nanotube is same as $\rho_A(T)$, i.e., $\rho_Z(T) = \rho_A(T)$. At temperatures lower than ω_B, on the other hand, the breathing mode does not contribute to the scattering and therefore the resistivity becomes smaller than that of an armchair nanotube with same radius, i.e., $\rho_Z(T) = \rho_A(T)B/(B+\mu) = \rho_A(T)(\lambda+\mu)/(\lambda+2\mu)$. Figure 45 shows calculated temperature dependence of the resistivity. Because of the small coupling constant g_2 the absolute value of the resistivity is much smaller than that in bulk 2D graphite dominated by much larger deformation-potential scattering. The resistivity of an armchair CN is same as that obtained previously [74] except for a difference in g_2.

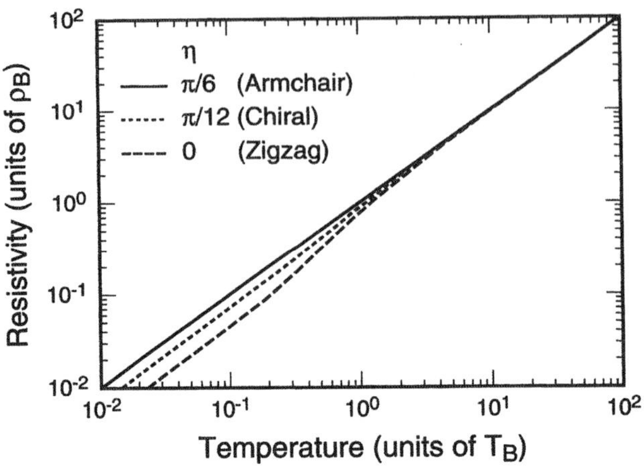

Fig. 45: (right) The resistivity of armchair (solid line) and zigzag (dotted line) nanotubes in units of $\rho_A(T_B)$ which is the resistivity of the armchair nanotube at $T=T_B$, and T_B denotes the temperature of the breathing mode, $T_B=\hbar\omega_B/k_B$.

7. Summary

Electronic and transport properties of carbon nanotubes have been discussed theoretically mainly based on a $\boldsymbol{k}\cdot\boldsymbol{p}$ scheme. The motion of electrons in carbon nanotubes is described by Weyl's equation for a massless neutrino with a helicity. This leads to interesting properties of nanotubes including Aharonov-Bohm effects on the band gap, the absence of backward scattering and the conductance quantization in the presence of scatterers with a potential range larger than the lattice constant, a conductance quantization in the presence of lattice vacancies, and power-law dependence of the conductance across a junction between nanotubes with different diameters. At high temperatures scattering by phonons starts to play some role, but is not important because conventional deformation potential coupling is absent and only much smaller coupling through bond-length change contributes to the scattering. Optical absorption is appreciable only for the light polarization parallel to the axis and almost all the intensity is transfered to excitons from continuum interband transitions due to the one-dimensional nature of nanotubes.

Acknowledgments

The author acknowledges the collaboration with H. Ajiki, T. Seri, T. Nakanishi, H. Matsumura, R. Saito, H. Suzuura, M. Igami, T. Yaguchi. This work was supported in part by Grants-in-Aid for Scientific Research, for Priority Area, Fullerene Network, and for COE Research (12CE2004 "Control of Electrons by Quantum Dot Structures and Its Application to Advanced Electronics") from Ministry of Education, Culture, Sports, Science, and Technology, Japan.

References

[1] S. Iijima, Nature (London), **354** ,56 (1991).

[2] S. Iijima, T. Ichihashi, and Y. Ando, Nature (London) **356**, 776 (1992).

[3] S. Iijima and T. Ichihashi, Nature (London) **363**, 603 (1993).

[4] D. S. Bethune, C. H. Kiang, M. S. de Vries, G. Gorman, R. Savoy, J. Vazquez, and R. Beyers, Nature (London) **363**, 605 (1993).

[5] N. Hamada, S. Sawada, and A. Oshiyama, Phys. Rev. Lett. **68**, 1579 (1992).

[6] J. W. Mintmire, B. I. Dunlap, and C. T. White, Phys. Rev. Lett. **68**, 631 (1992).

[7] R. Saito, M. Fujita, G. Dresselhaus, and M. S. Dresselhaus, Phys. Rev. B **46**, 1804 (1992).

[8] M. S. Dresselhaus, G. Dresselhaus, and R. Saito, Phys. Rev. B **45**, 6234 (1992).

[9] M. S. Dresselhaus, G. Dresselhaus, R. Saito, and P. C. Eklund: *Elementary Excitations in Solids*, ed. J. L. Birman, C. Sebenne and R. F. Wallis (Elsevier Science Publishers B. V., Amsterdam, 1992) p. 387.

[10] R. A. Jishi, M. S. Dresselhaus, and G. Dresselhaus, Phys. Rev. B **47**, 16671 (1993).

[11] K. Tanaka, K. Okahara, M. Okada and T. Yamabe, Chem. Phys. Lett. **191**, 469 (1992).

[12] Y. D. Gao and W. C. Herndon: Mol. Phys. **77**, 585 (1992).

[13] D. H. Robertson, D. W. Brenner, and J. W. Mintmire, Phys. Rev. B **45**, 12592 (1992).

[14] C. T. White, D. C. Robertson, and J. W. Mintmire, Phys. Rev. B **47**, 5485 (1993).

[15] H. Ajiki and T. Ando, J. Phys. Soc. Jpn. **62**, 1255 (1993).

[16] H. Ajiki and T. Ando, J. Phys. Soc. Jpn. **62**, 2470 (1993); [Errata, J. Phys. Soc. Jpn. **63**, 4267 (1994).]

[17] H. Ajiki and T. Ando, Physica B **201**, 349 (1994).

[18] H. Ajiki and T. Ando, J. Phys. Soc. Jpn. **64**, 4382 (1995).

[19] T. Ando, J. Phys. Soc. Jpn. **66**, 1066 (1997).

[20] N. A. Viet, H. Ajiki, and T. Ando, J. Phys. Soc. Jpn. **63**, 3036 (1994).

[21] H. Ajiki and T. Ando, in *The Physics of Semiconductors*, edited by D.J. Lockwood (World Scientific, Singapore, 1995), p. 2061.

[22] H. Ajiki and T. Ando, J. Phys. Soc. Jpn. **65**, 2976 (1996).

[23] H. Ajiki and T. Ando, Jpn. J. Appl. Phys. Suppl. **34-1**, 107 (1995).

[24] S. N. Song, X. K. Wang, R. P. H. Chang, and J. B. Ketterson, Phys. Rev. Lett. **72**, 697 (1994).

[25] J. E. Fischer, H. Dai, A. Thess, R. Lee, N. M. Hanjani, D. L. Dehaas, and R. E. Smalley, Phys. Rev. B **55**, R4921 (1997).

[26] M. Bockrath, D. H. Cobden, P. L. McEuen, N. G. Chopra, A. Zettl, A. Thess, and R. E. Smalley, Science **275**, 1922 (1997).

[27] L. Langer, V. Bayot, E. Grivei, J. -P. Issi, J. P. Heremans, C. H. Olk, L. Stockman, C. Van Haesendonck, and Y. Bruynseraede, Phys. Rev. Lett. **76**, 479 (1996).

[28] A. Yu. Kasumov, I. I. Khodos, P. M. Ajayan, and C. Colliex, Europhys. Lett. **34**, 429 (1996).

[29] T. W. Ebbesen, H. J. Lezec, H. Hiura, J. W. Bennett, H. F. Ghaemi, and T. Thio, Nature (London) **382**, 54 (1996).

[30] H. Dai, E.W. Wong, and C. M. Lieber, Science **272**, 523 (1996).

[31] A. Yu. Kasumov, H. Bouchiat, B. Reulet, O. Stephan, I. I. Khodos, Yu. B. Gorbatov, and C. Colliex, Europhys. Lett. **43**, 89 (1998).

[32] S. J. Tans, M. H. Devoret, H. Dai, A. Thess, R. E. Smalley, L. J. Geerligs, and C. Dekker, Nature (London) **386**, 474 (1997).

[33] S. J. Tans, R. M. Verschuren, and C. Dekker, Nature (London) **393**, 49 (1998).

[34] D. H. Cobden, M. Bockrath, P. L. McEuen, A. G. Rinzler, and R. E. Smalley, Phys. Rev. Lett. **81**, 681 (1998).

[35] S. J. Tans, M. H. Devoret, R. J. A. Groeneveld, and C. Dekker, Nature (London) **394**, 761 (1998).

[36] A. Bezryadin, A. R. M. Verschueren, S. J. Tans, and C. Dekker, Phys. Rev. Lett. **80**, 4036 (1998).

[37] J. Tersoff, Appl. Phys. Lett. **74**, 2122 (1999).

[38] M. P. Anantram, S. Datta, and Y.-Q. Xue, Phys. Rev. B **61**, 14219 (2000).

[39] K. -J. Kong, S. -W. Han, and J. -S. Ihm, Phys. Rev. B **60**, 6074 (1999).

[40] H. J. Choi, J. Ihm, Y. -G. Yoon, and S. G. Louie, Phys. Rev. B **60**, R14009 (1999).

[41] T. Nakanishi and T. Ando, J. Phys. Soc. Jpn. **69**, 2175 (2000).

[42] M. S. Dresselhaus, G. Dresselhaus, and P. C. Eklund, *Science of Fullerenes and Carbon Nanotubes*, (Academic Press 1996).

[43] T. W. Ebbesen, Physics Today **49** (1996) No. 6, p. 26.

[44] H. Ajiki and T. Ando, Solid State Commun. **102**, 135 (1997).

[45] R. Saito, G. Dresselhaus and M. S. Dresselhaus, *Physical Properties of Carbon Nanotubes*, (Imperial College Press 1998).

[46] C. Dekker, Phys. Today, **52**, 22 (1999).

[47] T. Ando, Semicond. Sci. Technol. **15**, R13 (2000).

[48] N. H. Shon and T. Ando, J. Phys. Soc. Jpn. **67**, 2421 (1998).

[49] W. -D. Tian and S. Datta, Phys. Rev. B **49**, 5097 (1994).

[50] T. Ando and T. Seri, J. Phys. Soc. Jpn. **66**, 3558 (1997).

[51] H. Ajiki and T. Ando, Physica B **216**, 358 (1996).

[52] R. Saito, G. Dresselhaus, and M. S. Dresselhaus, Phys. Rev. B **61**, 2981 (2000).

[53] R. Loudon, Am. J. Phys. **27**, 649 (1959).

[54] R. J. Elliot and R. Loudon, J. Phys. Chem. Solids **8**, 382 (1959); **15**, 196 (1960).

[55] J. W. Wildoer, L. C. Venema, A. G. Rinzler, R. E. Smalley, and C. Dekker, Nature (London) **391**, 59 (1998).

[56] A. Kasuya, M. Sugano, T. Maeda, Y. Saito, K. Tohji, H. Takahashi, Y. Sasaki, M. Fukushima, Y. Nishina, and C. Horie, Phys. Rev. B **57**, 4999 (1998).

[57] M. A. Pimenta, A. Marucci, S. A. Empedocles, M. G. Bawendi, E. B. Hanlon, A. M. Rao, P. C. Eklund, R. E. Smalley, G. Dresselhaus, and M. S. Dresselhaus, Phys. Rev. B **58**, R16016 (1998).

[58] H. Kataura, Y. Kumazawa, Y. Maniwa, I. Umezu, S. Suzuki, Y. Ohtsuka and Y. Achiba, Synth. Met. **103**, 2555 (1999).

[59] M. Ichida, S. Mizuno, Y. Tani, Y. Saito, and A. Nakamura, J. Phys. Soc. Jpn. **68**, 3131 (1999).

[60] T. Ando and T. Nakanishi, J. Phys. Soc. Jpn. **67**, 1704 (1998).

[61] Y. Ando, X. -L. Zhao, and M. Ohkohchi, Jpn. J. Appl. Phys. **37**, L61 (1998).

[62] M. V. Berry, Proc. Roy. Soc. London A392, 45 (1984).

[63] B. Simon, Phys. Rev. Lett. **51**, 2167 (1983).

[64] S. Hikami, A. I. Larkin, and Y. Nagaoka, Prog. Theor. Phys. **63**, 707 (1980).

[65] *Anderson Localization*, edited by Y. Nagaoka and H. Fukuyama (Springer, Berlin, 1982); *Localization, Interaction, and Transport Phenomena*, edited by B. Kramer, G. Bergmann, and Y. Bruynseraede (Springer, Berlin, 1984); P. A. Lee and T. V. Ramakrishnan: Rev. Mod. Phys. **57**, 287 (1985); *Anderson Localization*, edited by T. Ando and H. Fukuyama (Springer, Berlin, 1988).

[66] G. Bergmann, Phys. Rept. **107**, 1 (1984).

[67] F. Komori, S. Kobayashi, and W. Sasaki, J. Phys. Soc. Jpn. **51**, 3162 (1982).

[68] G. Bergmann: Phys. Rev. Lett. **48**, 1046 (1982).

[69] A. Bachtold, C. Strunk, J. P. Salvetat, J. M. Bonard, L. Forro, T. Nussbaumer, and C. Schöneberger, Nature (London) **397**, 673 (1999).

[70] A. Fujiwara, K. Tomiyama, H. Suematsu, M. Yumura, and K. Uchida, Phys. Rev. B **60**, 13492 (1999).

[71] P. L. McEuen, M. Bockrath, D. H. Cobden, Y. -G. Yoon, and S. G. Louie, Phys. Rev. Lett. **83**, 5098 (1999).

[72] A. Bachtold, M. S. Fuhrer, S. Plyasunov, M. Forero, E. H. Anderson, A. Zettl, and P. L. McEuen, Phys. Rev. Lett. **84**, 6082 (2000).

[73] S. Frank, P. Poncharal, Z. L. Wang, and W. A. de Heer, Science **280**, 1744 (1998).

[74] C. L. Kane, E. J. Mele, R. S. Lee, J. E. Fischer, P. Petit, H. Dai, A. Thess, R. E. Smalley, A. R. M. Verschueren, S. J. Tans, and C. Dekker, Europhys. Lett. **41**, 683 (1998).

[75] H. Suzuura and T. Ando, Physica E **6**, 864 (2000).

[76] H. Suzuura and T. Ando, Mol. Cryst. and Liq. Cryst. **340**, 731 (2000).

[77] H. A. Mizes and J. S. Foster, Science **244**, 559 (1989).

[78] O. Zhou, R. M. Fleming, D. W. Murphy, R. C. Haddon, A. P. Ramirez, and S. H. Glarum, Science **263**, 1744 (1994).

[79] S. Amelinckx, D. Bernaerts, X. B. Zhang, G. Van Tendeloo, and J. Van Landuyt, Science **267**, 1334 (1995).

[80] M. Fujita, K. Wakabayashi, K. Nakada and K. Kusakabe, J. Phys. Soc. Jpn. **65**, 1920 (1996).

[81] K. Nakada, M. Fujita, G. Dresselhaus, and M. S. Dresselhaus, Phys. Rev. B **54**, 17954 (1996).

[82] M. Fujita, M. Igami, and K. Nakada, J. Phys. Soc. Jpn. **66**, 1864 (1997).

[83] L. Chico, L. X. Benedict, S. G. Louie, and M. L. Cohen, Phys. Rev. B **54**, 2600 (1996).

[84] M. Igami, T. Nakanishi, and T. Ando, J. Phys. Soc. Jpn. **68**, 716 (1999).

[85] M. Igami, T. Nakanishi, and T. Ando, J. Phys. Soc. Jpn. **68**, 3146 (1999).

[86] T. Ando, T. Nakanishi, and M. Igami, J. Phys. Soc. Jpn. **68**, 3994 (1999).

[87] M. Igami, T. Nakanishi, and T. Ando, J. Phys. Soc. Jpn. **70**, 481 (2001).

[88] T. Kostyrko, M. Bartkowiak, and G. D. Mahan, Phys. Rev. B **59**, 3241 (1999).

[89] C. T. White and T. N. Todorov, Nature (London) **393**, 240 (1998).

[90] M. P. Anantram and T. R. Govindan, Phys. Rev. B **58**, 4882 (1998).

[91] S. Roche and R. Saito, Phys. Rev. B **59**, 5242 (1999).

[92] K. Harigaya, Phys. Rev. B **60**, 1452 (1999).

[93] T. Kostyrko and M. Bartkowiak, Phys. Rev. B **60**, 10735 (1999).

[94] J. Han, M. P. Anantram, R. L. Jaffe, J. Kong, and H. Dai, Phys. Rev. B **57**, 14983 (1998).

[95] B. I. Dunlap, Phys. Rev. B **46**, 1933 (1992).

[96] B. I. Dunlap, Phys. Rev. B **49**, 5643 (1994).

[97] R. Saito, G. Dresselhaus, and M. S. Dresselhaus, Phys. Rev. B **53**, 2044 (1996).

[98] L. Chico, V. H. Crespi, L. X. Benedict, S. G. Louie, and M. L. Cohen, Phys. Rev. Lett. **76**, 971 (1996).

[99] R. Tamura and M. Tsukada, Solid State Commun.,101,601,1997.

[100] R. Tamura and M. Tsukada, Phys. Rev. B **55**, 4991 (1997).

[101] R. Tamura and M. Tsukada, Z. Phys. D **40**, 432 (1997).

[102] V. Meunier, L. Henrard, and Ph. Lambin, Phys. Rev. B **57**, 2586 (1998).

[103] Z. Yao, H. W. C. Postma, L. Balents, and C. Dekker, Nature (London) **402**, 273 (1999).

[104] M. Menon and D. Srivastava, Phys. Rev. Lett. **79**, 4453 (1997).

[105] J. Li, C. Papadopoulos, and J. Xu, Nature (London) **402**, 253 (1999).

[106] G. Treboux, P. Lapstun, Z. Wu, and K. Silverbrook, J. Phys. Chem. B **47**, 8671 (1999).

[107] G. Treboux, J. Phys. Chem. B **47**, 10381 (1999).

[108] A. J. Stone and D. J. Wales: Chem. Phys. Lett. **128**, 501 (1986).

[109] H. Terrones and M. Terrones, Fullerene Sci. Technol. **4**, 517 (1996).

[110] K. Kusakabe, K. Wakabayashi, M. Igami, K. Nakada, and M. Fujita, Mol. Cryst. Liq. Cryst. **305**, 445 (1997).

[111] H. J. Choi and J. Ihm, Phys. Rev. B **59**, 2267 (1999).

[112] H. Matsumura and T. Ando, J. Phys. Soc. Jpn. **70**, 2657 (2001).

[113] S. Itoh and S. Ihara, Phys. Rev. B **48**, 8323 (1993).

[114] V. Meunier, Ph. Lambin, and A. A. Lucas, Phys. Rev. B **57**, 14886 (1998).

[115] M. -F. Lin, J. Phys. Soc. Jpn. **67**, 1094 (1998).

[116] M. F. Lin, Physica B **269**, 43 (1999).

[117] B. I. Dunlap, Phys. Rev. B **50**, 8134 (1994).

[118] V. Ivanov, J. B. Nagy, Ph. Lambin, A. A. Lucas, X. B. Zhang, X. F. Zhang, D. Bernaerts, G. van Tendeloo, S. Amelinckx, and J. van Lunduyt, Chem. Phys. Lett. **223**, 329 (1994).

[119] S. Ihara and S. Itoh, Carbon **33**, 931 (1995).

[120] K. Akagi, R. Tamura, M. Tsukada, S. Itoh, and S. Ihara, Phys. Rev. Lett. **74**, 2307 (1995).

[121] S. Iijima, M. Yudasaka, R. Yamada, S. Bandow, K. Suenaga, F. Kokai, and K. Takahashi, Chem. Phys. Lett. **309**, 165 (1999).

[122] H. Matsumura and T. Ando, J. Phys. Soc. Jpn.,67,3542,1998.

[123] H. Matsumura and T. Ando, Mol. Crys. Liq. Crys. **340**, 725 (2000).

[124] R. Tamura and M. Tsukada, Phys. Rev. B **58**, 8120 (1998).

[125] R. Saito, T. Takeya, and T. Kimura, G. Dresselhaus, and M. S. Dresselhaus, Phys. Rev. B **57**, 4145 (1998).

[126] See for example, W. A. Harrison, *Electronic Structure and the Properties of Solids* (W.H. Freeman and Company, San Francisco, 1980).

Chapter 2

Vertical diatomic artificial quantum dot molecules

D. G. Austing[a†], S. Sasaki[a], K. Muraki[a], Y. Tokura[a], K. Ono[b]
S. Tarucha[a,b], M. Barranco[c], A. Emperador[c], M. Pi[c], F. Garcias[d]

[a]*NTT Basic Research Laboratories, NTT Corporation, 3-1 Morinosato Wakamiya, Atsugi, Kanagawa, 243-0198, Japan*

[b]*Departament of Physics and ERATO Mesoscopic Correlation Project, University of Tokyo, 7-3-1 Hongo, Bunkyo-ku, Tokyo, 113-0033, Japan*

[c]*Departament ECM, Facultat de Fsica, Universitat de Barcelona, E-08028 Barcelona, Spain*

[d]*Departament de Fsica, Facultat de Cincies, Universitat de les Illes Balears, E-07071 Palma de Mallorca, Spain*

[†]*Also at: Institute for Microstructural Sciences, M23A, National Research Council, Montreal Road, Ottawa Ontario K1A 0R6, Canada,*
E-mail: guy.austing@nrc.ca

Abstract

Circular vertically coupled semiconductor double quantum dots can be employed to study the filling of electrons in quantum dot artificial molecules. When the dots are quantum mechanically strongly coupled, the electronic states in the system are substantially delocalized, and the Coulomb diamonds and the addition energy spectra of the artificial molecule resemble those of a single quantum dot artificial atom in the few-electron limit. When the dots are quantum mechanically weakly coupled, the electronic states in the system are substantially localized on one dot or the other, although the dots can be electrostatically coupled, and this leads to a pairing of the counductance peaks in the several-electron regime. We also describe more generally the dissociation of the few-electron artificial molecules at 0 T as a function of interdot distance from the strong coupling limit to the weak coupling limit. Slight mismatch unintentionally introduced in the fabrication of the artificial molecules from materials with nominally identical constituent quantum wells is responsible for electron localization as the interdot coupling becomes weaker. This offsets the energy levels in the quantum dots by up to 2 meV, and this plays a crucial role in the appearance of the addition energy spectra as a function of coupling strength particularly in the weak coupling limit.

1. Introduction

Semiconductor quantum dots (QD's) are widely considered as artificial atoms, and are uniquely suited to study fundamental electron-electron interactions and quantum effects [1]. We have recently reported atomic-like properties of artificial semiconductor atoms by measuring conductance (Coulomb) oscillations in high quality disk-shaped vertical quantum dots containing a tunable number of electrons starting from zero [2]. A 'shell' structure marked by 'magic' numbers in the addition energy spectrum, a pairing of conductance peaks in the presence of a magnetic field applied parallel to the tunneling current due to spin degeneracy, and modifications in line with Hund's first rule can all be observed. Knowledge of the attributes of a single quantum dot is invaluable for understanding single electron phenomena in more complex quantum dot systems.

There are many analogies with 'natural' atoms. One of the most appealing is the capability of forming molecules. There is much current interest in semiconductor systems composed of two quantum dots, particularly for possible solid-state quantum computing, as they constitute basic qubits. Indeed, systems composed of two QD's, artificial quantum molecules (QM's), coupled either laterally or vertically, have recently been investigated experimentally [3,4] and theoretically [5–8].

In this article we outline how vertically coupled disk-shaped dots can be employed to study the filling of electrons in artificial semiconductor molecules. When the dots are quantum mechanically strongly coupled the electronic states in the system are essentially delocalized. On the other hand, when the dots are quantum mechanically weakly coupled, the electronic states in the system are usually localized. Nevertheless, the weakly coupled dots are still coupled electrostatically, and this can lead to a pairing of conductance peaks.

The direct observation of a systematic change in the addition energy spectra for few-electron (number of electrons, $N < 13$) QM's as a function of interdot coupling, has not been reported before, and calculations of QM properties widely assume a priori that the constituent QD's are identical [5–7]. Our special transistors incorporating QM's [9] made by vertically coupling two well-defined and highly symmetry QD's [2] are ideally suited to observe the former and test the latter. We stress that our quantum dot molecules are different from those reported recently [3,4]. The coupling strength can be tuned *in-situ*, a highly desirable attribute, in planar (lateral) double quantum dot transistors, but usually only a many-electron QM can be realized [10]. Single-particle states can be observed in small ungated vertical triple barrier structure devices, but this kind of device usually is not designed to accumulate electrons one-by-one at zero bias (equilibrium condition). In another type of vertical QM, the two QD's are actually coupled laterally, and the coupling strength is a strong function of N. There are also semiconductor QM's based on self-assembled quantum dots. Either one probes electrically a large ensemble of such QM's, so there is inhomogeneous broadening, or one studies optically a single QM. So far, N is restricted to just a few electrons (and holes), or the coupling has only been varied over a limited effective range.

Few-electron artificial quantum molecules in the vertical geometry exhibit a particularly rich behavior [9], and a number of theoretical investigations of these vertical QM's, often assuming the two QD's are identical, have also been reported [5–8]. In practice, there is a small unavoidable mismatch (offset) between the sets of energy levels in the two constituent QD's which cannot be neglected especially when quantum mechanical coupling and electrostatic coupling between the dots is weak. We later show that the mismatch energy, 2δ, is typically 0.5 to 2 meV. The degree of electron localization-delocalization in a QM system significantly depends on how large 2δ is relative to the quantum mechanical coupling energy, $\Delta_{\rm SAS}$, and this is reflected dramatically in the addition energy spectrum. Spectra from new and realistic model calculations by Local Spin Density Functional Theory (LSDFT) for double dot structures [11,12], with and without mismatch, can now be compared to the experimental spectra to shed new light on how the addition energy spectra evolve with $\Delta_{\rm SAS}$ and 2δ. This allows us to evaluate whether our QM's are homonuclear-like or heteronuclear-like.

2. The vertical artificial molecule transistor

The molecules we study are formed by vertically coupling, quantum mechanically and electrostatically, two QD's which individually can show clear atomic-like features [2,9]. This quantum dot molecule can be realized in the vertical geometry, as illustrated schematically in Fig. 1, by placing a single gate around a sub-micron cylindrical mesa incorporating a $GaAs/Al_{0.2}Ga_{0.8}As/In_{0.05}Ga_{0.95}As$ triple barrier structure (TBS) specially designed to accumulate electrons in the linear transport regime. In a simple picture, each dot (dot 1 and dot 2) in our molecule can be thought of as a circular disk. The thickness of the disk (~10 nm) is typically ten times smaller than the effective diameter (~100 nm) in the few-electron limit. The thickness is well defined because of heterostructure nature of the TBS tunneling barriers. The diameter is determined by the depletion region spreading from the side-wall of the mesa, the extent of which is regulated by the action of the Schottky gate.

Our QM devices, also shown schematically in Fig. 2 (a), are fabricated from TBS's with nominally identical quantum wells of width 12 nm, and the outer barriers are typically about 7 to 8 nm wide. Figure 2 (b) shows a scanning electron micrograph of a typical mesa after gate metal deposition. The two vertically coupled QD's

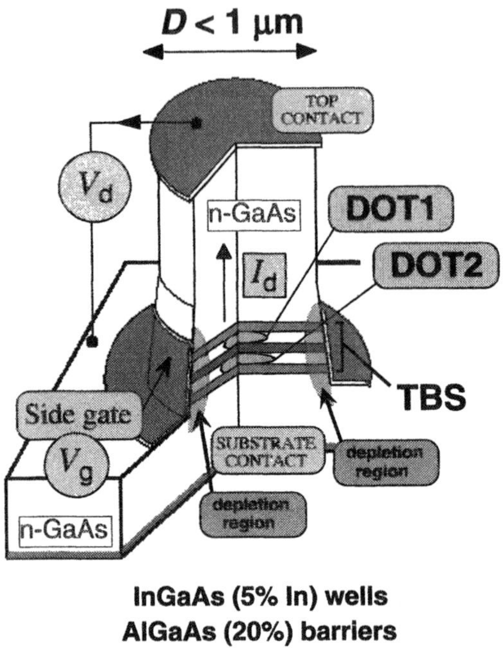

Fig. 1: Schematic section through our single electron transistor device incorporating two vertically coupled quantum dots.

are located inside the circular mesa of geometric diameter $D < 1\mu$m. The TBS starting material, and the processing recipe are described fully elsewhere [9]. The processing involves a special two stage etch to form a mesa with undercut prior to deposition of the gate metal. Note that the side-wall of the mesa is not perfectly vertical. The base of the mesa is actually slightly wider than the top of the mesa (see Fig. 1). Current I_d flows through the two coupled QD's, separated by the central barrier of thickness b, in response to bias voltage V_d applied between the substrate contact and grounded top contact, and voltage on the single surrounding side gate V_g. By measuring the properties of the current (conductance) oscillations and the Coulomb diamonds discussed later, we are able to identify attributes of quantum dot molecules. Our single electron transistors are ideal for studying single electron tunneling and single electron charging phenomena. The transistor structures are cooled to about 300 mK or less and no magnetic field is applied.

Referring to Fig. 2 (c), it is convenient now to state that the artificial molecule is well modeled by combining a radial circularly symmetric harmonic oscillator potential (with realistic 5 meV lateral confinement energy) with a double quantum well potential in the vertical (z) direction [11,12]. The wells are of width w (=12 nm) and depths $V_0 \pm \delta$ ($V_0 = 225$ meV $\gg \delta$). A barrier height of about 225 meV is realistic for the actual barriers in our starting material. δ is included as a simple means to induce the QM to change from being homonuclear-like to heteronuclear-like. Quantum mechanical coupling in the z-direction gives rise to bonding ($|\mathrm{B}\rangle$) and anti-bonding

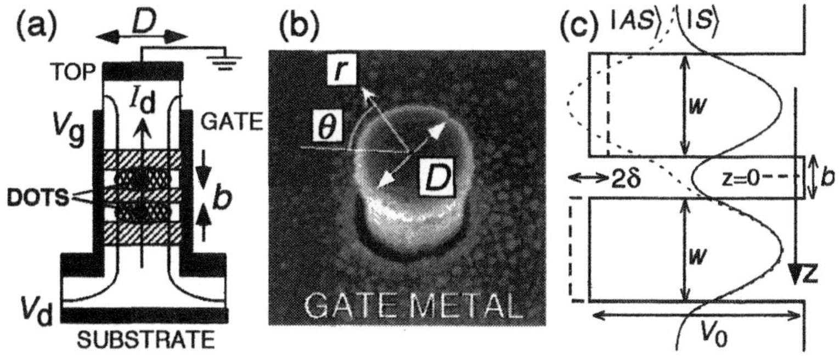

Fig. 2: Schematic diagrams of (a) mesa containing two vertically coupled quantum dots and (c) double quantum well structure, and (b) scanning electron micrograph of a typical circular mesa just after the gate metal has been deposited.

($|AB\rangle$) states that would ideally be shared 50:50 between the constituent QD's if $2\delta = 0$ meV. Note that for a perfectly symmetric system, it is also widespread to refer to the bonding (anti-bonding) states, as symmetric (anti-symmetric) states. The unperturbed symmetric and anti-symmetric states are marked ($|S\rangle$) and ($|AS\rangle$) respectively in Fig. 2 (c).

3. Control of the coupling between the two dots in the artificial molecule

By changing the thickness of the central $Al_{0.2}Ga_{0.8}As$ barrier, b, we can control how strongly the two QD's are coupled. For the materials we typically use, the energy splitting between the bonding and anti-bonding sets of single particle (sp) molecular states, Δ_{SAS}, can be varied from about 3.5 meV for b=2.5 nm (strong quantum mechanical coupling) to about 0.1 meV for b=7.5 nm (weak quantum mechanical coupling) [9]. This is expected to have a dramatic effect on the electronic properties of vertical QM's [6–8,11,12].

Figure 3 shows how Δ_{SAS} varies with b based on a simple one-dimensional flat band model calculation with a material-dependent effective mass. The triple barrier potential is assumed to be perfectly symmetric. Thick marks along the lower axis mark b values for six different TBS's we study, namely 2.5, 3.2, 4.0, 4.7, 6.0, 7.5 nm. Strong (weak) quantum mechanical coupling means $\Delta_{SAS} \gtrsim \hbar\omega$ ($\Delta_{SAS} \ll \hbar\omega$) where $\hbar\omega$ is the bare lateral confinement energy in the few-electron limit ($N \to 0$). $\hbar\omega$ can be ~5 meV in a single QD structure [2,13]. However, $\hbar\omega$ can be ~30% lower in a double QD structure. The difference is because the exact lateral depletion potential profile in the vicinity of the QD and the larger QM is not quite the same [14].

Figure 4 shows a simple cartoon indicating how the important characteristic energies, Δ_{SAS}, $E_{\text{electrostatic}}$, E_{charging}, of a diatomic quantum dot molecule change with central barrier thickness, b. $E_{\text{electrostatic}}$ is the classical electrostatic charging energy, i.e., the minimum energy (when 2δ=0 meV) to place the next electron on one dot

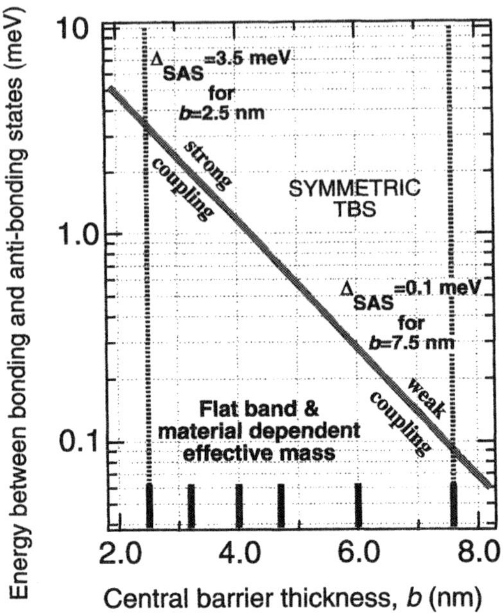

Fig. 3: Calculated value of Δ_{SAS} as a function of central barrier thickness, b, according to a simple model and assuming a symmetric triple barrier potential. See text for more details.

if the previous electron was added to the other dot, assuming the two dots are 'distinct'. $E_{charging}$ is the classical charging energy for placing sequentially electrons one-by-one on just 'one dot'. For strong and intermediate coupling, quantum mechanical coupling is dominant ($\Delta_{SAS} \gg E_{electrostatic}$), and bonding and anti-bonding states are well separated (experimentally resolvable). If coupling is very strong the QM behaves like a single QD in the few-electron limit where only bonding states are initially populated. $E_{electrostatic}$ decreases with b but not as strongly as Δ_{SAS}. Thus, for weak coupling, electrostatic coupling is dominant ($E_{electrostatic} \gg \Delta_{SAS}$), and the QM takes on the characteristics of two separate QD's (particularly if the dots are not perfectly identical). Clearly, the competition between the two mechanisms as b is varied is expected to have a profound effect on the transport properties of the two dot system. Note $E_{charging}$ changes also with b. In the strong coupling limit it is approximately half the value in the weak coupling limit because the QM behaves like just 'one dot' with an effective electronic volume double that of the constituent dots. Finally, as we discuss in detail later, the mismatch energy (offset), 2δ, is shown as an irregular region because it can vary for material-to-material and device-to-device. Naively, one should expect that the unintentional offset is less important for smaller b ($\Delta_{SAS} \gtrsim 2\delta$). On the other it should become more important for larger b ($2\delta \gg \Delta_{SAS}$, $E_{electrostatic}$). We later demonstrate this to be true.

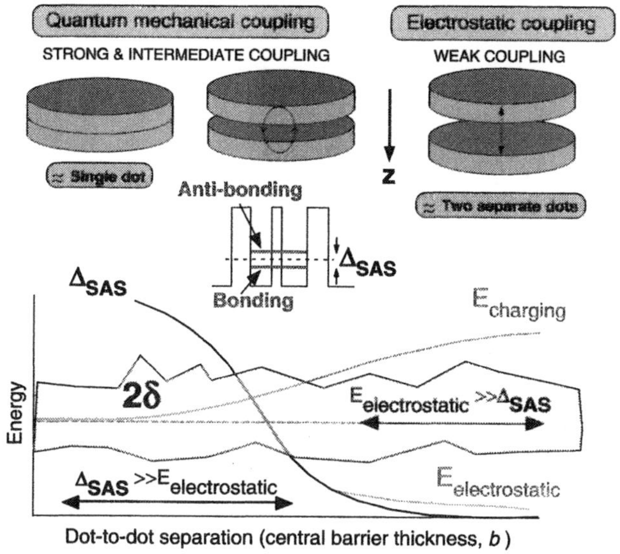

Fig. 4: Simple cartoon showing how the characteristic energies, Δ_{SAS}, $E_{electrostatic}$, $E_{charging}$, of a diatomic quantum dot molecule change with central barrier thickness, b. For strong and intermediate coupling, quantum mechanical coupling is dominant, and bonding and anti-bonding states are well separated. If coupling is very strong the QM behaves like a single QD in the few-electron limit. For weak coupling, electrostatic coupling is dominant, and the QM takes on the characteristics of two separate QD's. The mismatch energy (offset), 2δ, is shown as an irregular region because it can vary for material-to-material and device-to-device.

4. General behavior in the strong and weak coupling limits

An *addition energy* spectrum is a simple and convenient way to characterize the energy required to add one-by-one electrons to a QD or QM system. It provides information about the effective lateral confinement and the Coulomb energies (as well as 2δ). The latter contains direct, exchange and correlation contributions. Nonetheless, it does require careful interpretation. Experimentally, addition energy spectra can be deduced straightforwardly and accurately from the relative spacings between Coulomb oscillation peaks (I_d measured as a function of V_g for $V_d \to 0$ V), or absolutely from the half-widths of the associated Coulomb diamonds (I_d measured in the plane $V_g - V_d$) [2,13]).

Figure 5 shows Coulomb diamonds up to N=14 for a $D = 0.56\mu$m mesa containing two strongly coupled QD's (b=2.5 nm) at 0 T. In the lower panel, regions of black, grey, and white, respectively, represent positive current, zero current, and negative current. The relative size of the diamonds (width along V_g or V_d axis) reflects clearly the underlying shell structure. The half-widths of the diamonds along the V_d axis directly give $\Delta_2(N)$, the change in electro-chemical potential, as a function of number of electrons in the two dot system, N, and this is the quantity plotted

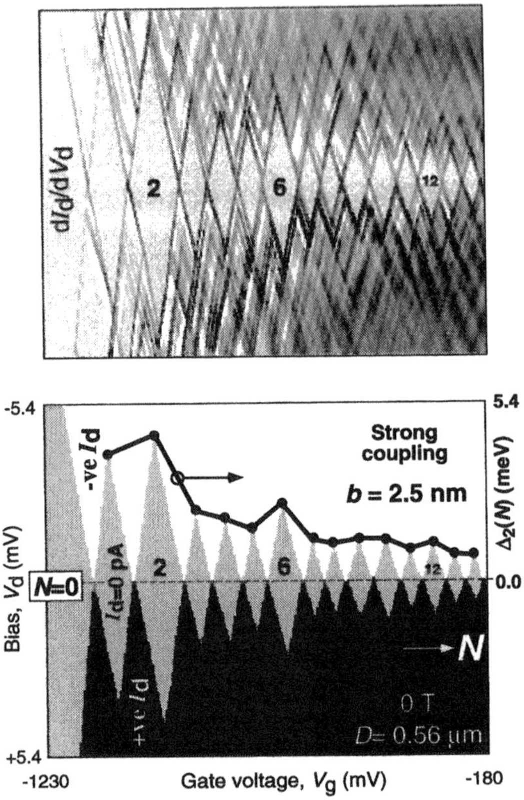

Fig. 5: Coulomb diamonds for a $D = 0.56\mu$m mesa containing two strongly coupled QD's (b=2.5 nm) at 0 T in the bias-gate voltage ($V_d - V_g$) plane up to N=14. The relative size of the diamonds reflects the underlying shell structure, and the half-widths of the diamonds here directly give $\Delta_2(N)$, the change in electro-chemical potential, plotted in the addition energy spectrum. In the lower panel, regions of black, grey, and white, respectively, represent positive, zero, and negative I_d. In the upper panel, exactly the same data set is used except the more familiar dI_d/dV_d is plotted instead of I_d. Black, grey, and white, respectively, represent positive, zero, and negative values of dI_d/dV_d.

in the addition energy spectrum. In the upper panel, exactly the same data used to generate the lower panel is shown except dI_d/dV_d is plotted instead of I_d. Black, grey, and white, respectively, represent positive, zero, and negative values of dI_d/dV_d. This representation is good for investigating not just the N-electron ground states, but also the excited states. Excited state spectroscopy in QD and QM systems is discussed elsewhere [13,15].

The Coulomb diamonds in Fig. 5 are well-formed, highly regular, and symmetric with respect to the bias polarity, so they look very much like those seen for single QD's [13]. By regular and symmetric, we mean that the sides of neighboring diamonds are defined by just two sets of effectively parallel lines: one set of lines has

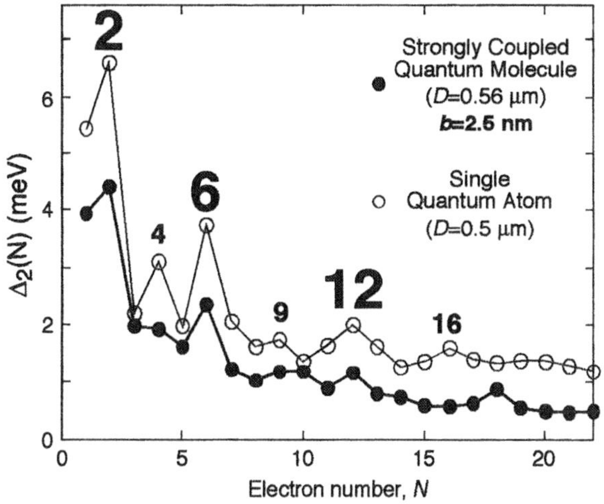

Fig. 6: Addition energy (change in electro-chemical potential), $\Delta_2(N)$, for the $D = 0.56\mu$m quantum mechanically strongly coupled (b=2.5 nm) double dot transistor, and a $D = 0.5\mu$m single dot transistor [2]. For a single disk-shaped artificial atom 'magic' numbers 2, 6, 12, mark the complete filling of a shell, and 4, 9, 16, mark the half filling of a shell (Hund's first rule related) [2,17].

positive dV_g/dV_d, while the other set has negative dV_g/dV_d. The electronic states responsible must therefore be delocalized over the whole QM system. This is what one would expect if well-developed bonding and anti-bonding states are present. Notice that the N=2 and 6 Coulomb diamonds are unusually large compared to the adjoining diamonds. Actually, this is what one would expect for electrons being added to just bonding-states for N up to at least 6. Then the peaks in $\Delta_2(N)$ at N=2 and 6 can be explained by the complete filling of the first and second shell of bonding Fock-Darwin like states [2]. The magnetic field dependence up to 4 T of at least the first few peaks (not shown here) is also consistent with this interpretation [14].

Figure 6 shows together $\Delta_2(N)$ up to N=22 for the same b=2.5 nm QM, and a $D = 0.5\mu$m single dot transistor [2]. For the single QD artificial atom, magic number 2, 6, 12, ... marking the complete filling of the first, second, and third shells, and 4, 9, 16, ... marking half-shell filling (Hund's first rule related), respectively are immediately apparent from the principle and secondary peaks in the spectra [2]. For the strongly coupled QM, principle peaks at 2 and 6 are very clear (12 less so). The secondary peak at N=4 is noticeably weaker. The reason for this is not completely understood, but could be due to small ($\sim$ 10%) deviations from perfect circular symmetry of the QD disks [16,17]. In line with earlier comments, $\Delta_2(N)$ is generally less for the QM than the QD: $\Delta_2(1)$ is approximately 30% lower, and $\Delta_2(N > 15)$ is approximately 50% lower [14]. For $N > 6$ deviations between the QD and QM spectra are indicative that in the QM, one electron eventually enters the lowest anti-bonding Fock-Darwin like state [18]. The point where this actually

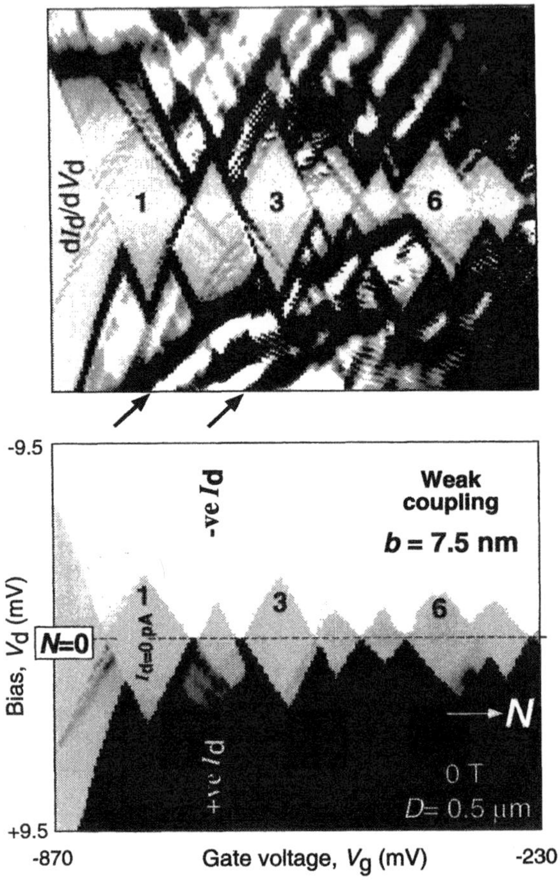

Fig. 7: Coulomb diamonds for a $D = 0.5\mu$m mesa containing two weakly coupled QD's (b=7.5 nm) at 0 T in the bias-gate voltage ($V_d - V_g$) plane up to N=7. As in Fig. 5, the relative size of the now somewhat distorted, asymmetric and apparently poorly formed diamonds still reflects the underlying shell structure [see addition energy spectrum in Fig. 9 (b)]. In the lower panel, regions of black, grey, and white, respectively, represent positive, zero, and negative I_d. In the upper panel, exactly the same data set is used except the more familiar dI_d/dV_d is plotted instead of I_d. Black, grey, and white, respectively, represent positive, zero, and negative values of dI_d/dV_d. Note in addition to the distorted diamonds (more kite-like in shape), there are resonance lines cutting across the diamonds in forward bias (indicated by arrows), which are absent in Fig. 5.

occurs is hard to predict but is probably $N \gtrsim 12$, but certainly the lower states are all bonding states.

Figure 7 shows Coulomb diamonds up to N=7 for a $D = 0.5\mu$m mesa containing two weakly coupled QD's (b=7.5 nm) at 0 T. In the lower panel, regions of black, grey, and white, respectively, represent positive current, zero current, and negative current. As in Fig. 5, the relative size of the now somewhat distorted, asymmetric and apparently poorly formed diamonds still reflects the underlying shell structure. The addition energy spectrum for this QM is shown in Fig. 9 (b). In the upper panel, exactly the same data used to generate the lower panel is shown except dI_d/dV_d is plotted instead of I_d. Black, grey, and white, respectively, represent positive, zero, and negative values of dI_d/dV_d.

The Coulomb diamonds in Fig. 7 are apparently less well-formed, highly irregular, and certainly asymmetric with respect to the bias polarity, so they look very different to those for single QD's [13], and indeed those for the strongly coupled QM's (see Fig. 5). By less well-formed, we mean that the onset of current flow in the vicinity of the border of each diamond is not so steep as that for the strongly coupled QM. The grey scale in the lower panel is set such that if the absolute current is ~500 fA off from "zero current" the color saturates at black or white. Weak structure apparent near the borders of the diamonds is not noise but represents real current flow on the order of ~100 fA or less. Interesting spin-related and co-tunneling physics associated with these low current features, particularly for N=2, will be discussed in detail elsewhere [15]. By irregular and asymmetric, we mean that the sides of neighboring diamonds are not defined by just two sets of effectively parallel lines, as is the case for the single QD's or the strongly coupled QM's. Close inspection of Fig. 7 reveals two sets of almost parallel lines with positive but different dV_g/dV_d, and two other sets of parallel lines with negative but different dV_g/dV_d. In fact, most of the Coulomb diamonds have a shape that is more kite-like than diamond-like. Note also that in addition to the distorted diamonds (kites), there are resonance lines (black and white stripes) cutting across the diamonds (kites) in forward bias (indicated by arrows in upper panel), which are clearly absent in the upper panel of Fig. 5. These resonance lines, and other similar lines running out to several 100 mV's in both bias directions (not shown) are related to zero-dimensional zero-dimensional (0D-0D) resonant tunneling of electrons through the individual dot states (resonance width ~0.3 meV). This too is discussed in detail elsewhere [15]. Taking these observations together, the in total four sets of lines with different dV_g/dV_d defining the diamonds (kites), and the presence of 0D-0D resonant tunneling at finite bias is clear evidence that the electronic states responsible must be substantially localized on one dot or the other. This is what one would expect if bonding and anti-bonding states are built-up substantially from states of just dot 1 or just dot 2. Generally, we can say that non-resonant processes are largely responsible for the filling of the two separate dots as we run along the V_g axis ($V_d \sim 0$ V). Notice that the N=1 and N=3 Coulomb diamonds (kites) are unusually large compared to the adjoining diamonds (kites). Naively, these magic numbers are somewhat surprising, and certainly they are different from the magic numbers of both the single QD and the strongly coupled QM for small-N. These unexpected magic numbers, and the clear asymmetry with respect to bias polarity in Fig. 7 of both the diamonds (kites) and the 0D-0D resonances are direct evidence that some key attribute of the vertical QM system, particularly important in the weak coupling regime, has been overlooked. Actually, the general behavior in Fig. 7 is what one

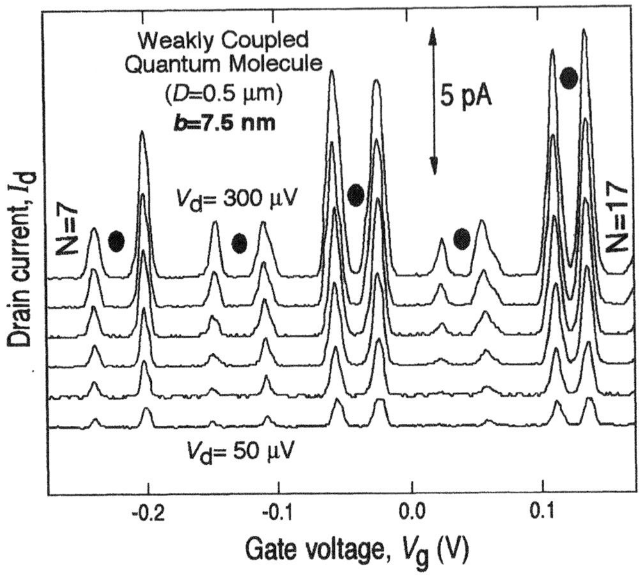

Fig. 8: Excitation voltage dependence from $V_d = 50\mu$V to 300μV of the $I_d - V_g$ characteristic for the $D = 0.5\mu$m quantum mechanically weakly coupled (b=7.5 nm) double dot transistor showing five consecutive pairs of current peaks (Coulomb oscillations) from N=7 to 17. Each pair is marked by '•'. This pairing is discussed in the text.

would expect for a QM system that is not perfectly symmetric, i.e., the QM is heteronuclear-like rather than homonuclear-like. The attribute in question is some slight offset in the energy levels of the two nominally identical constituent dots. This mismatch can affect the appearance of the diamonds (kites) as well as the shape of $\Delta_2(N)$. We will shortly investigate in detail the influence of mismatch (offset) as b is varied from 2.5 to 7.5 nm, and quantify the energy scale of the mismatch energy (2δ).

Finally, in this section, we show in Fig. 8 the excitation voltage dependence from $V_d = 50\mu$V to 300μV of the $I_d - V_g$ characteristic for the same $D = 0.5\mu$m quantum mechanically weakly coupled (b=7.5 nm) double dot transistor. On entering the several-electron regime ($N > 6$), it appears that at 0 T the dots are filled alternately. Five consecutive pairs of current peaks (Coulomb oscillations) from N=7 to 17 are evident (each pair is marked by '•'). We presume that the pairing arises from electrostatic coupling between the dots [19]. From the related Coulomb diamonds (kites) (not shown in Fig. 7), we can estimate the energy splitting between the peaks belonging to each pair. This energy splitting of ~0.7 meV is a measure of $E_{\text{electrostatic}}$, and does not appear very sensitive to N. The unusual pairing and resonant enhancement of conductance peaks with magnetic field will be discussed elsewhere [20]. Note that in Fig. 8, odd-N peak spacings are larger than even-N peak spacings. We speculate that this surprising pattern, rather than the more intuitively expected opposite pattern [see Fig. 9 (c)], is due to mismatch (offset) and its complexity, but is not well understood [18].

5. Mismatch and its effect on electron localization

In this section we present experimental and theoretical addition energy spectra characterizing the dissociation of slightly asymmetric vertical diatomic QM's on going from the strong to the weak coupling limits that correspond to small and large interdot distances, b, respectively. We also show that spectra calculated for symmetric diatomic QM's only resemble those actually observed when the coupling is strong. The interpretation of our experimental results is based on the application of local-spin density-functional theory (LSDFT) [11,21,22]. It follows the development of the method thoroughly described in Ref. [11], which includes finite thickness effects of the dots, and uses a relaxation method to solve the partial differential equations arising from a high order discretization of the Kohn-Sham equations on a spatial mesh in cylindrical coordinates [23]. Axial symmetry is imposed, and the exchange-correlation energy has been taken from Perdew and Zunger [21]. To analyze the experiments we have modeled the QM by two axially symmetric QD's. The QM is confined in the radial direction by a harmonic oscillator potential $m\omega^2 r^2/2$ of strength $\hbar\omega = 5$ meV (a realistic lateral confinement energy for a single QD in the few electron limit [2,18]), and in the axial (z) direction by a double quantum well structure whose wells are of same width w, and have depths $V_0 \pm \delta$, with $\delta \ll V_0$ [24]. Figure 2 (c) schematically shows the double quantum well structure and its unperturbed bonding and anti-bonding single particle (sp) wavefunctions. If δ is set to zero, the artificial molecule is symmetric ('homonuclear' diatomic QM); otherwise, it is asymmetric ('heteronuclear' diatomic QM). In the LSDFT calculations here, δ is 0 meV, or it is set to a realistic value of 0.5 or 1 meV [26]. In the homonuclear case, $\Delta_{SAS}$is well reproduced by the law $\Delta_{SAS}(b) = \Delta_0 \exp(-b/b_0)$ with $b_0 = 1.68$ nm, and $\Delta_0 = 19.1$ meV. Note that this law gives very slightly different values of $\Delta_{SAS}(b)$ compared to those shown in Fig. 3 because the details of the calculation are slightly different. It is easy to check that in the weak coupling limit ($2\delta > \Delta_{SAS}(b) \rightarrow 0$ meV), 2δ is approximately the energy splitting between the bonding and anti-bonding sp states which would be almost degenerate if δ is 0 meV. For this reason we call the mismatch (offset) the quantity 2δ. We stress that the common assumption that the two dots are perfectly identical (meaning perfect alignment of states in dot 1 and dot 2, or equivalently identical electron densities in dot 1 and dot 2) is too idealistic although certainly it is computationally convenient. In reality the two dots will not be perfectly identical (meaning small misalignment of states in dot 1 and dot 2, or equivalently slight difference between electron densities in dot 1 and dot 2).

Figure 9 (a) shows calculated addition energy spectra, $\Delta_2(N) = U(N+1) - 2U(N) + U(N-1)$, for homonuclear QM's ($2\delta$=0 meV) for a sequence of realistic values of b conveniently normalized as $\Delta_2(N)/\Delta_2(2)$. $U(N)$ is the total energy of the N-electron system. As we have noted $\Delta_2(N)$ can reveal a wealth of information about the energy required to put an extra electron into a QD or QM system [2,17].

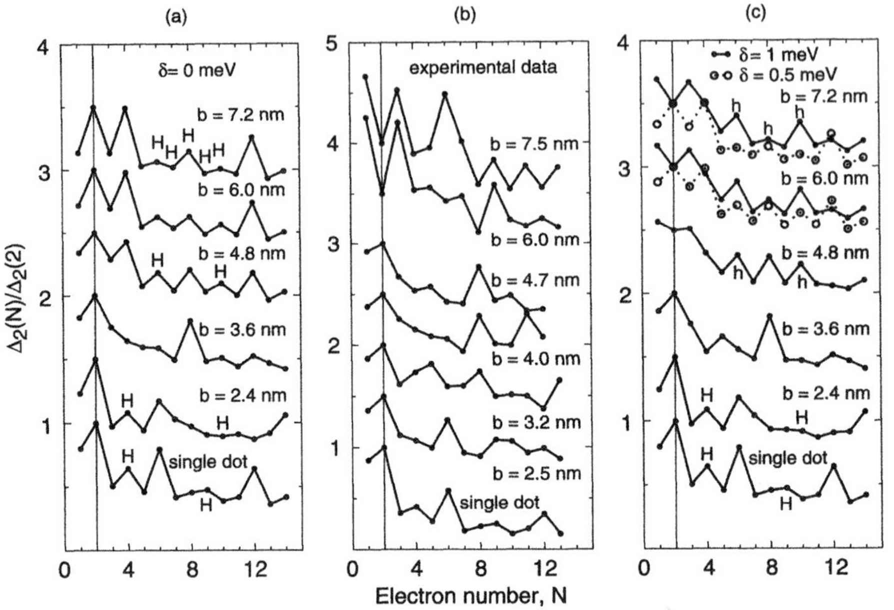

Fig. 9: (a) Calculated $\Delta_2(N)/\Delta_2(2)$ for homonuclear QM's with different interdot distances, b. Also shown is the calculated reference spectrum for a single QD. (b) Experimental QM addition energy spectra, $\Delta_2(N)/\Delta_2(2)$, for several interdot distances between 2.5 and 7.5 nm. Also shown is an experimental reference spectrum for a single QD [17]. (c) Same as panel (a) but for heteronuclear QM's obtained using a 2δ=2 meV mismatch (dotted lines for b=6.0 and 7.2 nm are for 2δ=1 meV). In each panel the curves have been vertically offset so that at N=2 they are equally separated by 0.5 units for clarity. All traces in panels (a) and (c) except 3.6 and 6.0 nm: H(h) marks cases where we could clearly identify Hund's first rule like filling within single dot, or bonding or anti-bonding states (constituent dot states).

Clearly the spectral features are very sensitive to b. For small b ($\Delta_{\mathrm{SAS}} \gtrsim \hbar\omega$), the calculated spectrum of a few electron strongly coupled QM is rather similar to that of a single QD, at least for $N < 7$. In this calculation only $|\mathrm{B}\rangle$ states are occupied for $N \lesssim 7$ [18]. At intermediate dot separation, the spectral pattern changes and becomes more complex. $|\mathrm{AB}\rangle$ states can now be populated at smaller N. However, a simpler picture emerges at larger interdot distances when the molecule is about to dissociate. For example, at b=7.2 nm strong peaks at N=2, 4, 12, and a weaker peak at N=8 appear that can be easily interpreted from the peaks appearing in the single QD spectrum. The peaks at N=4 and 12 in the QM are a consequence of symmetric dissociation into two closed shell (magic) N=2 and 6 constituent QD's respectively (i.e., $N = 4 \to 2+2, N = 12 \to 6+6$), whereas the peak at N=8 corresponds to the dissociation of the QM into two identical stable QD's holding four electrons each filled according to Hund's first rule to give maximal spin [2,7]. The QM peak at N=2 is related to the localization of one electron on each constituent dot, the two-electron state being a spin-singlet QM configuration.

Since the modeled QM is homonuclear, each single particle (sp) wave function is shared 50%-50% between the two constituent QD's. Electrons are completely delocalized in the strong coupling limit. As b increases, $\Delta_{\rm SAS}$ decreases and eventually symmetric, $|{\rm S}\rangle$, and anti-symmetric, $|{\rm AS}\rangle$, sp molecular states become quasi-degenerate. Electron localization can thus be achieved combining these states as $(|{\rm S}\rangle \pm |{\rm AS}\rangle)/2^{\frac{1}{2}}$.

We conclude from Fig. 9 (a) that the fingerprint of a dissociating few-electron homonuclear diatomic QM is the appearance of peaks in $\Delta_2(N)$ at N=2, 4, 8 and 12 [7]. This is a robust statement, as it stems from the well understood shell structure of a single QD. Nonetheless, we will now argue that our vertical QM's are slightly heteronuclear ($V_0 \gg \delta > 0$ meV), and that particularly in the weak coupling limit, the observed addition energy spectra for real QM devices are not well explained if we assume that the QM's are homonuclear. If we compare Fig. 9 (a) with the experimental spectra shown in Fig. 9 (b), we are led to conclude that the experimental devices are not homonuclear, but heteronuclear QM's.

The exact mechanism is not fully understood, but the origin of the mismatch is the difficulty in fabricating two perfectly identical constituent QD's in the QM's discussed here, even though all the starting materials incorporate two nominally identical quantum wells. This mismatch can clearly influence the degree of delocalization-localization, and the consequences will depend on how big 2δ is in relation to $\Delta_{\rm SAS}$ [8,25]. Elsewhere we will discuss how the effective value of $\Delta_{\rm SAS}$ is measured, and how the mismatch is determined for all values of b [26]. We merely note here that 2δ is typically 0.5 to 2 meV [this is consistent with the theoretical data in Fig. 9 (c)], and nearly always with the upper QD (dot 1 nearest top contact of mesa in Fig. 1) states at higher energy than the corresponding lower QD states.

Figure 9 (b) shows experimental spectra, also normalized as $\Delta_2(N)/\Delta_2(2)$, for several QM's with b between 2.5 and 7.5 nm, deduced accurately, as discussed above (see Figs. 5 and 7), from peak spacings between conductance (Coulomb) oscillations ($I_d - V_g$) measured by applying an arbitrarily small bias ($V_d < 100\mu$V). Likewise, also shown is a reference spectrum for a single QD, which shows the familiar shell structure for an artificial atom with peaks at N=2, 6 and 12 [17]. Note this single QD is different from the single QD whose $\Delta_2(N)$ spectrum is shown in Fig. 6, although the two spectra are practically identical. The diameters of the mesas all lie in the range of 0.5 to 0.6 μm. While all mesas are circular, we can not exclude the possibility that the QM's and QD's inside the mesas may actually be slightly non-circular ($\sim$ 10% deviation is typical), and that the confining potential is not perfectly parabolic as N increases [16,17]. The experimental QM spectra evolve in a complex manner as $\Delta_{\rm SAS}$ is systematically reduced, but we emphasize the following key observations: i) The spectrum for the most strongly coupled QM (b=2.5 nm) resembles that of the QD up to the third shell (N=12) (see Fig. 6), and indeed looks somewhat like the calculated homonuclear QM spectrum when the coupling is strongest. ii) For intermediate coupling (b=3.2 to 4.7 nm), the QM spectra are quite different from the QD spectrum, and a fairly noticeable peak appears at N=8. iii) For weaker coupling (b=6.0 and 7.5 nm) the spectra are different again, with

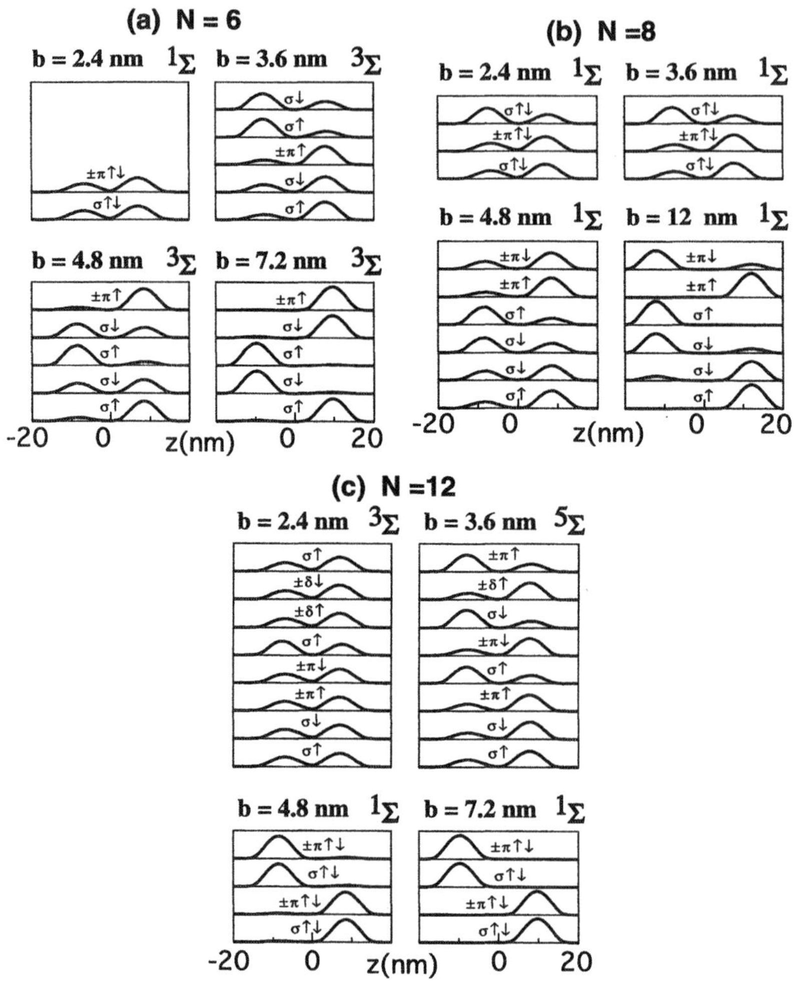

Fig. 10: Calculated probability distribution functions, $P(z)$ in arbitrary units, as a function of z for the heteronuclear $N = 6, N = 8$, and $N = 12$ QM's (a), (b), and (c) respectively, using a 2δ=2 meV mismatch.

prominent peaks at N=1 and 3 (also see Fig. 7) which cannot be explained if the QM is truely homonuclear.

We confirmed the slightly asymmetric heteronuclear character of the QM's by performing LSDFT calculations with a 2δ=2 meV mismatch. The normalized theoretical spectra are displayed in Fig. 9 (c) for the same b values used to generate the spectra in Fig. 9 (a). For b=6.0 and 7.2 nm, spectra for a 1 meV mismatch are also given. One-to-one comparison between theory and experiment of absolute values is not helpful, because the QM's (QD's) actually behave in a very complex way [18]. In particular, 2δ can vary from device-to-device, as well as material-to-material, and probably it decreases with N [26]. Nonetheless, the overall agreement between theory and experiment of the general spectral shape is quite good, indicating the crucial role played by mismatch. The most strongly coupled QM spectrum is still similar to the single QD spectrum. Crucially, however, the appearance of the spectra in the weak coupling limit for small N values is now correctly given (see clear peaks at N=1 and 3), as well as the evolution with b of the peak appearing at N=8 for intermediate coupling. Clearly mismatch and electron localization become relatively more important as Δ_{SAS} is decreased. A comparison between panels (a) and (c) of Fig. 9 reveals that for smaller values of b ($\lesssim$ 4.8 nm), for a reasonable choice of parameters (ω, δ), mismatch does not produce sizeable effects. The reason is that the electrons are still rather delocalized, and distributed fairly evenly between the two dots. Exceptions to this substantial delocalization may arise only when both the constituent single QD states are magic, as discussed below, at intermediate coupling. For larger interdot distances, mismatch induces electron localization. The manner in which it happens is determined by the balance between interdot and intradot Coulomb repulsion, and by the degree of mismatch between the single particle energy levels, and so is difficult to predict except in some trivial cases for certain model parameters (ω, δ). For example, a large mismatch compared to $\hbar\omega$ will cause the QD of depth $V_0 - \delta$, to eventually 'go away empty'.

Finally, still assuming perfect coherency, a deeper theoretical understanding of heteronuclear QM dissociation can now be gained from the analysis of the evolution with b of the sp molecular wave functions. For each single particle (sp) wave function, $\phi_{n\ell\sigma}(r,z,\theta) = u_{n\ell\sigma}(r,z)\exp(-i\ell\theta)\chi_\sigma$ we introduce a z-probability distribution function defined as,

$$P(z) \equiv 2\pi \int_0^\infty drr\,[u(r,z)]^2 \,.$$

Figure 10 shows $P(z)$ for (a) N=6, (b) N=8, and (c) N=12 (deeper well always in the $z > 0$ region), each for several values of b. States are labeled as $\sigma, \pm\pi, \pm\delta, \ldots$, depending on the $\ell = 0, \pm1, \pm2, \ldots$ sp angular momentum, and $\uparrow, \downarrow$ indicate the spins. In each sub-panel, the probability functions are plotted, ordered from bottom to top, according to the increasing energies of the orbitals. For each b, the third component of the total spin and total orbital angular momentum of the ground state are also indicated by the standard spectroscopic notation ${}^{2S_z+1}|L_z|$ with $\Sigma, \Pi, \Delta, \ldots$, denoting $|L_Z|$= 0, 1, 2, ... We conclude that: i) QM's dissociate more easily at smaller values of b, if they yield magic number constituent QD's, as is the case for $N = 12 \to 6 + 6$ for b=4.8 nm (c) or $N = 4 \to 2 + 2$ (not shown), for example. ii) Particularly for intermediate values of b, not all orbitals contribute equally to the QM bonding, i.e., the degree of hybridization is not the same for all QD sp orbitals. See for example the π and σ states in the b=4.8 nm panel of (a). iii) At larger b, dissociation can lead to Hund's first rule like filling in one of the QD's and full shell filling in the other dot. See for example the b=7.2 nm panel in (a) for N=6, which dissociates into 2+4. The same happens for the N=10 QM, which dissociates into 4+6 (not shown). In other cases, dissociation leads to Hund's first rule like filling in each of the QD's, as shown in the b=12 nm panel of (b) for N=8, which breaks into 4+4. In close analogy with natural molecules, atomic nuclei, or multiply charged simple metal clusters [27], homo and heteronuclear QM's choose preferred energetically favorable dissociation channels yielding the most stable QD configurations. iv) Some configurations are extremely difficult to disentangle: even at very large b, there can still be orbitals contributing to the QM bonding. A good example of this is the N=8 QM for b=12 nm (b).

6. Summary

In conclusion, we have shown that gated sub-micron vertical triple barrier structures are ideal for studying complex and interesting properties of coupled quantum dots, i.e., quantum dot molecules. As a function of central barrier thickness, we can alter the degree of coupling between the two dots, and the nature of the dominant coupling mechanism. For quantum mechanically strongly coupled dots, the lower electronic states are bonding-like and largely delocalized over the entire system, and the attributes of the molecule resemble those of a single dot. For quantum mechanically weakly coupled dots, the electronic states of the system are mostly localized on one dot or the other, nevertheless electrostatic coupling is most likely responsible for a pairing of several consecutive conductance peaks in the several-electron regime.

One of our key findings is that the experimental addition energy spectra only resemble those calculated for symmetric homonuclear QM's when the coupling is strong. For intermediate and weak coupling, however, noticeable differences appear between spectra of real QM devices and spectra calculated assuming two identical QD's. Peaks at N=1 and 3 are observed rather than predicted peaks at N=2 and 4, for example, for b=6.0 and 7.5 nm. This is a signature that the constituent dot energy levels are offset, and that electron localization becomes relatively more important as Δ_{SAS} is decreased. This is confirmed by looking at calculated spectra for slightly asymmetric heteronuclear diatomic QM's (2δ=1 or 2 meV), which correctly recover the N=1 and 3 peaks for weak coupling.

Acknowledgements

This work has been performed under grants PB98-1247 and PB98-0124 from DGESIC, and 2000SGR-00024 from Generalitat of Catalunya, and partly funded by NEDO program (NTDP-98). We are very grateful for the assistance of T. Honda with processing the samples.

References

[1] M.A. Kastner, Phys. Today 46, No. 1, 24 (1993); R.C. Ashoori, Nature **379**, 413 (1996).

[2] S. Tarucha et al., Phys. Rev. Lett. **77**, 3613 (1996).

[3] F.R. Waugh et al., Phys. Rev. Lett. **775**, 705 (1995); T. Schmidt et al., Phys. Rev. Lett. **78**, 1544 (1997); G. Schedelbeck et al., Science **278**, 1792 (1997); R.H. Blick et al., Phys. Rev. Lett. **80**, 4032 (1998); M. Brodsky et al., Phys. Rev. Lett. **85**, 2356 (2000).

[4] A. Lorke and R.J. Luyken, Physica B **256-258**, 424 (1998); M. Bayer et al., Science **291**, 451 (2001).

[5] C. Yannouleas and U. Landman, Phys. Rev. Lett **82**, 5325 (1999); A. Wensauer et al., Phys. Rev. B **62**, 2605 (2000).

[6] J.J. Palacios and P. Hawrylak, Phys. Rev. B **51**, 1769 (1995); J. Hu et al., Phys. Rev. B **54**, 8616 (1996); J.H. Oh et al., Phys. Rev. B **53**, R13264 (1996); H. Tamura, Physica B **249-251**, 210 (1998); Y. Asano, Phys. Rev. B **58**, 1414 (1998); M. Rontani et al., Solid State Commun. **112**, 151 (1999).

[7] B. Partoens and F.M. Peeters, Phys. Rev. Lett. **84**, 4433 (2000).

[8] O. Mayrock et al., Phys. Rev. B **56**, 15760 (1997); Y. Tokura et al., J. Phys. Condens. Matt. **11**, 6023 (1999); Y. Tokura et al., Physica E **6**, 676 (2000); G. Burkard et al., Phys. Rev. B **62**, 2581 (2000).

[9] D.G. Austing et al., Physica B **249-251**, 206 (1998); D.G. Austing et al., Semicond. Sci. Technol. **11**, 388 (1996); D.G. Austing et al., Jpn. J. Appl. Phys. **34**, 1320 (1995).

[10] Few-electron lateral double quantum dot molecules have only recently been realized. A. Sachrajda, private communication.

[11] M. Pi et al., Phys. Rev B. **63**, 115316 (2001).

[12] M. Pi et al., Phys. Rev. Lett. **87**, 066801 (2001).

[13] L.P. Kouwenhoven et al., Science **278**, 1788 (1997).

[14] S. Amaha et al., Solid State Commun. **119**, 183 (2001).

[15] K. Ono et al., submitted to Science (2001).

[16] D.G. Austing et al., Phys. Rev. B **60**, 11514 (1999).

[17] P. Matagne et al., submitted to Phys. Rev. B (2001).

[18] In our QD's the effective confinement energy actually decreases with N as discussed by S. Tarucha et al., Appl. Phys. A **71**, 367 (2000). Additionally, the effective confinement energy in our QM's can actually be up to half that of the QD's [14]. Both effects are not well reproduced by any existing calculation. Because of these two effects, population

of anti-bonding states in real QM's can start at higher N than suggested by the calculations for strong coupling, and the filling sequence and observed spectral shape for $N > 6$ can be sensitively modified when the coupling is weak.

[19] G. Klimeck et al., Phys. Rev B. **50**, 2316 (1994).

[20] D.G. Austing et al., unpublished (2001).

[21] J.P. Perdew and A. Zunger, Phys. Rev. B **23**, 5048 (1981).

[22] M. Stopa, Phys. Rev. B **54**, 13767 (1996); M. Koskinen et al., Phys. Rev. Lett. **79**, 1389 (1997); I.H. Lee et al., Phys. Rev. B **57**, 9035 (1998); R.N. Barnett and U. Landman, Phys. Rev. B **48**, 2081 (1993); K. Hirose and N.S. Wingreen, Phys. Rev. B **59**, 4604 (1999).

[23] Self-interaction corrections [21] have not been included. We have checked [11] that they do not play an important role in the calculated addition spectra, see Fig. 12 of this reference.

[24] We have taken for the dielectric constant and the electron effective mass values corresponding to GaAs, i.e., ε=12.4 and m^*=0.067.

[25] K. Muraki et al., Solid State Commun. **112**, 625 (1999); T.H. Oosterkamp et al., Nature **395**, 873 (1998).

[26] S. Sasaki et al., unpublished (2001); K. Ono et al., unpublished (2001); D.G. Austing et al., unpublished (2001).

[27] M. Weissbluth, *Atoms and Molecules* (Academic Press, New York, 1978); P. Ring and P. Schuck, *The Nuclear Many-Body Problem* (Springer-Verlag, Berlin 1980); U. Näher et al., Phys. Rep. **285**, 245 (1997); C. Yannouleas et al., *Metal Clusters* (Wiley, New York 1999), W. Ekardt, Editor, p. 145.

Chapter 3

Optical spectroscopy of self-assembled quantum dots

David Mowbray* and Jonathan Finley

Department of Physics and Astronomy, University of Sheffield, Sheffield S3 7RH, U.K.
**E-mail: D.Mowbray@sheffield.ac.uk*

Abstract

Self-assembled quantum dots provide high optical quality structures suitable for electro-optical device applications and the study of physics in a zero-dimensional semiconductor system. In this article we describe a number of optical spectroscopic studies of In(Ga)As self-assembled quantum dots grown in a GaAs matrix. Studies of both single dots and large dot ensembles are discussed. We demonstrate that it is possible to obtain information concerning the confined electronic states, carrier transport processes and the physical structure of the dots. In addition the influences of Coulomb and exchange interactions between multiple carriers confined within a dot are studied.

1. Introduction

The electronics industry has witnessed a continuous and rapid decrease in circuit feature size, driven by requirements for increased complexity and faster operation speeds. With the present rate of decrease, device sizes will soon enter the regime where quantum mechanical effects become important, in many cases detrimental to conventional device operation. However the physics of semiconductor nanostructures is being actively investigated, as such structures are likely to form the basis of new and novel electronic and electro-optical devices, or improved conventional devices. Quantum wells, which form two-dimensional nanostructures, are already used in visble injection lasers [1] and form the critical component of quantum cascade lasers [2]. More recently, prototypes of a number of devices based on quantum dots (zero dimensional nanostructures) have been demonstrated, including low threshold current [3] and increased temperature stability [4] lasers, single photon sources [5] and detectors [6], normal incidence, far infrared photodetectors [7] and optical memories [8].

Whilst the study and commercial application of quantum wells represents a fairly mature field [9], the area of quantum dot physics and devices is relatively new, having for a long time been hindered by the lack of suitable structures. Quantum dots suitable for optical studies and electro-optical device applications must satisfy a number of requirements, including deep confining potentials, small size to ensure energy level spacings significantly greater than the room temperature thermal energy, confinement of both electrons and holes, high optical quantum efficiency with a low density of non-radiative recombination channels, high areal density, good size and shape uniformity and the ability to be incorporated in the intrinsic region of a *p-i-n* structure, allowing the electrical injection and extraction of carriers. In addition, compatability with existing epitaxial growth techniques is desirable. Quantum dots based on the self assembly technique [10] satisfy all these requirements, with the possible exception of uniformity.

Self-assembled quantum dots form spontaneously during the epitaxial growth of two semiconductors having very different lattice constants. The most extensively studied system consists of InAs dots grown within a GaAs matrix. Starting with a GaAs substrate, InAs is deposited using the epitaxial techniques of molecular beam epitaxy (MBE) or metal organic vapour phase epitaxy (MOVPE). Because of the 7% lattice mismatch between InAs and GaAs, the InAs is initially deposited as a highly strained, two-dimensional layer. The strain energy in this layer rapidly builds up with increasing layer thickness, resulting in a transformation, after the deposition of approximately one atomic layer, to three-dimensional growth in the form of nanometer size islands [11]. These islands form the quantum dots, which sit on the original thin, two-dimensional layer, known as the wetting layer. Although the surface area, and hence surface energy, is increased by this two-dimensional–three-dimensional growth transition, the InAs in the islands starts to relax back to its bulk value, reducing the strain energy [10]. Because this relaxation is elastic in nature, no misfit dislocations are formed and the dots have a high optical quantum efficiency. After growth the dots are generally overgrown with GaAs.

Figure 1 (a) shows a cross-sectional transmission electron micrograph (TEM) of an InAs quantum dot. Self assembled InAs dots have typical base lengths of ~10–50 nm, heights ~5–20 nm and densities ~10^9–10^{11}cm^{-2}. Of direct relevance to the present article is the shape of the dots. Despite extensive structural studies, no broad concensus as to their precise shape exists. Reported or assumed shapes include pyramids [12], truncated pyramids [13], truncated pyramids with octagonal bases [14], lenses [15] and cones [16]. It remains possible that the shape of self-assembled dots may be a function of the growth technique and growth conditions, and may change when the dots are overgrown.

Considerable information concerning the electronic structure of self assembled dots has been obtained from optical studies [10], which utilize a range of spectroscopic techniques. These studies fall into two main groups; studies of large dot ensembles (~10^6–10^8 dots) and studies of single dots. The former has the advantage of relative experimental simplicity, but the results are complicated by inhomogeneous broadening of the spectral features. The latter overcomes the problem of studying a large collection of slightly different dots, but at the expense of greater experimental complexity. In this article we will describe examples of the information that can be obtained from both ensemble and single dot studies. Photocurrent spectroscopy of large ensembles is used to study carrier transport processes and to deduce dot structural parameters. Single dot photoluminescence spectroscopy is used to study coulomb interactions and exchange effects in dots containing multiple electrons and holes.

2. Photocurrent spectrsocopy of quantum dot ensembles

In this section the application of interband photocurrent spectroscopy to determine the absorption spectra of InAs self-assembled quantum dots and to study the effects on these spectra of large electric fields is described. By studying the intensity of the photocurrent as a function of both electric field and temperature, the mechanisms responsible for the escape of carriers from the dots can be identified. The interband transitions exhibit a strong quantum confined Stark shift which is asymmetrical about zero-field. By comparing this behavior with the results of a theoretical model it is possible to deduce information concerning the shape and composition of the dots.

Nominal InAs self-assembled quantum dots were grown by molecular beam epitaxy on (001) GaAs substrates at a temperature of 500°C. The dots were deposited at 0.01 monolayers per second (ML/s), which results in dots of areal density $\sim 1.5 \times 10^9$ cm^{-2}, base size 18 nm and height 8.5 nm, as determined from transmission electron microscopy studies. The asymmetrically shaped dots [see Fig. 1 (a)] are formed on a ~1 ML thick wetting layer and have their apex oriented along the growth direction. Single layers of dots were grown within the intrinsic region of both *p-i-n* and *n-i-p* structures, which allow fields up to 300 kV/cm to be applied either parallel or anti-parallel to the growth direction. Applying a reverse bias to

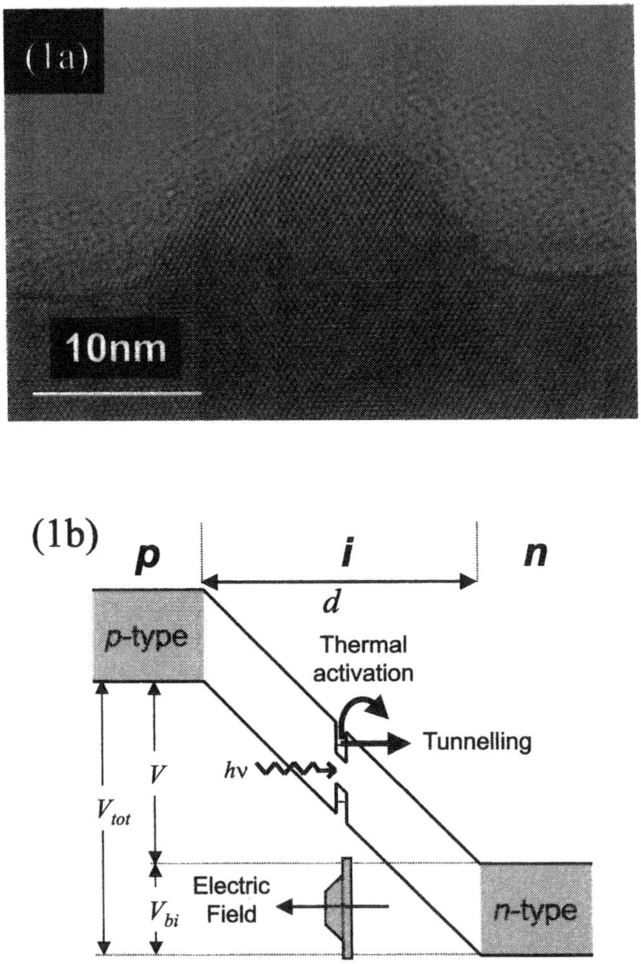

Fig. 1: (a) Cross-sectional transmission electron micrograph of a nominal InAs self-assembled quantum dot. The growth of this structure was terminated after the growth of the dots (no GaAs overgrowth). (b) Schematic band diagram of a GaAs *p-i-n* structure with a single layer of quantum dots grown within the intrinsic region.

a *p-i-n* structure (*p* region at the surface) results in an electric field (F) pointing from the substrate to the surface [see Fig. 1 (b)]. For an *n-i-p* structure the field direction is reversed. Hence by growing nominally identical dots in *n-i-p* and *p-i-n* structures the effects of fields between $\sim\pm 300$kV/cm can be studied. The total electric field is given by the equation $F = (V + V_{bi})/d$, where V is the externally applied voltage, V_{bi} is the built-in junction voltage ($\approx$1.5 V) and $d(= 0.3\mu\text{m})$ is the intrinsic region width. Photocurrent spectra were measured over the temperature range 10 to 300 K using 400 μm diameter mesa devices with optical access, annular contacts. The mesas contain $\sim 2 \times 10^6$ dots. Very low intensity, monochromated white light ($\sim$ 3mW/cm^2, bandwidth $\approx$8 meV) from a tungsten-halogen lamp was

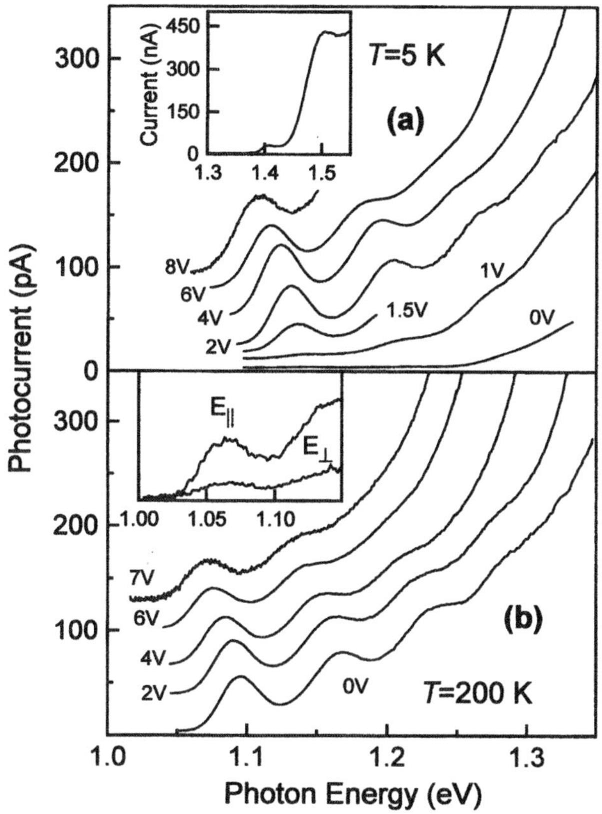

Fig. 2: Photocurrent spectra as a function of applied reverse bias for a single layer of quantum dots. (a) shows spectra recorded for a sample temperature of 5K (b) is for a sample temperature of 200K. The upper inset shows a spectrum to higher spectral energy showing absorption into the wetting layer and bulk GaAs. The lower inset shows polarized photocurrent spectra for inplane propagating light in a waveguide structure. The lowest energy quantum dot transition is strongly polarized for the incident electric field vector along the growth direction.

used for excitation and the photocurrent was detected using lock-in techniques (f_{mod} $\sim$200 Hz), allowing very low ($\sim$1 pA) photocurrents to be detected. The low incident optical power results in extremely low dot carrier occupancies ($\ll 1$) and hence the influence of many carrier effects on the optical spectra can be neglected [17].

Figure 2 shows photocurrent spectra, recorded as a function of applied reverse bias from 0 to 8V (total electric field F from 50 to 320kV/cm) for a *p-i-n* sample at temperatures of 5 and 200K. At low temperatures and fields, carriers are unable to escape from the dots before radiative recombination occurs, and a photocurrent signal is only observed from the InAs wetting layer (1.41 eV) and the GaAs barriers (1.5 eV) (see upper inset to Fig. 2). However, as the field is increased the photocreated carriers escape from the dots by quantum mechanical tunneling and four well-defined

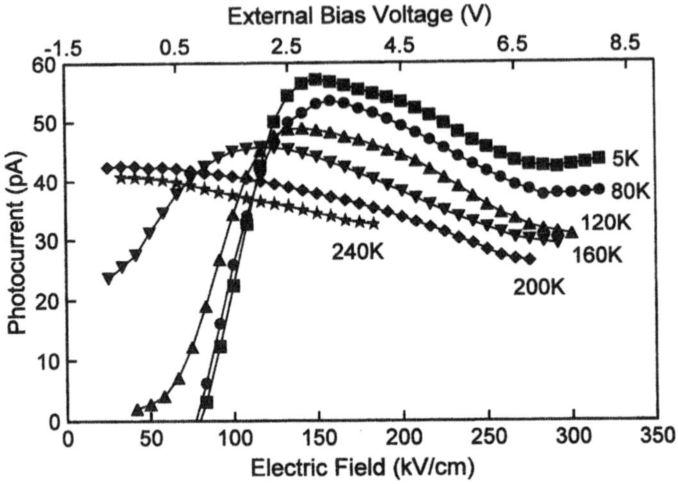

Fig. 3: The intensity of the quantum dot ground state transition plotted as a function of the applied electric field and for temperatures in the range 5 to 240K.

features in the energy range 1.1 to 1.3 eV (at 5 K) are observed. These features arise from inter-band quantum dot transitions between confined electron and hole states. A similar behavior is observed at 200 K, although the quantum dot transitions are observed at increasingly lower fields as the temperature is raised, consistent with a transition to carrier escape by thermal excitation at elevated temperatures. This behavior is more clearly seen in Fig. 3 where the intensity of the lowest energy quantum dot transition is plotted as a function of electric field and for a range of temperatures [18]. At low temperatures the photocurrent intensity exhibits a sharp onset at a field ~80kV/cm, reflecting the switching on of tunnelling carrier escape from the dots. The escape rate associated with this process will vary rapidly with applied field, and tunnelling escape will dominate when this escape rate becomes faster than the radiative recombination rate (~1ns [19]). With increasing temperature the onset of the photocurrent intensity becomes weaker and by 200K the photocurrent intensity is approximately independent of temperature. This behavior indicates that at high temperatures carrier escape from the dots is dominated by thermal activation, a field independent mechanism. The slight decrease in the photocurrent intensity at high electric fields, and all temperatures, reflects a decrease in the transition oscillator strength as the electron and hole wavefunctions are pulled apart by the applied electric field.

With increasing field, and for all temperatures, all the quantum dot transitions shift strongly to lower energy (by 30 meV at 8 V (≡300 kV/cm)). This behavior is a result of the quantum confined Stark effect, which has been extensively studied in higher dimensionality systems [20]. The ground state transition energy for nominally identical dots in *p-i-n* and *n-i-p* [21] samples, at 200K, is plotted as a function of field in Fig. 4. The transition energy exhibits a significant asymmetry about zero field,

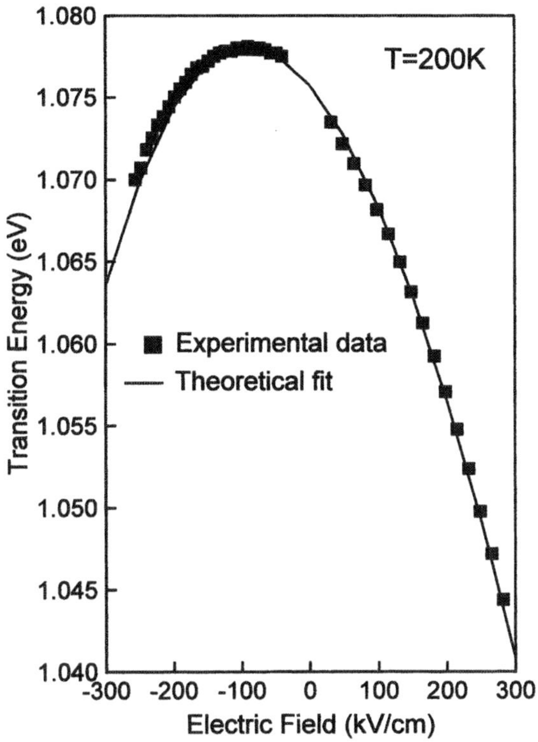

Fig. 4: The quantum dot ground state transition energy plotted as a function of the total electric field. Positive and negative fields were obtained by measuring two different samples, one a *p-i-n* structure the other a *n-i-p* structure. The solid line is the theoretical fit to the experimental data using parameters given in the text.

with the maximum transition energy occurring for a non-zero field of −90 kV/cm. This asymmetry implies that the self-assembled quantum dots have a permanent dipole moment (p), arising from a zero field spatial separation of the electron and hole wave functions along the growth axis (the field direction in the present experimental geometry).

The field dependence of the transition energy (E) in Fig. 4 can be well described by the equation $E = E_0 + pF + \beta F^2$, where E_0 is the transition energy at zero field. The second term arises from the non-zero dipole moment (p), and the third term (β) arises from polarization of the dots in the applied field (the quantum confined Stark effect). By fitting the above equation to the experimental data in Fig. 4, a value of $p = (7 \pm 2) \times 10^{-29}\mathrm{cm}^{-1}$ is determined, corresponding to an electron-hole separation of $r = 4.0 \pm 1$ Å, obtained from $p = er$.

The maximum transition energy in Fig. 4 occurs for a negative field, corresponding to a field direction from the apex to the base of the dots. For this field direction the electron is attracted to the apex (hole attracted to the base) of the dots. This result implies that the electron charge density distribution lies closer to the base than that of the hole at F=0, with the resultant dipole pointing from base to apex. A permanent dipole moment for self assembled InAs quantum dots is predicted from theoretical modelling, due to the non-uniform quantum dot shape along the growth axis [22]. However, the sign of the dipole moment deduced from the present measurements (hole above electron) is opposite to that predicted by previous theoretical studies of pure InAs dots. For example, the sophisticated theoretical modelling of Refs. [23] and [24] both predict a hole wavefunction which is localized toward the base of the dots, below that of the electron. This alignment, which occurs for pure InAs dots and for any shape for which the lateral dot size decreases from base to apex (e.g., the pyramidal shape used in the models of Refs. [25-29]) results from the strain-induced form of the valence band edge profile [25] and the ratio of the electron and hole effective masses along the growth direction ($m^*_{hh} \gg m^*_e$).

To determine the dot structure necessary to reverse the relative alignment of the electron and hole wavefunctions, the quantum dot shape, size and compositional dependence of the permanent dipole moment (p) and the quadratic field coefficient (β) was calculated using the envelope function method, with the electrons and holes treated with separate one-band Hamiltonians [13]. Although strain will mix the light and heavy hole valence bands, the results of an 8-band **k.p** model calculation indicate that the lowest confined hole state is predominantly (~90%) heavy hole-like [30]. This is confirmed by experimental measurements. The lower inset to Fig. 2 shows polarized photocurrent spectra for light propagating in the plane of the dots [31]. The ground state transition is found to be strongly polarized for the electric field vector along the growth axis (TE mode), consistent with a heavy hole character [32]. These theoretical and experimental results both indicate a ground state with a predominant heavy hole character and hence support the use of a one-band model [33]. The strain distribution for a given dot shape was obtained using a Green's function technique which provides an analytical expression in the form of a Fourier series for the strain tensor [34]. The band gaps and offsets were calculated using model solid theory [35], including hydrostatic strain effects; the heavy-hole Hamiltonian included the spatial variation of the biaxial strain deformation potential and the directional dependence of the heavy-hole mass. Carrier effective masses, determined using 3-band **k.p** theory, and band offsets were assumed to vary linearly with composition.

Initially calculations were performed for pure InAs, pyramidal dots [36]. The results obtained were found to be in good agreement with previous theoretical calculations, based on more sophisticated models, [22-29] with the hole wavefunction always located below that of the electron. This alignment, which is opposite to that determined experimentally for the present dots, therefore appears to be a universal result for InAs pyramidal dots. To reverse the calculated electron and hole alignment it was found to be necessary to alter the assumed dot structure in two ways; a graded $In_xGa_{1-x}As$ composition, with x increasing from base to apex (the holes tend to be localized in the region with the largest In composition) is required and it is also necessary to severely truncate the pyramidal shape (strain effects localize the hole strongly below the electron until the truncation factor is greater than ≈0.6 [36]). Neither of these effects alone is sufficient to reverse the sign of the dipole, both must be used in combination.

The continuous line in Fig. 4 shows the best fit to the experimental data. This is achieved using a pyramid of base length 15.5 nm, height 22 nm, of which the top 75% is truncated to give an actual dot height of 5.5 nm, and an In mole fraction which varies linearly from 50% at the base to 100% at the (truncated) top surface. These parameters give a good fit to the experimental data. Although other combinations of size, shape and composition may give a similar quality of fit [36], the present shape represents a good approximation to that obtained from structural measurements [see Fig. 1 (a)]. In addition, the structural parameters deduced from the fit will be dependent on the model used, with more sophisticated models expected to give slightly different parameters [23]. However the main conclusion, namely that a non-zero and non-uniform Ga composition and a truncated shape are required to give the correct vertical alignment of the electron and hole wavefunctions, is a general result.

Evidence for non-pyramidal shaped dots and the presence of Ga in nominal InAs dots has recently been obtained from a number of structural measurements. Joyce et al. [37] used scanning tunnelling microscopy (STM) to compare the total volume of the dots with that of the deposited InAs. For high growth temperatures, $\sim$ 500°C, similar to that used to grow the dots studied in the present work, the total volume of the dots was found to be greater than that of the deposited InAs. This behavior can only be explained if Ga from the GaAs matrix which surrounds the dots, diffuses into the dots either during or after their growth. Liu et al. [38] studied the shape and composition of $In_{0.5}Ga_{0.5}As$ dots using cross-sectional STM. The dots had a trapezoidal (truncated pyramid) shape with an In rich core in the form of an inverted-triangle shape. The composition grading required to explain the sign of the dipole moment observed in the present work (a higher Ga concentration at the dot base) has been observed by Grandjean et al. in STM studies of $In_{0.3}Ga_{0.7}As$ quantum dots [39]. This grading is attributed to In segregation effects during growth. Finally, Kegel et al. [40] studied the composition profile of nominal InAs dots using surface-sensitive x-ray diffraction. The composition was found to vary continuously from GaAs at the base of the dots to InAs at the top; a gradient of sign consistent with that deduced from the present optical measurements.

The experimental results described in this section demonstrate that optical spectroscopy of self-assembled quantum dot ensembles is capable of providing important information concerning the electronic and structural properties of the dots. Despite the large inhomogeneous linewidth ($\sim$30 meV), photocurrent spectroscopy is a powerful tool because, in the present case, the Stark shifts are comparable to, or exceed the linewidth. A further notable feature of photocurrent spectroscopy is its high sensitivity. The absorption of a single layer of quantum dots is very low, making direct absorption spectroscopy very difficult. Warburton et al. [17] were able to measure the absorption of a single layer of quantum dots but their measurements, which gave absorptions of $\sim 1 \times 10^{-4}$ for the ground state transition, required the use of a state-of-the-art Fourier transform spectrometer and integration times of a few hours. In contrast, the present photocurrent measurements use relatively simple

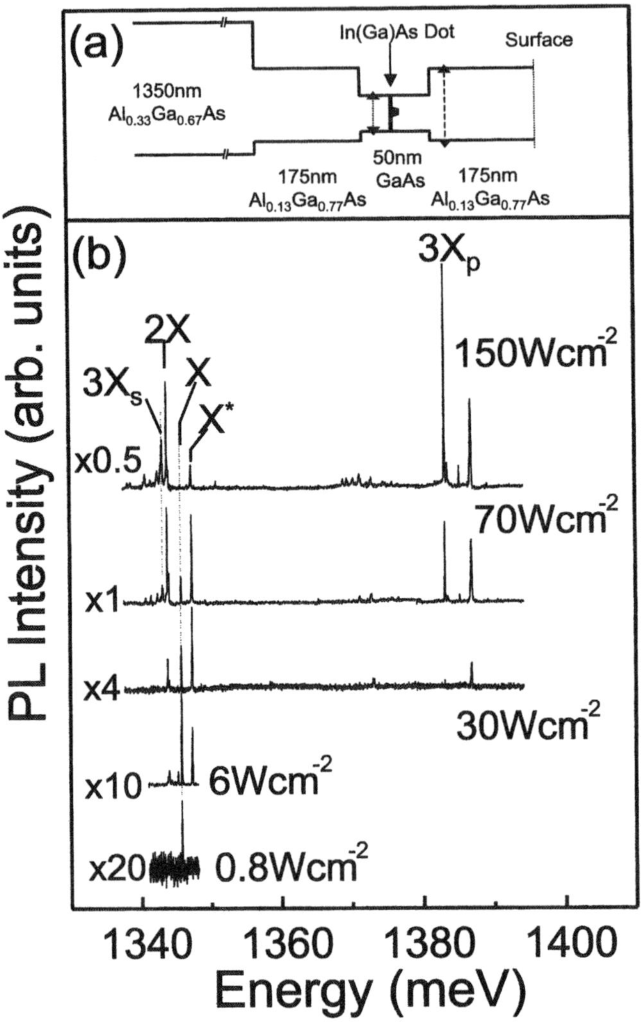

Fig. 5: (a) Schematic band diagram of a sample used to study multiple excitons in a single quantum dot. (b) Photoluminescence spectra recorded as a function of incident laser power, and hence average exciton occupancy, for a single quantum dot. The two groups of emission lines correspond to exciton recombination processes in the ground (s-shell) and first excited (p-shell) states of the dot.

equipment and spectra with excellent signal-to-noise can be acquired in approximately five minutes. This difference is a consequence of the fact that photocurrent spectroscopy, unlike absorption spectroscopy, is a background-less technique. Photocurrent spectroscopy can also be used to determine absolute absorption strengths. Under certain conditions (high electric fields, high temperature or a combination of both – see Fig. 3) all the photoexcited carriers escape from the dots before recombining and the dot absorption strength (A) can be determined from the magnitude of the photocurrent (I) and the relationship $I = APe/h\nu$ where P is the total incident optical power at frequency ν. For a single layer of quantum dots [41] a value of $A = (2 \pm 0.6) \times 10^{-4}$ is obtained for the normal incidence absorption of the ground state transition. That this low value, which is in good agreement with the value obtained from the direct absorption measurements of Warburton et al. [17], can be determined from spectra acquired in only a few minutes demonstrates the sensitivity of the photocurrent technique.

3. Single dot spectroscopy

The previous section has shown that, despite the inhomogeneous broadening, measurements of dot ensembles are capable of providing meaningful results. This is possible when the effects being studied occur on an energy scale comparable to, or greater than the inhomogeneous line-width. However effects resulting from the interaction between multiple carriers confined within a dot are predicted to occur on an energy scale of the order of only a few meV [42]. Such effects will therefore be obscured by the inhomogeneous broadening, which has typical values $\sim 20 - 30$ meV. Many carrier effects must hence be studied using single dot spectroscopy. In the following sections we describe the use of single dot spectroscopy to study the behavior of multiple exciton complexes and excitons in charged dots.

4. Multiple excitons

In this section a study of the optical properties of a single dot containing one, two or more excitons is described. The sample investigated consisted of a single layer of MBE grown InAs quantum dots deposited in the centre of a 50 nm GaAs layer. The band structure of this device is shown in Fig. 5 (a). Two $Al_{0.13}Ga_{0.87}As$ layers were grown on either side of the GaAs layer and a $Al_{0.33}Ga_{0.67}As$ layer was grown between this double heterostructure and the GaAs substrate. Following growth the structure was rapidly thermally annealed (300s at 750°C) to blue shift the low-temperature quantum dot emission to $\sim$1330 meV, allowing it to be measured by high sensitivity Si-based detectors. To permit single dot spectroscopy arrays of widely spaced $\sim$100 and $\sim$200 nm diameter mesas were formed using electron beam lithography followed by plasma etching. Mesas exhibiting a single optically active dot, displaying only a single emission line in the limit of very low laser excitation powers, were used for detailed studies of multiple exciton complexes. Single dot spectroscopy was performed for a sample temperature of 10K using a large numerical aperture microscope objective to produce a sub-micron size, focussed laser spot. The objective position and focussing was achieved using piezo-electric actuators. PL was excited using light from a titanium-sapphire laser and was dispersed and detected using a double monochromator and multi-channel charge coupled device (CCD) detector respectively. By varying the incident laser power the exciton occupancy of the dot could be varied in a controllable manner [43-45].

Figure 5 (b) shows spectra from a single dot as a function of incident laser power P_{ex} (and hence exciton number N_X). For these spectra the excitation energy is 1520

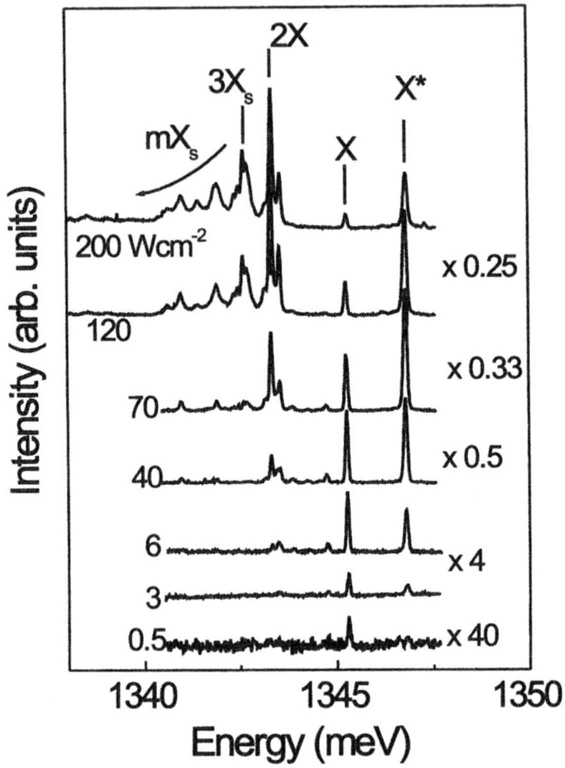

Fig. 6: Photoluminescence spectra of the s-shell emission for a single quantum dot.

meV, close to the band edge of GaAs. The PL spectra consist of two groups of lines, separated by ~40 meV. The highest energy group is not observed until the laser power reaches a certain level, and is hence attributed to the recombination of carriers in the first excited state of the dots (the p-shell). Such recombination is not expected until the dot ground state (s-shell) is fully occupied, preventing further carrier relaxation into this state. By this argument the lower energy group of lines is attributed to excitonic processes involving the recombination of carriers from the dot ground state.

The s-shell recombination is shown in more detail in Fig. 6. At the lowest laser power the spectra consist of a single narrow line (X) (full-width-half-maximum $< 40\mu$eV, resolution limited). This line is attributed to single exciton recombination. With increasing laser power additional lines are observed to both higher (X^*) and lower ($2X$) energy than X. The dominant lower energy line, $2X$, is attributed to biexciton recombination; the recombination of a single exciton in a dot initially occupied by two excitons. The ~2 meV red spectral shift of $2X$ with respect to X is a result of the additional Coulomb interactions between the four particles of the biexciton. With further increase in power, additional lines are observed below $2X$.

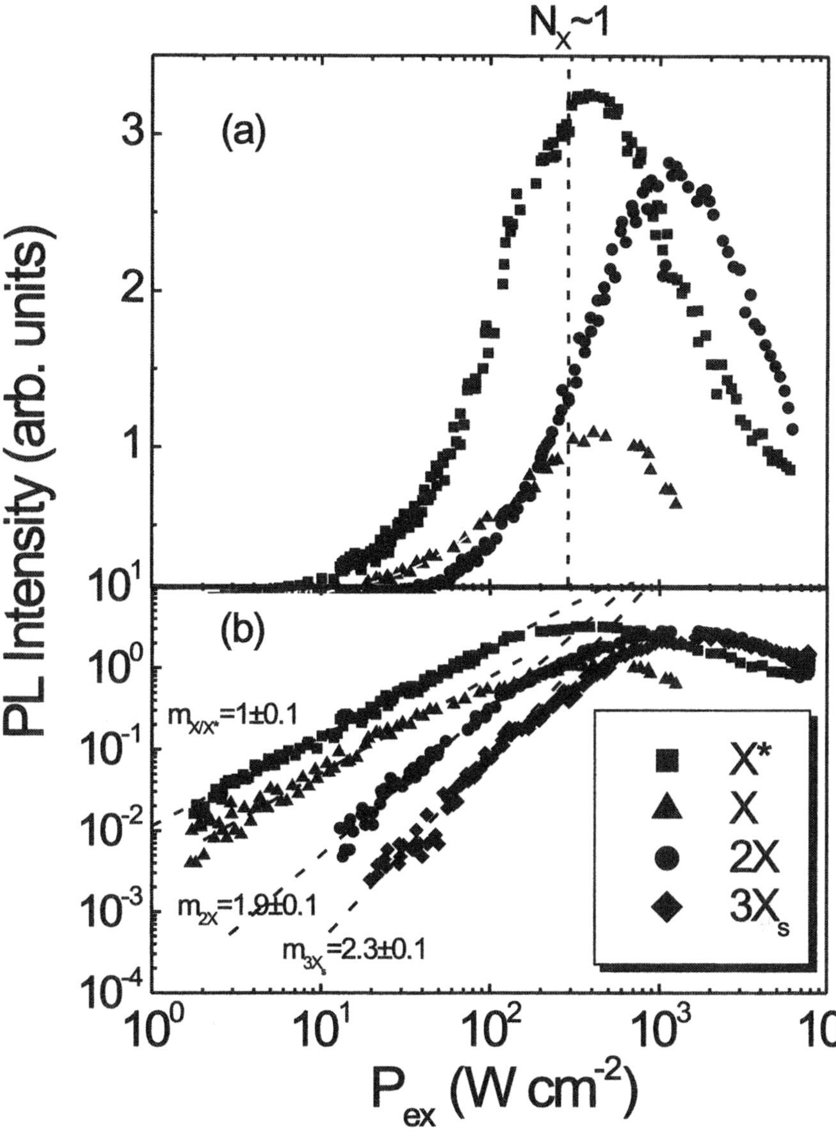

Fig. 7: Intensities of the lowest order exciton emission lines plotted as a function of incident laser power. (a) shows a linear-log plot, (b) shows a log-log plot.

These features arise from multi-exciton recombination processes (the recombination of a single s-shell exciton in a dot initially occupied by > 2 excitons) with the energy of the recombining exciton being perturbed by the other excitons. With increasing laser power the centre of gravity of the emission shifts to lower energy, in a similar manner to band gap renormalization observed in higher dimensionality systems [46].

To confirm the identification of the different emission lines observed in the spectra of Fig. 6, the dependence of their intensities on excitation power was measured. Intensities as a function of incident laser power for lines X, X^*, $2X$ and $3X$ are plotted in both a semi-log and log-log form in Fig. 7. Single exciton recombination (X) involves the creation of a single electron-hole pair and hence the intensity of this process should scale linearly with power. In Fig. 7 (b), X exhibits a unity gradient on the log-log plot. In contrast, biexciton recombination involves the creation of two electrons and two holes and hence should scale quadratically with power. In Fig. 7 (b) the intensity of $2X$ exhibits a gradient of two, confirming its identification. Higher order exciton lines should exhibit an even stronger dependence on power. The $3X$ line however exhibits a gradient of only 2.3, suggesting that for high exciton occupancies significant carrier escape from the dot occurs.

For a given laser power the spectra of Fig. 5 and 6 exhibit a number of different recombination lines, reflecting the statistical nature of carrier capture by the dot. Over the integration time used to record the spectra, the exciton occupancy of the dot will fluctuate, resulting in the appearance of more than one recombination process in the spectra. The intensity of each emission line initially increases with increasing laser power but eventually reaches a maximum before decreasing [see the log-linear plot of Fig. 7 (a)]. This behavior reflects the increasing average exciton occupancy of the dot with increasing power. For example, the probability of the dot containing exactly two excitons, and hence giving the biexciton line, initially increases with increasing power when the average exciton occupancy is less than two. However at high powers the average exciton occupancy will be much greater than two. In this case the probability of a fluctuation in the exciton occupancy resulting in two excitons is small, and decreases with further increase in power. The intensity of the biexciton line will hence decrease at high powers.

The line labelled X^* in Fig. 6, occurring 1.5 meV above X, exhibits unity gradient in the log-log plot of Fig. 7 (b), consistent with a single exciton. X^* is attributed to a single charged exciton, which is created when the dot captures unequal numbers of electrons and holes. This identification is supported by the absence of the X^* line when the laser energy is reduced to give excitation directly into the dot. In this case equal numbers of electrons and holes are created in the dot and hence charged excitons can not be formed. Although the present measurements do not allow the precise nature of X^* (positive (X^+) or negative (X^-) exciton) to be deduced, a comparison with the results of the next section suggest that it is X^+.

5. Charged excitons

In this section the influence of excess electrons or holes on the properties of both excitons and biexcitons is studied. By growing the quantum dots in a charge tuneable structure it is possible to form both positive and negatively charged excitons [47, 48] and biexcitons. Charge tuneable structures were formed by incorporating a single layer of $In_{0.5}Ga_{0.5}As$ self-assembled quantum dots within either n-type (for electron loading of the dot) or p-type (for hole loading) GaAs-AlGaAs metal-insulator-semiconductor (MIS) Schottky gated structures. The quantum dot layer consisted of

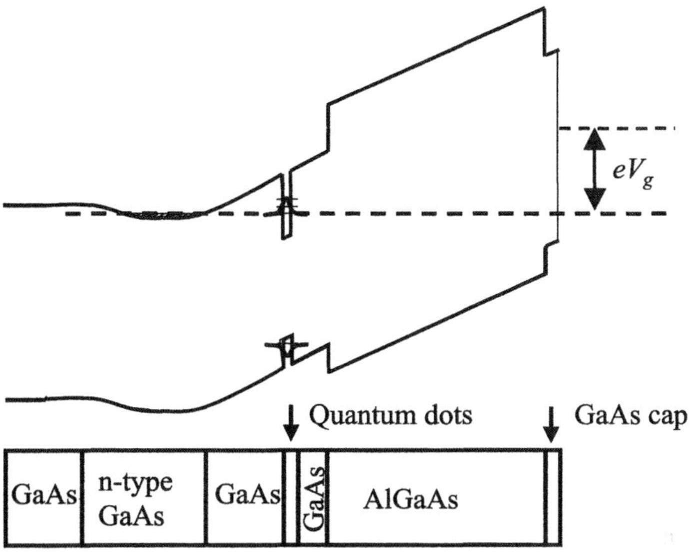

Fig. 8: Schematic band structure of an n-type metal insulator semiconductor (MIS) GaAs-AlGaAs structure used to give controllable electron loading of a single quantum dot.

6ML of $In_{0.5}Ga_{0.5}As$ deposited at 530°C [49]. The nominal layer sequence and band structure of an n-type structure is shown in Fig. 8. After growth, ohmic contacts were established to the doped layer and a Ti(5 nm)-Au(300 nm) Schottky gate was formed on the surface. The design of the structure allows the sequential charging of the dots by varying the voltage between the Schottky gate and doped contact (V_g) [50]. This alters the position of the confined dot states with respect to the Fermi energy of the system, which is defined by a reservoir of carriers produce by the doping. As the state are pushed below the Fermi energy they become occupied by carriers tunnelling from the reservoir. Due to quantum confinement and Coulomb blockade effects [51] it is possible to sequentially load the dot states with carriers. The charging state of the dot was determined from capacitance-voltage measurements of large ensembles [50]. Single quantum dots were probed through sub micron apertures opened in the opaque Schottky gate using electron beam lithography and dry etching. Luminescence measurements were performed at T~10K using the micro-PL set up described in the previous section. Optical excitation intensities were selected such that single exciton recombination dominated the PL spectra.

The evolution of the single dot ground state emission with excess electron number (N_e) is summarized by the gray-scale image of Fig. 9 (a). Spectra obtained for specific N_e are shown in Fig. 9 (b) for comparison For an n-type MIS structure N_e increases with decreasing reverse bias voltage V_g. For large negative V_g, the dot is uncharged and the spectrum is dominated by emission from the neutral exciton (X^0), with much weaker biexciton emission ($2X^0$) observed ~2 meV to lower energy. Photocurrent measurements performed for $V_g < -1$V (dots fully depleted of electrons) support the

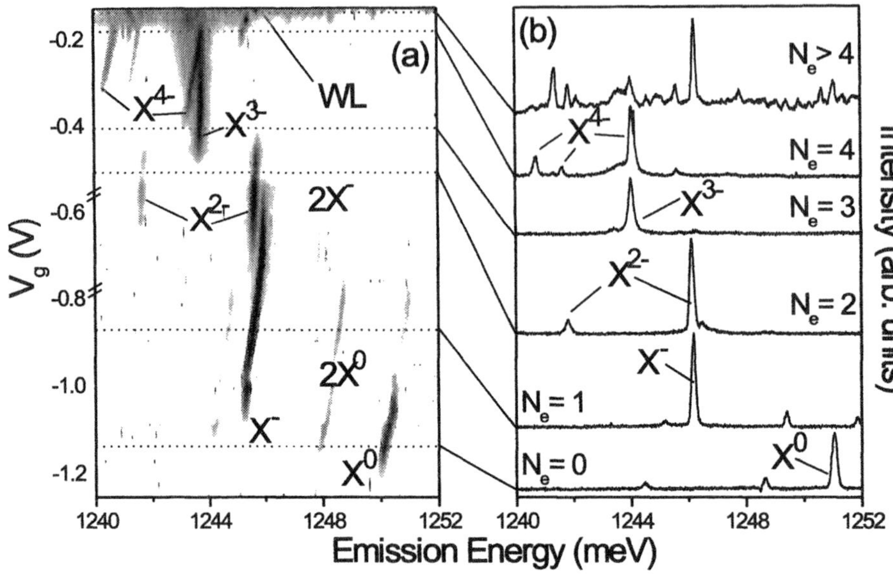

Fig. 9: (a) A gray-scale plot of the emission from a n-type charge tuneable single quantum dot recorded as a function of gate voltage and hence excess electron number. (b) Representative photoluminescence spectra.

identification of X^0, revealing an absorption feature which evolves with decreasing V_g into X^0 as observed in PL. As V_g is reduced further, the PL spectra undergo a series of pronounced changes as a result of the controlled addition of electrons into the dot. For charging with a single excess electron ($N_e = 1$), the X^0 emission line is quenched ($V_g = -1.175$V), being replaced by a new line (X^-) which is red-shifted from X^0 by -5.5 ± 0.7 meV. X^- is attributed to the recombination of a negatively charged exciton. The sign of the spectral shift between X^0 and X^- indicates that electron-hole attraction dominates over electron-electron repulsion in the three particle (2e+h) configuration of X^-. This arises as a consequence of the different lateral spatial extents of the electron (l_e) and hole (l_h) wavefunctions, with the red-shift of X^- with respect to X^0 requiring that $l_h < l_e$; the holes being more strongly localized than the electrons [47]. This is physically reasonable given the larger hole effective mass. Seven different dots with emission energies spanning the range $1260 - 1350$ meV (representing the inhomogeneous broadening of the dot ensemble) were studied. All seven dots exhibited a red shift of X^- with respect to X^0, with the size of the shift being almost constant, varying only weakly in the range $4.9 - 5.8$ meV. This behavior indicates a weak sensitivity of the $X^- - X^0$ separation on the detailed dot parameters.

At $V_g \sim -0.6$, X^- disappears and is replaced by an emission doublet (X_a^{2-} and X_b^{2-}) separated by 4.1 ± 0.5 meV. As discussed by Warburton et al. [48], this doublet structure arises from the two energetically different configurations which are possible following the recombination of an exciton in a dot with two excess electrons. The two

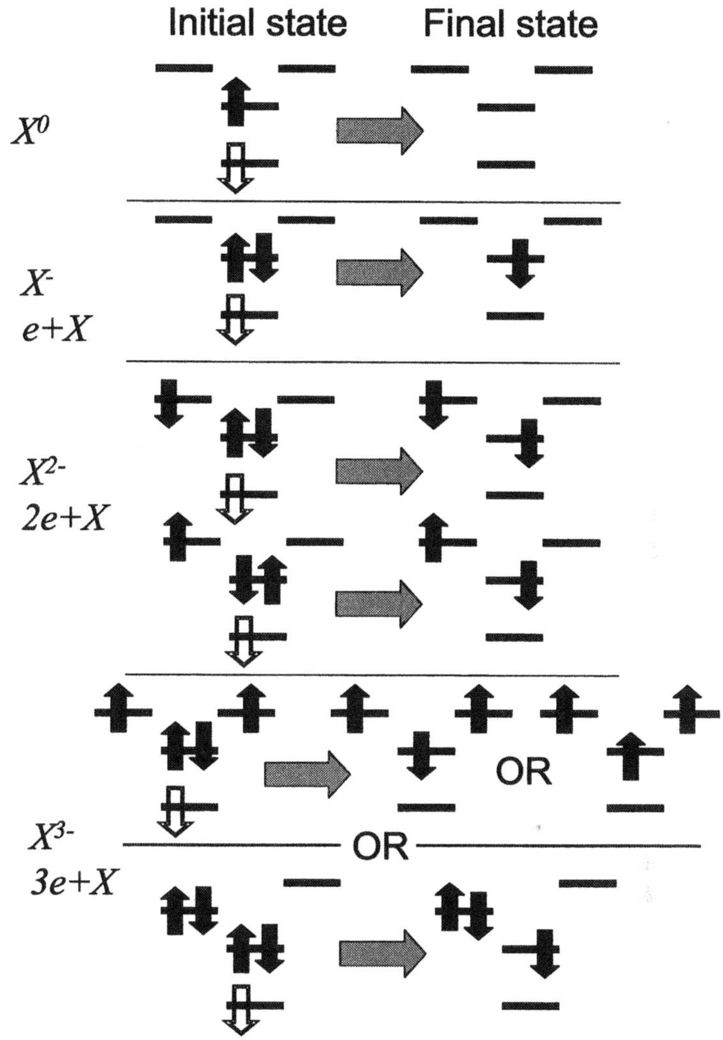

Fig. 10: Initial and final carrier configurations for the experimentally observed negatively charged excitons. For X^{3-}, two possible configurations are shown corresponding to degenerate and non-degenerate p-state levels.

remaining electrons reside in the ground and first excited states and may have either parallel (total spin S=1, X_a^{2-} triplet state) or antiparallel (total spin S=0, X_b^{2-}, singlet state) spins. These two configurations are split by the exchange interaction between the two electrons, with the splitting being equal to twice the exchange energy (Fig. 10).

For a gate voltage of ~ -0.4V the dot is charged with an additional electron ($N_e = 3$) and the $N_e = 2$ doublet (X_a^{2-} and X_b^{2-}) is replaced with a single line X^{3-}. The presence of only a single dominant emission line for the triply-charged exciton

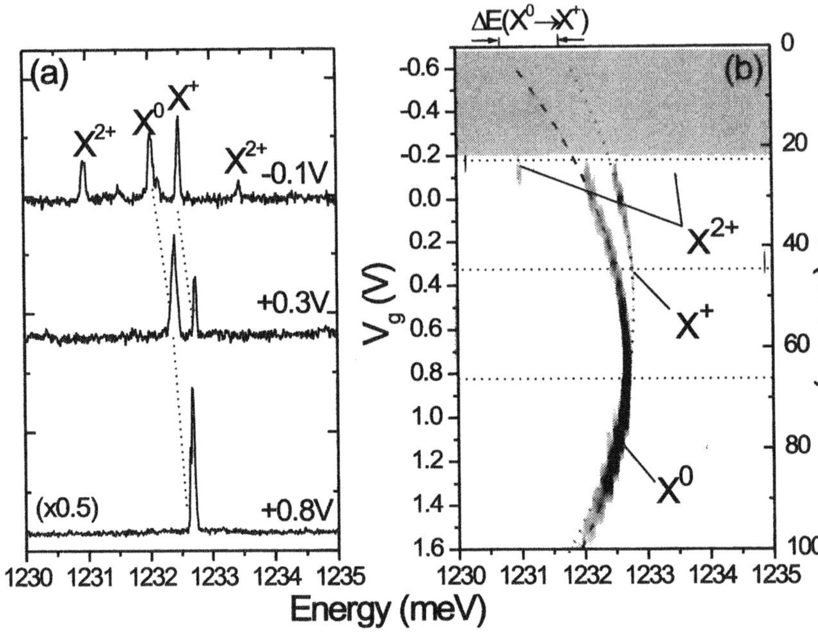

Fig. 11: (a) A grey scale plot of the emission from a p-type charge tuneable single quantum dot recorded as a function of gate voltage and hence excess hole number. (b) Representative photoluminescence spectra.

is somewhat surprising as two, energetically distinct, final configurations have been predicted theoretically for systems which possess perfect cylindrical symmetry [52]. This is a consequence of the Hund's rule filling of the first excited p-like state, which results in parallel spin electrons (see Fig. 10). The appearance of only a single line for X^{3-} suggests a lifting of the degeneracy of the two p-state sub-orbitals such that the lowest energy configuration for X^{3-} consists of two anti-parallel electrons in the same sub-orbital. In this case only one energetically distinct final state exists, consistent with the experimental observation. The degeneracy of the p-levels may be lifted by a number of mechanisms, including interaction with higher d-levels [53] or the inequivalence of the [110] and $[\bar{1}10]$ crystallographic directions [54].

For further decrease in the gate voltage the $N_e = 4$ charging state is reached and the X^{3-} single emission line is replaced with an emission multiplet, indicating a number of energetically distinct final configurations. At even lower gate voltages ($V_g \sim -0.16$V) a strong PL background appears, accompanied by a broadening of the sharp emission lines. This behavior probably reflects the filling of wetting layer states and the perturbation of the dot states by carrier fluctuations in these two-dimensional states. This explanation is supported by capacitance-voltage measurements which show a rapidly increasing capacitance signal for $V_g > \sim -0.2$V as the wetting layer is filled.

The spectra of Fig. 9 show a weak biexciton feature ($2X^0$) in addition to the much stronger exciton feature. The observation of the former results from the statistical nature of the dot photoexcited carrier capture. The energy of the biexciton recombination is unaffected by the first charging event ($N_e = 0 \rightarrow N_e = 1$) because for $2X^0$ the s-shell is already completely filled with two electrons. Hence the first charging event for the biexciton occurs when the voltage is sufficient to inject an electron into the p-shell. This occurs for $V_g \sim -0.7V$ and results in $2X^0$ being replaced with the negatively charged biexciton emission ($2X^-$) which is red-shifted by 0.9 meV with respect to $2X^0$. The energy shift between $2X^0$ and $2X^-$ (0.9 meV) is significantly smaller than between X^0 and X^- (5.8 meV). This arises since for $2X^-$ the interaction of the additional electron with the four particle carrier system ($2e + 2h$) in the dot is strongly reduced due to the completely filled s-shell. This behavior is a direct manifestation of shell filling phenomena for quantum dots.

Figure 11 shows spectra and a gray-scale plot of the emission from a single dot grown within a p-type MIS structure. For this structure increasing negative gate bias results in an increasing number of excess holes (N_h). For large forward biases a single line is observed; attributed to the recombination of the single, charge neutral exciton (X^0). With decreasing forward bias voltage this emission is Stark shifted [55] and at $V_g \sim +0.5V$ an additional feature appears (X^+) which is blue shifted with respect to X^0 by 1.0 ± 0.2 meV [56]. X^+ is attributed to the recombination of the single, positively charged exciton. In contrast to X^-, X^+ is blue shifted with respect to the uncharged exciton, consistent with $l_h > l_e$ and a more localized hole wavefunction. With further decrease in the gate voltage a doublet structure appears (X^{2+}) with components on either side of X^0 and X^+. These features are attributed to the double positively charged exciton with, as is the case for X^{2-}, the two final configurations split by the exchange interaction between the two remaining carriers.

The form of the spectra for the n- and p-type structures shows an important difference. For the p-type structure each spectra exhibits features representing many different charged states. In contrast, except near the transition voltages, the spectra for the n-type structure contain features due to only one charged state. This difference reflects the fact that non-equilibrium carrier configurations are possible in the p-type sample as a result of the long hole tunnelling time through the 25 nm thick barrier. For the n-type sample the smaller electron effective mass results in faster tunnelling times, keeping the system close to equilibrium.

Finally, we briefly describe a series of magneto-optical measurements of the different negatively charge exciton states. For these measurements PL was studied as a function of magnetic field in the Faraday geometry (B$\|$ z, z-growth axis). In the weak interaction regime the effect of the B-field on the $J = \pm 1$ components of the ground state exciton consists of a linear Zeeman splitting $\left(\Delta E_{\text{Zeeman}} = g_X \mu_B B\right)$ of the spin states and a diamagnetic shift of their centre of gravity ($\Delta E_{\text{dia}} \approx \gamma_2 B^2$). For charged excitons, the overall behavior will be determined by the difference of the total magnetic interaction energy ($\Delta E_{\text{Zeeman}} + \Delta E_{\text{dia}}$) between the initial and final states.

Figure 12 shows PL spectra, recorded as a function of applied field, for both the neutral (X^0) and triply, negatively charged (X^{3-}) excitons. In addition the g-factor of the spin splitting and the diamagnetic shift coefficient are plotted as a function of

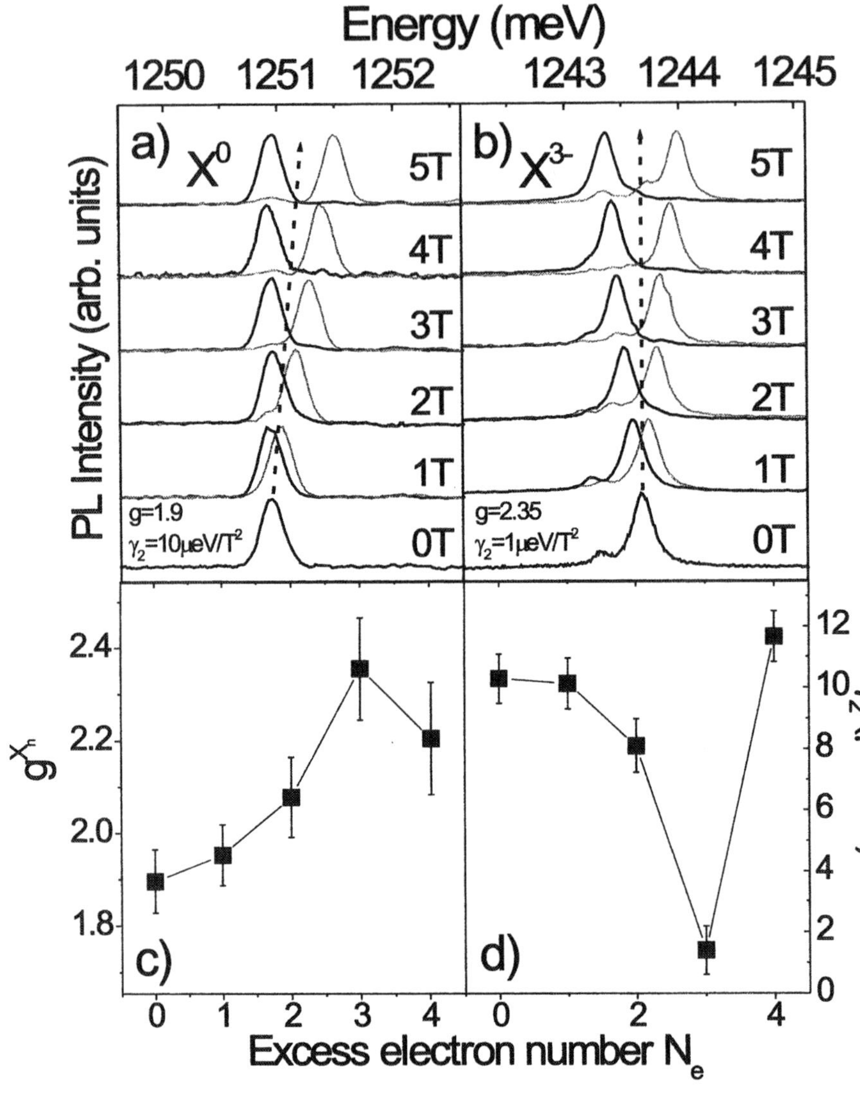

Fig. 12: Circularly polarized photoluminescence spectra for the neutral exciton (a) and triply negatively charged exciton (b) recorded as a function of magnetic field (Faraday configuration) for a single quantum dot. (c) The spin splitting g-factor plotted as a function of excess electron number. (d) The diamagnetic shift parameter plotted as a function of excess electron number.

excess electron number, N_e. X^0 exhibits a linear spin splitting and diamagnetic shift characterized by $g_{X^0} = 1.90 \pm 0.1$ and $\gamma_2(X^0) = 10.3 \pm 0.7 \mu eV T^{-2}$ respectively. These values are typical for the presently investigated quantum dots. The singly charged exciton exhibits a very similar g-factor ($g_{X^-} = 1.93 \pm 0.1$) to X^0, reflecting an almost identical ΔE_{Zeeman} for X^0 and X^-. X^0 is composed of one electron and one hole in the initial state, which both contribute to ΔE_{Zeeman}, but has no magnetically active particles in the final, vacuum state. For X^-, the two electrons in the initial state have antiparallel spins (total spin of zero) which gives no Zeeman splitting. Thus only a hole splitting is present in the initial state of X^- but the remaining electron in the final state also gives a splitting. The overall Zeeman splitting of X^- should hence be identical to that of X^0, in agreement with the experimental observation. Whereas higher charged states should also exhibit the same Zeeman splitting as X^0 the experimental data plotted in Fig. 12 (c) shows a significant increase of the g-factor with increasing N_e. The reason for this behavior is not fully understood but may indicate a perturbation of the carrier wavefunctions with increasing N_e, causing the exciton to sample different regions of the dot. For dots of non-uniform $In_xGa_{1-x}As$ composition this would alter the g-factor which, for $In_xGa_{1-x}As$, is a strong function of x.

The diamagnetic shift is governed by the combined effects of lateral confinement and carrier-carrier Coulomb interactions [57]. Within the experimental accuracy the diamagnetic coefficients for X^0 and X^- are identical, with a value of $10.1 \pm 0.7 \mu eVT^{-2}$. This observation indicates that the Coulomb interaction and correlation effects for X^- provide only a modest perturbation of the exciton structure. This is consistent with the exciton binding energy (~20 meV [28]) being much larger than $\Delta E(X^0 \to X^-) \sim 5$ meV and the non-interacting nature of the final single electron state. In contrast, further addition of electrons produces a very pronounced reduction of the diamagnetic shift for $N_e \geq 2$ [Fig. 12 (d)], with, for X^{3-} the diamagnetic shift practically vanishing (diamagnetic coefficient of $1.0 \pm 0.7 \mu eVT^{-2}$). This observation indicates almost identical shifts for the initial (two p-electrons, two s-electrons and one hole) and final (two p electrons and one s electron) states. The reason for this coincidence is unclear; a full understanding requiring a detailed knowledge of the lateral spatial extent of the multi-carrier wavefunctions.

6. Conclusions

The application of ensemble and single dot optical spectroscopy to the study of the structural and electronic properties of In(Ga)As self-assembled quantum dots has been described. Self-assembled quantum dots provide high optical quality systems suitable for electro-optical device applications and permit the study of physical processes in zero-dimensional semiconductor structures. The optical spectroscopic studies provide information on the confined electronic states of the dots, the dot physical structure, carrier transport mechanisms and the nature of carrier-carrier interactions.

Acknowledgements

The authors would like to thank the following for contributions made to the work described in this article. M. S. Skolnick, I. E. Itskevich, P. W. Fry, A. D. Ashmore, A. Lematre, R. Oulton, A. I. Tartakovskii and L. R. Wilson for help with the experiments and interpretation of the results. J. A. Barker, E. P. O'Reilly and P. A. Maksym for performing the theoretical calculations. M. Hopkinson and M. J. Steer for the growth of samples. J. C. Clark and G. Hill for sample processing. M. Al-Khafaji and A. G. Cullis for structural analysis of the quantum dots.

This work was supported by the United Kingdom Engineering and Physical Sciences Research Council (UK-EPSRC).

References

[1] L. A. Coldren and S. W. Corzine, *Diode Lasers and Photonic Integrated Circuits* (Wiley, Chichester, 1995).

[2] J. Faist, F. Capasso, D. L. Sivco, C. Sirtori, A. L. Hutchinson and A. Y. Cho, Science **264**, 553 (1994).

[3] O. B. Shchekin, G. Park, D. L. Huffaker, Q. W. Mo and D. G. Deppe, IEEE Photonics Technol. Lett. **12**, 1120 (2000).

[4] H. Chen, Z. Zou, O. B. Shchekin and D. G. Deppe, Electron. Lett. **36**, 1703 (2000).

[5] P. Michler, A. Kiraz, C. Becher, W. V. Schoenfeld, P. M. Petroff, L. D. Zhang, E. Hu and A. Imamoglu, Science 290, 2282 (2000).

[6] A. J. Shields, M. P. O'Sullivan, I. Farrer, D. A. Ritchie, R. A. Hogg, M. L. Leadbeater, C. E. Norman and M. Pepper, Appl. Phys. Lett. **76**, 3673 (2000).

[7] S. -W Lee, K. Hirakawa and Y. Shimada, Physica E **7**, 499 (2000).

[8] J. J. Finley, M. Skalitz, M. Arzberger, A. Zrenner, G. Bohm and G. Abstreiter, Appl. Phys. Lett. **73**, 2618 (1998).

[9] J. H. Davies in *The Physics of Low-Dimensional Semiconductors* (Cambridge, 1998).

[10] D. Bimberg, M. Grundmann and N. N. Ledentsov, in *Quantum Dot Heterostructures* Wiley, Chichester (1999)

[11] D. Leonard, K. Pond and P. M. Petroff, Phys. Rev. B **50**, 11687 (1994).

[12] S. Rumimov, P. Werner, K. Scheerschmidt, J. Heydenreich, U. Richter, N. N. Ledentsov, M. Grundmann, D. Bimberg, V. M. Ustinov, A. Yu Egorov, P. S. Kop'ev and Zh. I. Alferov, Phys. Rev. B **51**, 14766 (1995).

[13] P. W. Fry, I. E. Itskevich, D. J. Mowbray, M. S. Skolnick, J. J. Finley, J. A. Barker, E. P. O'Reilly, L. R. Wilson, I. A. Larkin, P. A. Maksym, M. Hopkinson, M. Al-Khafaji, J. P. R. David, A. G. Cullis, G. Hill and J. C. Clark, Phys. Rev. Lett. **84**, 733 (2000).

[14] K. Zhang, Ch. Heyn, W. Hansen, Th. Schmidt and J. Falta, Appl. Phys. Lett. **76**, 2229 (2000).

[15] X. Z. Liao, J. Zou, X. F. Duan, D. J. H. Cockayne, R. Leon and C. Lobo, Phys. Rev. B **58**, R4235 (1998).

[16] J.-Y. Marzin and G. Bastard, Solid State Commun. **92**, 437 (1994).

[17] R. J. Warburton, C. S. Durr, K. Karrai, J. P. Kotthaus, G. Medeiros-Ribeiro, P. M. Petroff, Phys. Rev. Lett. **79**, 5282, (1997).

[18] P. W. Fry, I. E. Itskevich, S. R. Parnell, J. J. Finley, L. R. Wilson, K. L. Schumacher, D. J. Mowbray, M. S. Skolnick, M. Al-Khafaji, A. G. Cullis, M. Hopkinson, J. C. Clark and G. Hill, Phys. Rev. B **62**, 16784 (2000).

[19] P. D. Buckle, P. Dawson, S. A. Hall, X. Chen, M. J. Steer, D. J. Mowbray, M. S. Skolnick and M. Hopkinson, J. Appl. Phys. **86**, 2555 (1999).

[20] D. A. B. Miller, D. S. Chemla, T. C. Damen, A. C. Gossard, W. Wiegmann, T. H. Wood and C. A. Burrus, Phys. Rev. B **32**, 1043 (1985).

[21] A small shift of 7 meV to higher energy has been applied to the results of the *p-i-n* structure, in order to obtain a continuous variation of the peak positions between positive and negative electric fields. The *p-i-n* and *n-i-p* structures were grown consecutively in order to obtain the minimum possible run-to-run variation in dot parameters between samples. The observed energy difference of 7 meV corresponds to only a ~2.5% variation in dot base size.

[22] M. Grundmann, O. Stier and D. Bimberg, Phys. Rev. B **52**, 11969 (1995).

[23] A. J. Williamson, L. W. Wang and A. Zunger, Phys. Rev. B **62**, 12963 (2000).

[24] M. Grundmann, O. Stier and D. Bimberg, Phys. Rev. B **52**, 11969 (1995).

[25] M. A. Cusack, P. R. Briddon and M. Jaros, Phys. Rev. B **54**, R2300 (1996).

[26] H. Jiang and J. Singh, Phys. Rev. B **71**, 3239 (1997).

[27] L.-W. Wang, J. Kim and A. Zunger, Phys. Rev. B **59**, 5678 (1999).

[28] O. Stier, M. Grundmann and D. Bimberg, Phys. Rev. B **59**, 5688 (1999).

[29] C. Pryor, Phys. Rev. B **57**, 7190 (1998).

[30] O. Stier Private communication.

[31] These spectra were recorded for dots grown in a laser structure where the optical waveguide allows the in-plane geometry to be accessed.

[32] G. Bastard, in *Wave Mechanics Applied to Semiconductor Heterostructures* Les Editions de Physique (Paris 1988).

[33] A one band model will be less suitable for calculating excited states where the amount of light hole admixture will be considerably greater.

[34] A. D. Andreev, J. R. Downes, D. A. Faux and E. P. O'Reilly, J. Appl. Phys. **86**, 297 (1999).

[35] M. P. M. C. Krijn, Semicond. Sci. Technol. **6**, 27 (1991).

[36] J. A. Barker and E. P. O'Reilly, Phys. Rev. B **61**, 13840 (2000).

[37] P. B. Joyce, T. J. Krzyzewski, G. R. Bell, B. A. Joyce and T. S. Jones, Phys. Rev. B **58**, R15981 (1998).

[38] N. Liu, J. Tersoff, O. Baklenov, Al. Holmes Jr. and C. K. Shih, Phys. Rev. Lett. **84**, 334 (2000).

[39] N. Grandjean, J. Massies and O. Tottereau, Phys. Rev. B **55**, R10189 (1997).

[40] I. Kegel, T. H. Metzger, A. Lorke, J. Peisl, J. Stangl, G. Bauer, J. M. Garca and P. M. Petroff, Phys. Rev. Lett. **85**, 1694 (2000).

[41] These dots are grown under slightly different conditions to those described previously and as a result have a higher areal density of 5×10^{10} cm^{-2}.

[42] L. Jacak, P. Hawrylak and A. Wjs, *Quantum Dots* Springer (Berlin 1998).

[43] M. Bayer, O. Stern, P. Hawrylak, S. Farfard and A. Forchel, Nature (London) **405**, 923 (2000).

[44] E. Dekel, D. Gershoni, E. Ehrenfreund, J. M. Garcia and P. M. Petroff, Phys. Rev. B **62**, 11038 (2000).

[45] A. Hartmann, Y. Ducommun, E. Kapon, U. Hohenester and E. Molinari, Phys. Rev. Lett. **84**, 5648 (2000).

[46] C. Delalande, G. Bastard, J. Orgonasi, J. A. Brum, H. W. Liu, M. Voos, G. Weimann and W. Schlapp, Phys. Rev. Lett. **59**, 2690 (1987).

[47] F. Findeis, M. Baier, A. Zrenner, M. Bichler, G. Abstreiter, U. Hohenester and E. Molinari, Phys. Rev. B **63**, R 121309 (2001).

[48] R. Warburton, C. Schäflein, D. Haft, F. Bickel, A. Lorke, K. Karrai, J. M. Garcia, W. Schönfeld and P. M. Petroff, Nature (London) **405**, 926 (2000).

[49] Atomic force microscopy performed on similar uncapped quantum dots shows disk shaped dots with lateral (vertical) dimensions of 23 ± 7 nm (3 ± 1 nm)

[50] J. J. Finley, P. W. Fry, A. D. Ashmore, A. Lematre, A. I. Tartakovskii, R. Oulton, D. J. Mowbray, M. S. Skolnick, M. Hopkinson, P. D. Buckle and P. A. Maksym, Phys. Rev. B **63**, R161305 (2001).

[51] H. Drexler, D. Leonard, W. Hansen, J. P. Kotthaus and P. Petroff, Phys. Rev. Lett. **72**, 2252 (1994).

[52] A. Wojs and P. Hawrylak, Phys. Rev. B **55**, 13066 (1997).

[53] P. Hawrylak, G. A. Narvaez, M. Bayer, A. Forchel, Phys. Rev. Lett. **85**, 389 (2000).

[54] L. Wang, J. Kim and A. Zunger, Phys. Rev. B **59**, 5678 (1999).

[55] This Stark shift is asymmetrical about zero electric field and implies a permanent dipole moment of sign equivalent to that deduced for nominal InAs dots as described in a previous section.

[56] This is the value obtained by extrapolating the Stark shifts of X^0 and X^+ to zero electric field.

[57] S. N. Walck and T.L. Reinecke, Phys. Rev. B **57**, 9088, (1998)

Chapter 4

Generation of single photons using semiconductor quantum dots

A.J. Shields[a*], R. M. Stevenson[a], R. M. Thompson[a,b], Z. Yuan[a], and B. E. Kardynal[a]

[a] *Toshiba Research Europe Limited, 260 Cambridge Science Park, Milton Road, Cambridge CB4 0WE, UK*
**E-mail: andrew.shields@crl.toshiba.co.uk*

[b] *Cavendish Laboratory, University of Cambridge, Madingley Road, Cambridge CB3 0HE, UK*

Abstract

Applications in optical quantum information technology require the development of a new type of light source for which exactly one photon is emitted periodically. We review here recent progress in using semiconductor quantum dots as the active medium for generating both single photons, as well as photon pairs. Anti-bunching and single photon emission is observed for both optical and electrical injection of the recombining electrons and holes into the dots.

1. Introduction

Light sources typically display a statistical distribution in the number of emitted photons in a given time interval, which gives rise to shot noise in optical measurements. A source for which the emission time of each photon is completely random obeys Poissonian statistics. A central aim of quantum optics is the generation of light fields with suppressed photon number fluctuations. Ideally such a source would emit an exact number of photons at regularly spaced time intervals. This would be useful for making optical measurements with a noise level below the shot noise limit, or for new applications in quantum information technology, such as quantum communications [1] or photonic quantum computing [2].

In quantum cryptography, a cryptographic key can be formed between two parties using bits encoded upon single photons transmitted along an optical fibre or through free space [1]. By using single photons the sender and intended recipient are able to guarantee the security of their key, since quantum mechanics dictates that measurement by a third party will inevitably produce a detectable change to the encoded single photons.

In the absence of single photon source, practical demonstrations of quantum key distribution have used a pulsed laser diode, for which the light level is so strongly attenuated that the average number of photons per pulse $\mu \ll 1$. However, even for these very weak intensities, it is impossible to avoid occurrences of pulses containing two or more photons. Since a coherent source can be shown to also obey Poissonian statistics, the rate of multi-photon pulses is given by $(\mu^2/2)$. These multi-photon pulses pose a threat to the security of quantum cryptography, since they allow an eavesdropper to split off a photon to make a measurement, while allowing the others to be transmitted undisturbed to the intended recipient. They thereby provide an eavesdropper a means to determine at least part of the key while remaining undetected. Partial knowledge of the key would greatly reduce the computational resource needed to crack any subsequent encrypted communication.

Recently Brassard et al. [3] have shown that in order to guarantee *unconditional* security for the key, the photon detection rate at the end of the fibre must exceed the rate at which multi-photon pulses are created by the source. Since the signal rate is strongly reduced by the finite efficiency of the source (typically $\mu \sim 0.1$), the efficiency of the detector (typically $\eta \sim 0.1$) and loss in the fibre and connectors, it is practically impossible to guarantee unconditional security when using an attenuated laser as the photon source. Secure quantum key distribution therefore requires the development of a true single photon source.

The photon statistics of light signals can be studied via the normalized second-order correlation function, $g^{(2)}(\tau)$. It measures the correlation between the intensity of the light field with that after a delay τ. For a light source with random emission times, the second-order correlation function is flat, $g^{(2)}(\tau) = 1$. A source for which $g^{(2)}(\tau = 0) > 1$ is described as 'bunched' since there is an enhanced probability of two photons being emitted within a short time delay. On the other hand, the photons emitted by an 'anti-bunched' source, for which $g^{(2)}(\tau = 0) < 1$, tend to be separated in time. For an ideal single photon source, we expect $g^{(2)}(\tau = 0) = 0$, since two photons can not be emitted simultaneously.

The first proposal to create an anti-bunched source was based upon the resonance fluorescence of a single two-level atom, using the fact that the emission of a photon returns the atom to its ground state, after which it must be re-excited before a second photon can be emitted [4]. This was first demonstrated experimentally in the resonance fluorescence of a low density vapor of Na atoms [5]. This showed a dip in $g^{(2)}(\tau)$ around $\tau = 0$, although the behavior was complicated by the fact that emission was collected from several atoms. Later experiments on a single trapped Mg^+ ion showed a more ideal demonstration of anti-bunching [6]. In recent years, the photon emission statistics of a wide range of other quantized, two-level systems have been studied, such as single molecules [7-11], CdSe/ZnS nanocrystals [12] and nitrogen vacancy centers in diamond [13,14].

A quantum dot is often described as the semiconductor analogue of an atom, since the three-dimensional confinement of the electrons results in their energy spectrum consisting of a series of discrete lines. Each of these levels can accommodate just two electrons of different spin, due to the Pauli exclusion principle. It was proposed that a quantum dot formed by etching a lithographically defined pillar in a hetero-junction could be used to make a device for emitting single photons [15]. Subsequent experimental work demonstrated light emission from such a structure [16]. However, the small confinement energies of these relatively large quantum dots required milli-

Kelvin operating temperatures for the device, while it was not possible to verify its emission statistics.

An experimental breakthrough came with the use of InAs self-organized quantum dots, which are formed using a natural growth mode of strained layer semiconductors [17]. When InAs is deposited on GaAs it initially grows as a strained two-dimensional sheet, but beyond some critical thickness islands form in order to minimize the strain. Subsequent overgrowth of the islands leads to the encapsulation of quantum dots inside GaAs lattice. The advantage of forming quantum dots using this method is that the dot dimensions are much smaller than can be produced using lithography, leading to zero-dimensional properties at room temperature, while the non-radiative defects associated with etching can be avoided. These properties have led to the successful integration of InAs self-assembled quantum dots into a number of optoelectronic devices, such as high density optical memory [18] and a detector of single photons [19]. Using self-assembled quantum dots has allowed us [20-22] and several other groups [23-26] to observe photon anti-bunching and single-photon emission for a single, optically-excited InAs quantum dot. We summarize some of our results here along with a more detailed analysis.

A major advantage of self-assembled quantum dots is that they can be easily integrated inside a device structure, allowing the possibility of electrically injecting the recombining electrons and holes into the dots. This would avoid having to use a pump laser, as well as its intricate alignment with the quantum dot. We describe here a realization of an electrically injected single-photon source, consisting of a quantum dot within the intrinsic region of a *p-i-n* junction. Such a single-photon emitting diode has the advantages of being highly stable, robust, long-lived and potentially cheap to manufacture [27].

The rest of the article is arranged as follows. In the following section we describe the samples studied and the experimental techniques used. Section 3 reports a study of the photoluminescence spectra of individual InAs quantum dots, as well as their emission dynamics studied by time-resolved techniques. The photon statistics of the dot photoluminescence is studied in Sect. 4. Here we concentrate upon the form of the second-order correlation function for the different exciton complexes which can be confined in the dot. Cross-correlation measurements show that the exciton and bi-exciton transitions of the dot can be used to generate pairs of photons. Section 5 details our progress in realizing an electrically injected single-photon emission device. We report for the first time photon anti-bunching for a dc injection current, and regulated single-photon emission by pulsing the injection current. A detailed analysis of the exciton emission intensities and the second-order correlation function is presented in Sect. 6 and discussed in Sect. 7. Finally, we summarize the outlook for using quantum dots as a single-photon source.

2. Experimental techniques

The samples were grown by molecular beam epitaxy on a (100)-oriented undoped GaAs substrate. For the optically injected samples, the layers were not intentionally

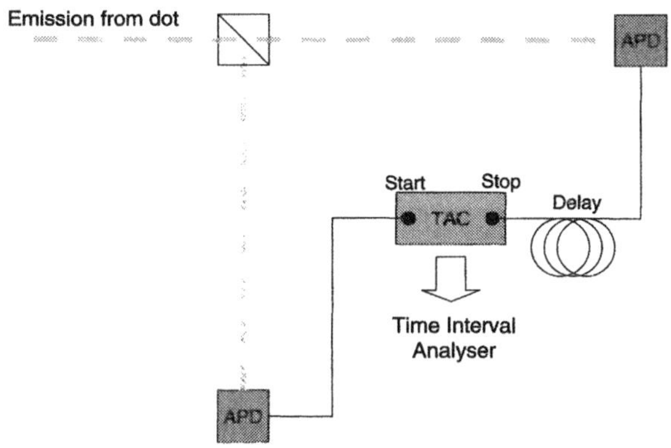

Fig. 1: Schematic of Hanbury-Brown and Twiss experimental arrangement used for correlation measurements.

doped, with the epitaxially grown layers consisting of 0.5 μm GaAs buffer layer, an InAs layer which self assembles into quantum dots and a 0.3 μm GaAs cap. The results presented here were recorded on a sample for which the nominal thickness of the InAs layer was 1.7 monolayers. Images recorded in an atomic force microscope of uncapped quantum dots grown under identical conditions revealed the dot density to vary from 2.5×10^{10} cm^{-2} at the center of the wafer, to $< 10^8$ cm^{-2} at the edge. This variation in dot density, which arises because the indium flux is higher at the center of the wafer, provides a convenient method of studying samples with different dot densities.

The wafers were prepared into an array of mesas with diameters ranging from 0.1 to 2 μm, using e-beam lithography and wet-etching. A 0.8 μm mesa containing a single quantum dot was selected for most of the experiments presented here. In other samples, the emission from a single dot was isolated by evaporating an opaque Al film on the wafer surface and etching sub-micron sized apertures in the film. Similar results were obtained from both types of structure.

The electrically injected single-photon emission experiments used a similar layer structure, except that the dots were placed inside the intrinsic region of a *p-i-n* junction. Further details of these samples and their processing is described in Sect. 5.

Photoluminescence was excited using a Ti-Sapphire laser of photon energy 1.55 eV, above the bandgap of the GaAs barrier layers. The laser was operated either continuous-wave (CW) or modelocked to produce 1 ps wide pulses at a repetition rate of about 77 MHz. The laser was focused to a $\sim 1\mu$m spot on the surface of the sample, mounted on the cold finger of a liquid He cryostat, by an infinity-corrected microscope objective lens, which also collimated the photoluminescence from the sample. The emitted light was dispersed by a grating spectrometer and detected using either a CCD or a photon-counting avalanche photodiode. The magnification

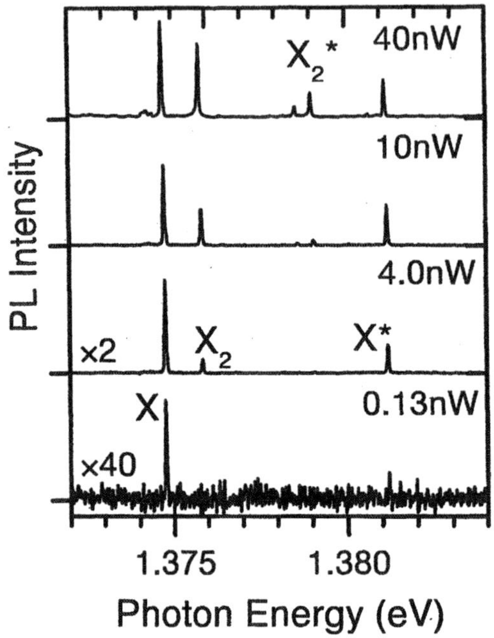

Fig. 2: Photoluminescence spectra of a single quantum dot under pulsed optical excitation, recorded for different laser excitation powers. The appearance of more lines can be seen as the power is increased.

was such that the image of the PL from a mesa was $\sim 100\mu$m, which allowed spatial filtering of the photoluminescence by adjusting the size of the spectrometer entrance slit. The spectral resolution of our system was $\sim 50\mu$eV, and the spatial resolution was $\sim 1\mu$m. A similar set-up was used to assess the electrically injected devices.

The statistics of the emitted luminescence was analyzed using the Hanbury-Brown and Twiss set-up, shown in Fig. 1, of a 50/50 beamsplitter and two-photon counting avalanche photodiodes. The time delay between subsequent clicks in the two single-photon detectors is measured repeatedly using time-correlated single-photon counting electronics. The distribution in the measured time delay is proportional to the normalized second-order correlation function $g^{(2)}(\tau)$ for time delays τ much less than the reciprocal of the average count rate in either detector, which is a good approximation for the time delays studied here.

3. Single quantum dot photoluminescence

3.1 Single dot spectra

The PL spectra of the mesa show a broad peak near 1.44 eV, which derives from the two-dimensional wetting layer on which the dots are formed, and several sharp lines to lower energy due to the quantum dot transitions. Figure 2 plots the quantum dot region of the PL spectra recorded for different average excitation powers of the 1 ps

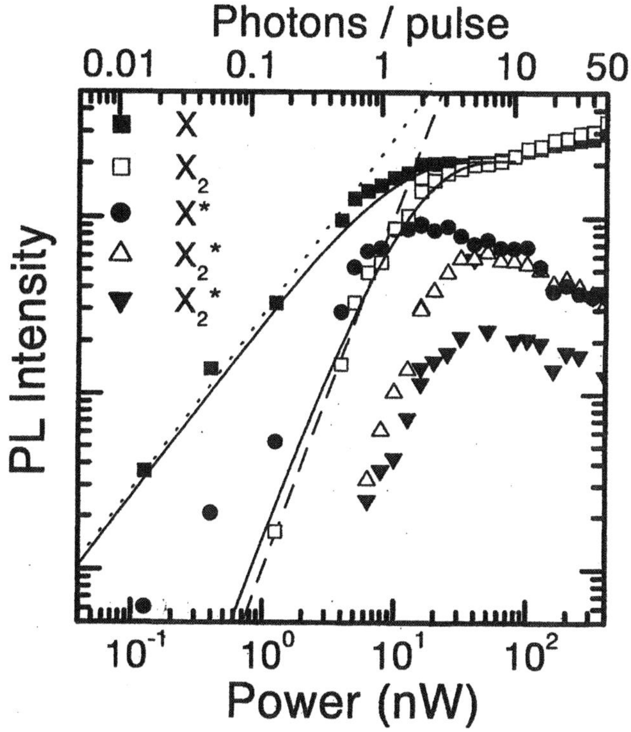

Fig. 3: Integrated intensities of PL lines as a function of pulsed laser excitation power. Open and solid triangles represent the higher and lower energy components of the X_2^* doublet. Dotted (dashed) line shows the gradient associated with linear (quadratic) power dependence. The solid lines show calculated power dependence of X and X_2 as a function of photons per pulse absorbed close to the dot.

pulsed laser. At the lowest laser power of 0.13 nW, two lines are seen: the stronger marked X at 1.3748 eV, and the weaker marked X^* at 1.3812 eV. As the power is increased to 4.0 nW, a third line (X_2) appears at an energy 1.1 meV higher than that of X. At 10 nW, a pair of lines X_2^* appear at 1.3786 eV and 1.3790 eV. All lines are saturated in intensity at the highest laser power shown of 40 nW, and no other lines are observed at any photon energy up to that of the wetting layer emission. We observed a qualitatively similar line structure for other mesas containing a single quantum dot.

The fact that no groups of lines appear to higher energy is strong evidence for the existence of only one pair of electron and hole levels in the quantum dot. Other PL studies on single quantum dots [28-31] have shown the appearance of excited state emission around 30-60 meV to higher energy of the lines seen at lowest powers. However, the InAs dots studied here have a smaller size, as demonstrated by their higher emission energy [28,31]. It is therefore reasonable to conclude that the energy-level spacing in our dots is much larger than found in other work, resulting in only a single confined electron or hole level.

To identify the exciton complexes responsible for the observed emission lines, we studied their integrated intensities as a function of laser power, as plotted in Fig. 3. The intensity of lines X and X^* show an approximately linear power dependence, with exponents of 0.94±0.04 and 1.09±0.09, and saturate at approximately the same power of 8 nW. Therefore X and X^* are attributed to quantum dot configurations containing only a single photo-excited electron-hole pair. The onset of emission from X_2 and X_2^* occurs at higher laser powers than for X and X^*, and the intensity of the emission increases with approximately quadratic power dependence, with fitted exponents of 1.95±0.12 and 2.02±0.15 respectively. This strongly suggests that X_2 and X_2^* are due to emission from the quantum dot containing two photo-excited electron hole-pairs. The fact that excited states are not observed rules out the possibility of attributing the emission to dots containing three or more excitons. In addition, the higher and lower energy components of the X_2^* pair have an almost constant intensity ratio. They are therefore regarded as a doublet from different configurations of the same multi-carrier complex.

At high excitation powers > 80 nW, there is a redistribution of the emission intensity from X^* and X_2^* towards X and X_2, demonstrating that emission from X and X_2 is a complementary process to emission from X^* and X_2^*. This is supported by the similar intensities at high laser power of X with X_2, and X^* with X_2^*. The line structure for X and X_2 is very different to that for $X^*$and X_2^*, which makes it difficult to attribute the emission to two different dots within the same mesa. This conclusion is also supported by emission from other mesas that seem to contain only a single quantum dot, which show qualitatively similar PL spectra and dependence on laser power, as that shown in Fig. 2. A more likely possibility is that the quantum dot may intermittently capture an excess carrier to allow formation of charged as well as neutral excitons, as has been suggested in Refs. 28, 30 and 31. The presence of charged biexciton emission (X_2^*) suggests that the quantum dot has either a second confined electron or hole level, although the absence of a tri-exciton transition rules out the possibility that both are bound. Transitions between ground and excited levels are parity forbidden. The redistribution of intensity from X and X_2 to X^* and X_2^* at high laser power suggests that X and X_2 derive from neutral excitons and X^* and X_2^* from charged excitons, since carriers photo-excited in the wetting layer will tend to neutralize charge trapped in the dots. The four features are thus attributed to the neutral (X) and charged exciton (X^*), and the neutral (X_2) and charged biexciton (X_2^*).

3.2 Time resolved photoluminescence

Figure 4 plots the temporal dependence of the emission from the different exciton complexes at different laser powers. At the lowest power of 3.8 nW, only emission from X is measured, displaying a single exponential decay with a lifetime of 1.36 ns, similar to that reported elsewhere [32]. The measured rise of the PL is limited by the response of the APD. As the power is increased to 12 nW, the two additional lines

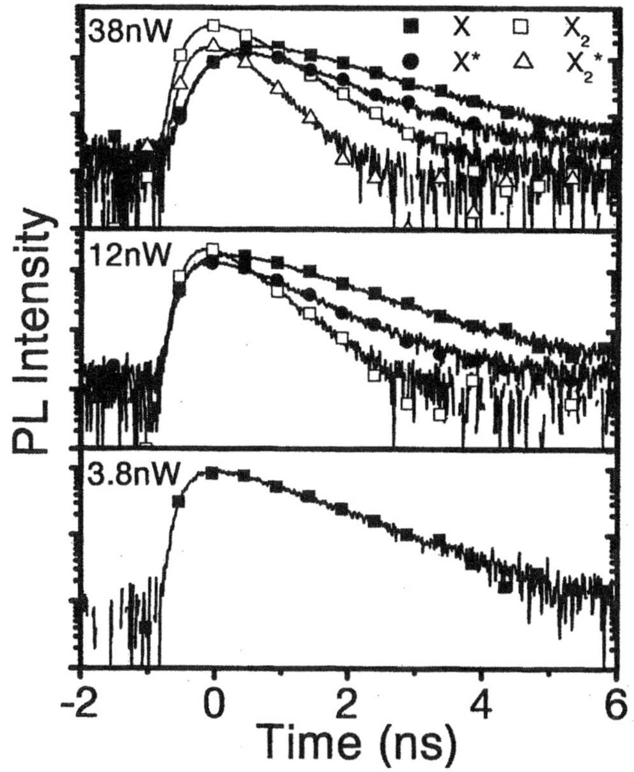

Fig. 4: Time resolved PL measured for each exciton complex at different laser excitation powers. Notice that each complex shows a distinct lifetime which is independent of the laser power. At the higher powers the emission of the single exciton (X) is delayed until after that of the biexciton (X_2). Similarly the charged exciton (X^*) is emitted after the charged biexciton (X_2^*).

X^* and X_2 can be measured, with decay times of 1.07±0.02 ns and 0.59±0.02 ns, respectively. The rise of these curves is again limited by the temporal resolution of the system. The PL due to X shows a similar decay time to that at lower power, but the peak intensity of the X PL is delayed by $\sim$ 0.23 ns relative to its position for the lowest laser power. This is attributed to the time delay associated with the radiative decay of X_2 into X for some of the pulses. At the highest power shown of 38 nW, all lines have reached their maximum intensity, and their temporal characteristics are found to be independent of laser power. In this case the maximum intensity of the X PL is shifted by 0.61 ns relative to that of X_2, as well as that of X for lower laser power. This is in excellent agreement with the measured radiative lifetime of X_2, of 0.59±0.02 ns, which in common with the lifetime of all exciton complexes studied, is found to be constant as function of power. Similarly, the peak of X^* is shifted by 0.44 ns, and is in agreement with the radiative lifetime of X_2^*, measured to be 0.52±0.05 ns.

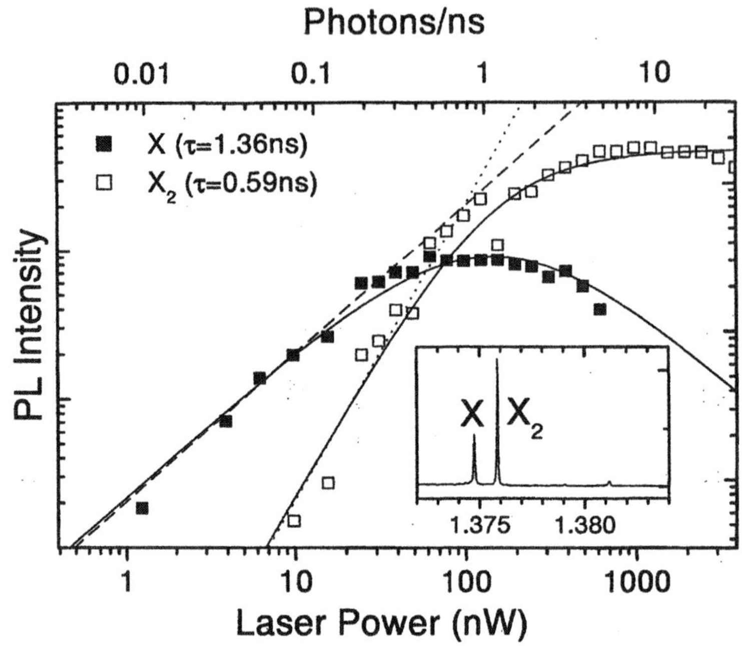

Fig. 5: Integrated intensities of PL lines as a function of power under CW laser excitation. Dotted (dashed) line shows the gradient associated with linear (quadratic) power dependence. The solid lines show calculated power dependence of X and X_2 as a function of the rate of absorbtion of photons close to the dot.

The time resolved PL demonstrates that the exciton emission follows that of biexciton, which is to be expected since the biexciton state decays radiatively into the single exciton ($X_2 \rightarrow X + \text{photon}$). Thus the temporal dependence confirms the assignment of these lines from the power dependence of their integrated intensities discussed above. Similarly, emission due to X^* follows that of X_2^*, as expected from the radiative decay of a charged biexciton state into a charged single exciton ($X_2^* \rightarrow X^* + \text{photon}$).

The radiative lifetime of X and X^* are determined to be 1.36±0.06 ns and 1.07±0.02 ns respectively, more than a factor of two longer than the corresponding biexciton. This is attributed to the two possible recombination paths for the two electron-hole pairs in the biexciton. In addition, X^* has a slightly shorter radiative lifetime than X, which may be due to the lack of a dark state for the charged exciton.

3.3 CW excitation

Photoluminescence was also collected after excitation of the mesa by a CW laser of similar energy. Such measurements revealed the same excitonic transitions, with identical photon energies (within experimental error) and similar linewidths as for pulsed laser excitation described above. Again the exciton line was found to have a

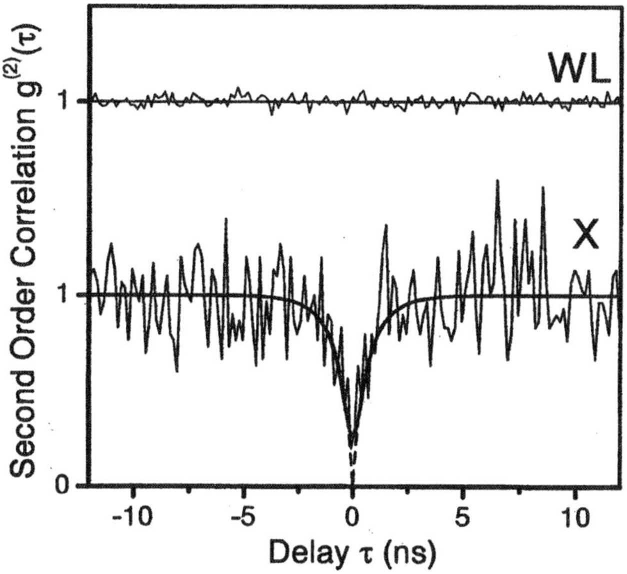

Fig. 6: Second-order correlation function measured under CW optical excitation of emission from the exciton (X). Smooth lines are calculated for the same experimental conditions. The top trace shows the measured and calculated correlation of emission from the wetting layer for comparison.

linear dependence upon laser power, while for the biexciton it was again approximately quadratic, as shown in Fig. 5.

The laser power dependence of the exciton intensities measured with pulsed and CW laser excitation differ at higher laser powers. This is because the exciton state of the dot can capture a second electron-hole pair for intense CW excitation before the emission of the exciton photon. For this reason the exciton emission weakens at the highest CW laser powers, while the maximum intensity of the biexciton is considerably stronger than that of the exciton for CW excitation.

4. Measurements of the photon statistics

The system was then arranged as in Fig. 1 to measure the second-order correlation function, $g^{(2)}(\tau)$, between photons emitted from the X, X^* and X_2 states. PL, excited with a relatively high laser power, so as to saturate the emission intensities, was spectrally filtered by the spectrometer so as to contain just one emission line. The second-order correlation function was measured for both continuous and pulsed laser excitation.

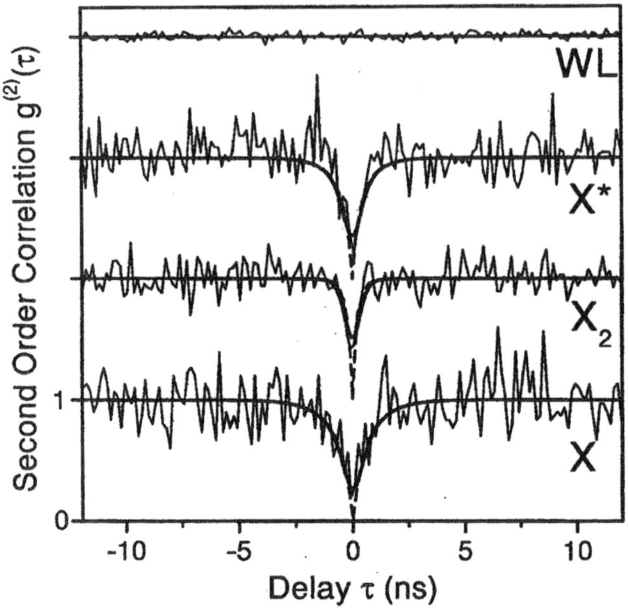

Fig. 7: Second-order correlation function measured under CW optical excitation for emission from the exciton (X), biexciton (X_2), and charged exciton (X^*). Smooth lines are calculated for the same experimental conditions. The top trace shows the measured and calculated correlation of wetting layer emission for comparison.

4.1 Photon anti-bunching in quantum dot emission

Figure 6 compares the second-order correlation function of the emission due to the exciton transition (X) of the quantum dot with that of the wetting layer after CW laser excitation. Notice that the wetting layer emission displays a flat correlation trace, which is the signature of a source displaying Poissonian statistics. In contrast, there is a clear dip in the correlation trace of the quantum dot around zero time delay, for which $g^{(2)}(0) = 0.20$. This anti-bunching behavior arises because after a photon is emitted, there is a finite time delay before the quantum dot captures a second photon and re-emits. It is direct evidence of the suppression of simultaneous emission of two photons by the quantum dot.

Figure 7 plots the second correlation function measured for three of the exciton complexes observed at high laser power. These traces were recorded by setting the spectrometer wavelength so that only emission due to either X, X_2 or X^* could pass. Notice that anti-bunching behavior is observed for all of the exciton complexes. If we consider the biexciton state of the dot, for instance, after emission of a biexciton photon so as to leave the exciton state, i.e., $X_2 \rightarrow X + \text{photon}$, there is a finite time delay before the biexciton state can be repopulated, allowing a second photon to be emitted at the biexciton photon energy. The anti-bunching dip is not so deep for the biexciton and charged exciton emission as the exciton; $g^{(2)}(0) = 0.43$ for X_2 and

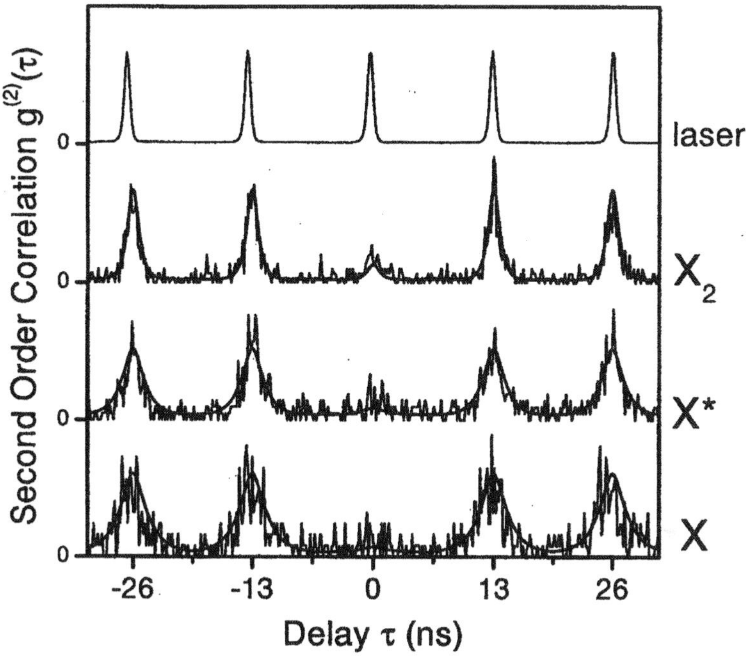

Fig. 8: Second-order correlation function measured under pulsed optical excitation for emission from the exciton (X), charged exciton (X^*), and biexciton (X_2). Smooth lines are calculated for the same experimental conditions. The top trace shows the measured correlation of an attenuated laser for comparison.

0.20 for X^*. However, as discussed later, the depth of the dip is limited (for all CW measurements) by the timing resolution of the Hanbury-Brown and Twiss set-up. The dip is thus more difficult to resolve for X_2 and X^*, for which the radiative lifetimes are shorter.

4.2 Single photon emission from a quantum dot

Figure 8 plots the second-order correlation function recorded for the X, X^*, and X_2 lines of the dot under *pulsed* laser excitation, as well as that measured for the laser itself. Each correlation trace consists of a series of peaks separated by the laser period of 13.0 ns. For the laser, the correlation peaks have roughly equal height, as expected for a coherent light source. In contrast, the emission from the quantum dot displays a strong suppression of the peak around zero time delay for each of the complexes. This is clear evidence for single photon emission from each of the exciton complexes.

The area of the zero time delay peak observed for the X transition, compared to those at finite delay, suggests the fraction of multi-photon pulses emitted to be just 6% of that of a coherent source of the same intensity. As discussed below, these multi-photon pulses derive from stray emission from the buffer, substrate and

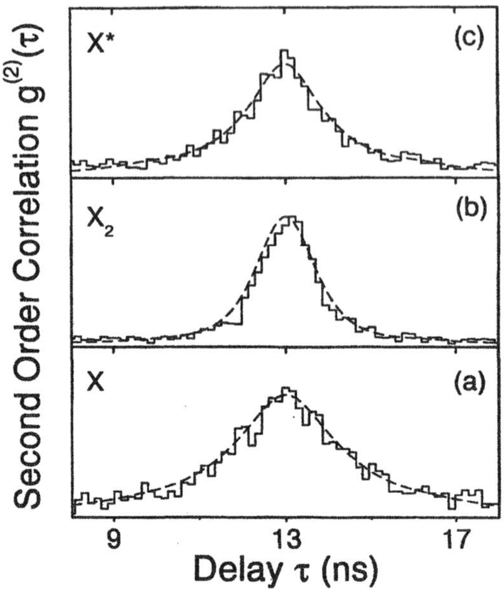

Fig. 9: Uncertainty in time between subsequent photon emission for the (a) exciton, (b) biexciton and (c) charged exciton.

wetting layer of the structure. Their number could be reduced by redesign of the sample structure or better spectral rejection.

Notice in Fig. 8, that the correlation peaks observed for X_2 appear to be significantly narrower than those for X. This can be seen more clearly in Fig. 9 which plots the average peak shape for exciton, biexciton, and charged exciton emission. The FWHM of these peaks demonstrates a reduction in the jitter between single photon emission events for the biexciton (1.5 ns) compared to the simple exciton (2.8 ns). This reduction in jitter derives from the shorter radiative decay time of the biexciton state, measured above to be 0.59 ns for X_2 compared to 1.36 ns for X.

There are potential advantages in designing single-photon emitting devices around biexcitonic emission from quantum dots rather than the single exciton transition. Since the radiative lifetime of the biexciton is shorter than that of the exciton, by at least a factor of two in these measurements, the maximum possible emission rate from the biexciton state can be higher. Another advantage is a reduction in the timing jitter associated with the uncertainty in the time between photons. This would allow the photon detector used in an application to be gated 'on' for a shorter time, thus reducing its dark count probability.

Since triexcitons cannot be confined in these small dots, the temporal evolution of the biexciton PL remains unchanged at higher excitation powers, and the jitter and maximum bit-rate are independent of power, unlike that for the single exciton for which we observe a delay at high laser power, or even for a biexciton in a dot with more than one pair of electron and hole levels. This means that the average

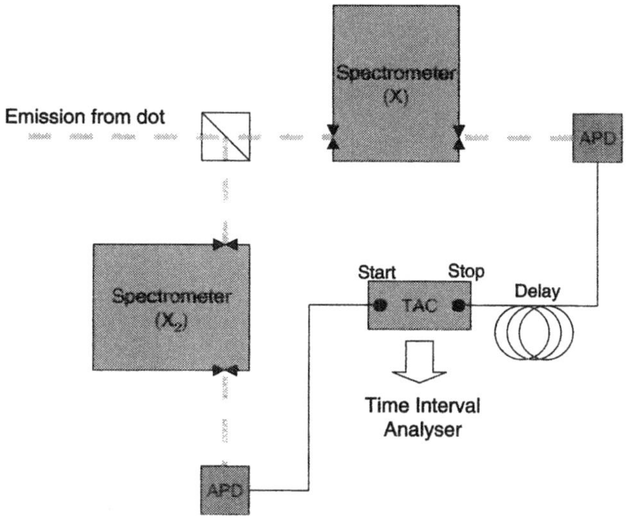

Fig. 10: Schematic of Hanbury-Brown and Twiss experimental arrangement used for cross correlation measurements between the exciton (X) and biexciton (X_2).

number of photons emitted by the device per pulse can be much closer to unity, as lower powers are not necessary to reduce the jitter. This has the effect of reducing the number of pulses that contain no photons, and thus increases the emission efficiency.

4.3 Cross-correlation measurements

We also measured the correlation between different exciton transitions emitted by the quantum dot [33]. This experiment was performed using the set-up of Fig. 10, where the spectrometers in the two arms of the beamsplitter were set to allow different excitonic transitions to pass. An electrical delay on the stop channel allows us to record negative as well as positive delays, i.e., to look at the case where the 'stop' occurs before the 'start'.

4.3.1 CW cross-correlation

Figure 11 shows the cross-correlation measured by using the biexciton X_2 transition to trigger the 'start' channel and the exciton X to supply the 'stop', recorded for CW laser excitation. The laser power was set so that the exciton and biexciton had roughly equal intensities. It can be seen that the cross-correlation shows evidence of anti-bunching at negative delays, and bunching at positive. Similar characteristics have been reported in Ref. 33.

This behavior can be understood by considering the radiative decay of the biexciton state of the dot. The time-resolved photoluminescence measurements showed the exciton photon to be emitted after the biexciton. Emission of a photon at the X_2 transition energy, prepares the dot in the exciton state. There is thus an enhanced probability of observing an X photon in the stop channel, after detecting an X_2 photon in the start, giving rise to the bunching behavior. The decay in $g^{(2)}(\tau)$ at positive delays is determined by the exciton lifetime for low laser powers and the exciton capture time at high powers. On the other hand, there is a suppressed chance of seeing an exciton photon before the biexciton, leading to the anti-bunching behavior for negative delays.

4.3.2 Pulsed cross-correlation

We repeated the same cross-correlation experiment by exciting the photoluminescence using a pulsed laser. A relatively high laser power was chosen so as to saturate the X and X_2 intensity in the spectrum. The cross-correlation shown in Fig. 12 displays a series of peaks separated by the laser period of 13.2 ns. In contrast to the single line correlation measurements, notice that the peaks at finite delay have an asymmetric lineshape, with the decay on the longer time delay side of the peak being slower than the rise at shorter delays. This derives from the difference in the lifetime of the X_2 and X states used for the start and stop channels. The zero delay peak is higher than the others and is much sharper on the negative delay side. This is again due to the fact that there is an enhanced probability of detecting a X photon after a X_2 photon, but greatly reduced chance of seeing the photons in the opposite time order. The area of the zero delay peak is similar to the other peaks, with a relative area of 1.09. The difference in this area comes from the certainty of the emission of an exciton photon following that of a biexciton. For other peaks, the two photons are emitted in different periods, and emission of the exciton photon is never certain and depends on the excitation power.

4.4 Polarized cross correlation measurements

The correlation of exciton emission with biexciton emission demonstrates the operation of the device as a photon pair emitter. This result would have great significance if the photon pair was polarization entangled, allowing an entangled photon pair source to be developed [34] for applications in quantum communications that is relatively simple, compact and cheap compared with the more usual practice of entangled pair generation through spontaneous parametric down conversion in non-linear crystals [35].

A statistical analysis can be made of the polarization of the first photon emitted at the biexciton energy, and of the second photon at the exciton energy, to determine the nature of the polarization relationship between them. The experimental system is shown in Fig. 13. In addition to the components present in the unpolarized cross-correlation system of Fig. 10, a linear polarization selecting beamsplitter is inserted after each spectrometer, along with two more APDs to detect the photons reflected from each of these polarization selectors. Thus, detectors T_1 and R_1 are triggered by photons at the energy of the first (biexciton) photon, that are transmitted or reflected at the polarization selector, corresponding to TM and TE polarized photons. Similarly, T_2 and R_2 are triggered by TM and TE photons at the energy of

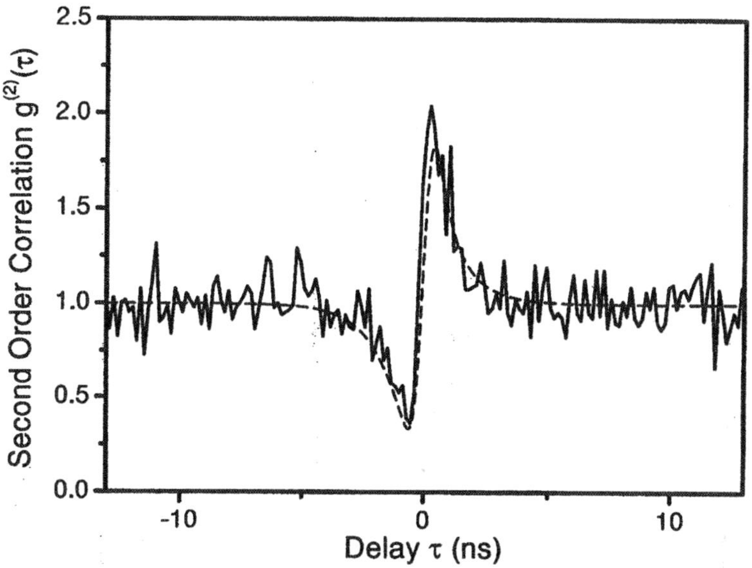

Fig. 11: Second-order correlation between emission from the biexciton (X_2) and exciton (X) under CW laser excitation. Dashed lines show expected correlation from calculations.

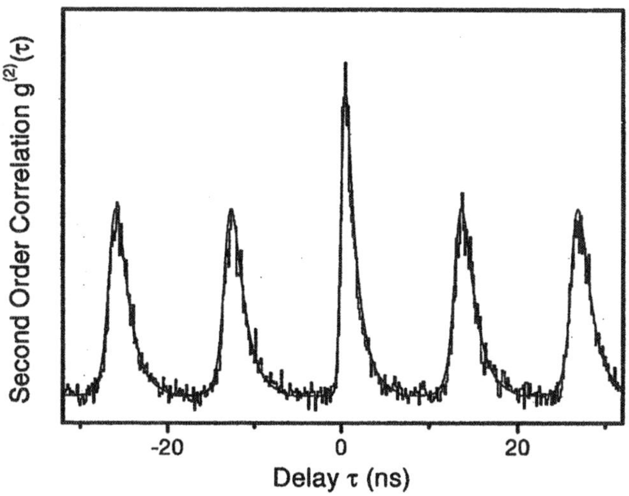

Fig. 12: Second-order correlation between emission from the biexciton (X_2) and exciton (X) under pulsed laser excitation. Smooth line shows the expected correlation trace from calculations.

the second (exciton) photon energy. In addition, a wave plate is inserted before the 50/50 beamsplitter. This wave plate is either a half or quarter wave plate, and is used to project the linear or circular polarization of the emitted photons to a linear polarization at a given angle to the polarization selectors.

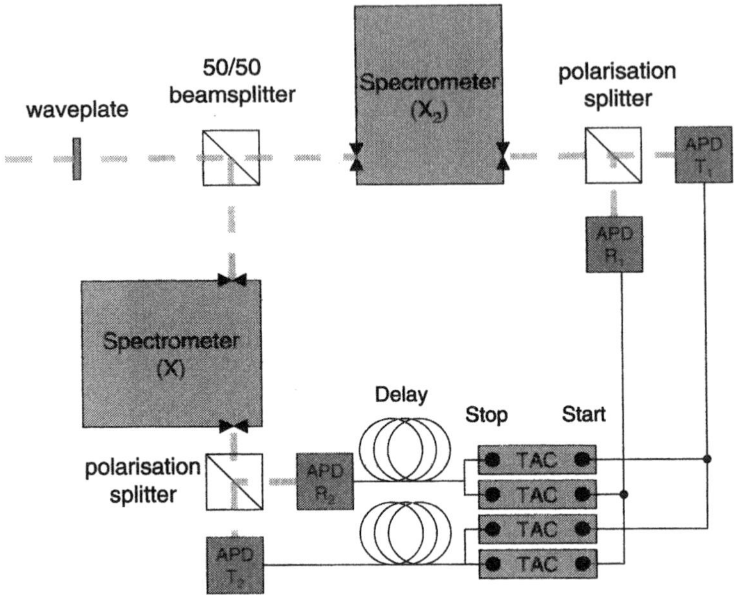

Fig. 13: Schematic of Hanbury-Brown and Twiss experimental arrangement used for polarization selective cross correlation measurements between the exciton (X) and biexciton (X_2).

Pulsed laser excitation was used as before, with the power set to 15 nW, so that the intensity of the biexciton line was comparable to that of the exciton. A half wave plate was used, and its angle was set so that there was no rotation of linear polarized light. The four possible polarization combinations of the second-order correlation between the biexciton and exciton photons were measured on four independent time interval analyzers. The results are shown on Fig. 14.

The second-order correlation of identically linear polarized pairs, shown by lines (a) and (d), are very similar to each other, and show a relatively strong peak at $\tau = 0$. The shape of this peak is similar to that measured using un-polarized detection shown in Fig. 12, and demonstrates the same suppression in the probability of detecting the exciton photon before the biexciton, characterized by the sharp rising edge. The second-order correlation of photons of opposite linear polarization, shown by lines (b) and (c), are also similar to each other, and have zero delay peaks that are similar in size to the other peaks, in sharp contrast to the pairs of the same polarization. The area of the central peaks is normalized to the average integrated area of the other peaks arising from pairs of photons in different laser periods. The resulting areas are 1.67±0.18 and 1.68±0.33 for the photons of the same polarization, and 0.92±0.14 and 0.88±0.23 for photons of opposite polarization. This shows that there is strong correlation in the polarization of the first and second photon emitted from the quantum dot, where 65% of the photon pairs contain photons of

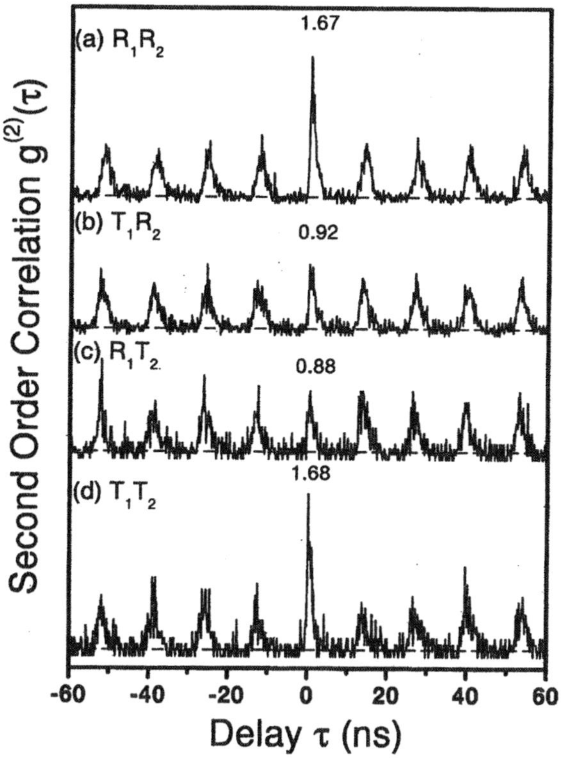

Fig. 14: Second-order correlation of biexciton photons with exciton photons, for different combinations of linear polarization. (a) and (d) show the correlation for pairs containing identically polarized photons, and (b) and (c) are for pairs containing oppositely polarized photons. Dashed lines show the CW background level for each trace.

the same linear polarization, while only 35% have opposite polarizations. This result is found to be independent of the polarization of the laser excitation, and the time integrated total emission is found to be un-polarized within experimental error.

The degree of polarization is however found to be strongly dependent on the rotation of the linear polarization relative to the polarization selectors. The correlation in the polarization of the photon pair reduces from a maximum close to no rotation of linear polarized light, to the case when no polarization dependence is resolvable beyond a rotation of the polarization by 20°. This is consistent with the expected behavior for correlated photon pairs, as the path selection by the polarization splitter becomes random as the orientation of the photon polarization approaches 45° to the vertical axis. It is also in contrast to the expected behavior of polarization entangled photon pairs, where the correlation is expected to be independent of the wave plate rotation. This is because the polarization of an entangled photon is not defined until measured, and the correlation between the first and second photon only depends on the relative angle between the two polarization analyzers, which is constant in these experiments.

When a quarter wave plate is used, we analyze the relationship between circular polarized photons emitted from the sample. We observe no polarization dependence in the second-order correlation, for any angle of rotation of the quarter wave plate, which shows that the circular component of the detected photons, including those components due to potential phase shifts in the measurement system, is small.

In irregularly shaped, or elongated quantum dots, we expect the degeneracy of the exciton emission line to be lifted, resulting in two linearly polarized emission lines in the spectra, separated by up to a few 100 μeV [36,37]. The result of the splitting of the exciton level is the formation of two distinct decay paths, which destroys the entanglement of the system. The polarization of the photons emitted are expected to be linear, the exciton photon is expected to be of the same polarization as the photon emitted by the parent biexciton, and the two distinct decay paths are expected to emit photons of opposite polarization [38]. Although we do not resolve such splitting in any of the dots we have studied, our maximum resolution of 50 μeV is still broader than the homogeneous linewidth of the quantum dot under study, determined from the lifetime to be around 0.5 μeV. In addition, the angle of maximum correlation in the polarization corresponds to emission linearly polarized along the [110] and [−110] axes, in agreement with studies that have shown elongation of quantum dots along the [−110] direction in particular [17]. We thus attribute the observed emission of linearly polarized pairs to an irregular dot shape.

The generally un-polarized emission of our quantum dot suggests that there is no preferential selection of a particular decay path, and the fact that a significant number of pairs are emitted in the opposite polarization suggests scattering of the exciton between the non-degenerate levels, or imperfect selection of the polarization in our measurement system. The scattering of linearly polarized laser light reflected off the sample surface into the orthogonal polarization is found to be less than 1%, and the polarization independence seen in the correlation of circularly polarized emission, suggests that exciton scattering may be the cause, although this must be happening on a timescale of similar order to the exciton lifetime, which is at least an order of magnitude faster than the spin scattering lifetime measured elsewhere [39].

In summary, we conclude that the photons emitted from the quantum dots studied here are strongly linearly polarized, with no circular component resolvable in these experiments, and that the time averaged emission is un-polarized. Strong correlation is found between a biexciton and an exciton photon of the same polarization, consistent with polarized pair emission of an anisotropic quantum dot. The polarization correlated pair are not entangled, due the distinguishable nature of the two decay paths. It may still be possible to generate polarization entangled photon pairs from a single quantum dot device, provided a suitably isotropic quantum dot can be found. However, the strong scattering in these results may also be present in an entangled photon source, which would reduce the degree of entanglement, and introduce errors into any quantum cryptography system of which it was a part.

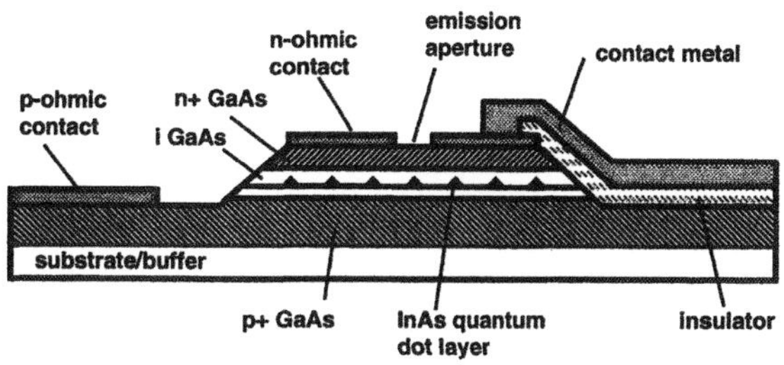

Fig. 15: Schematic of the single-photon emitting diode in cross-section.

5. Electrically injected single photon emission

One of the attractive features of using InAs quantum dots to generate single photons is the possibility of electrically injecting the electrons and holes and thereby dispensing with the pump laser and its elaborate alignment with the quantum dot. This section describes the observation of anti-bunching and single photon emission in the electroluminescence of quantum dots inside a *p-i-n* diode.

5.1 Device structure

The MBE grown layer structure for the electroluminescence measurements is very similar to that described previously for the PL experiments except that the quantum dot layer was placed in the intrinsic region of a *p-i-n* diode, as shown in Fig. 15. The wafers were prepared into mesas with lateral dimensions of 10×10 μm and Ohmic contacts were formed to the n and p-type regions. The emissive area was defined by an aperture in the opaque metal layers on the device surface. Figure 16 plots a current voltage characteristic recorded on a *p-i-n* diode, displaying nearly ideal behavior, with the injected current increasing rapidly around a forward bias of 1.5 V.

5.2 Electroluminescence spectra

The luminescence produced by the diode, for either optical or electrical injection, was collected in a similar manner to that described in the previous section. The same transitions, with similar linewidths, were observed in the photo- and electroluminescence spectra. Figure 17 plots electroluminescence spectra recorded on the diode at 5 K with different applied biases between the n- and p-type contacts. At low injection currents a sharp electroluminescence line is observed near 1.3942 eV. Since the intensity of this line increases approximately linearly with current, I (Fig. 18); it is ascribed to recombination of the single exciton (X). At higher injection currents, a

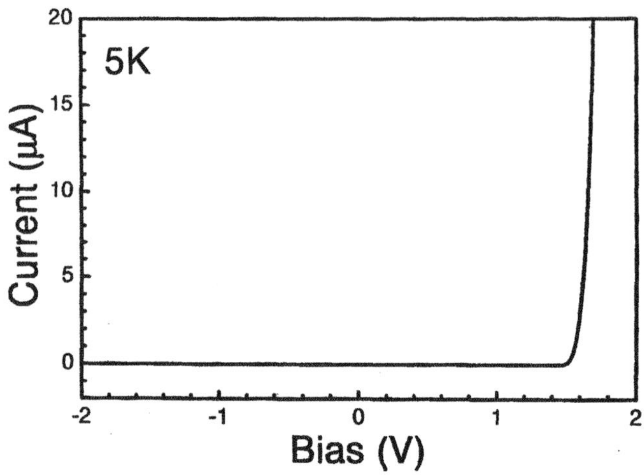

Fig. 16: Current voltage characteristic of the device shows ideal diode-like behavior.

second strong line (marked X_2) appears at higher energy. This line, which strengthens with current as $I^{2.0}$, is ascribed to the biexciton transition of the dot. Note that the strength of X drops for currents in excess of 5 μA, due to competition from the biexciton state. On the other hand, the biexciton intensity is seen to saturate at the highest currents, suggesting, as for the photoluminescence experiments, that tri- and higher order excitons cannot be excited in these dots.

Time resolved photoluminescence measurements on the same dot determined the exciton and biexciton lifetimes to be 1.02 and 0.47 ns, respectively. We also observe the X decay to be delayed relative to X_2 at high laser power, as one would expect, since the biexciton photon is emitted before the exciton [20].

5.3 Photon anti-bunching in electroluminescence

The second-order correlation function of the electroluminescence of the *p-i-n* diode was studied using a Hanbury-Brown and Twiss arrangement of Fig. 1. Figure 19 (i-iii) plots the correlation signal recorded for the single exciton electroluminescence with different injection currents. The dip in the correlation signal $g^{(2)}(\tau)$ at zero time delay, $\tau = 0$, due to anti-bunching, is clearly observed. The finite value of $g^{(2)}(0)$ in these traces derives mostly from the finite time resolution of the measurement system, as discussed later in the analysis section. This demonstrates that single photon emission can also be achieved for an electrically driven device. A very similar correlation trace was recorded by injecting the electron-hole pairs using continuous laser excitation, as in the previously described experiments. In contrast, emission from the two-dimensional wetting layer on which the dots form [Fig. 19 (v)] displays a flat correlation trace, as expected for Poissonian statistics.

We also measured the second-order correlation function of the biexciton electroluminescence line of the dot at a diode current of 6 μA [Fig. 19 (iv)]. We find

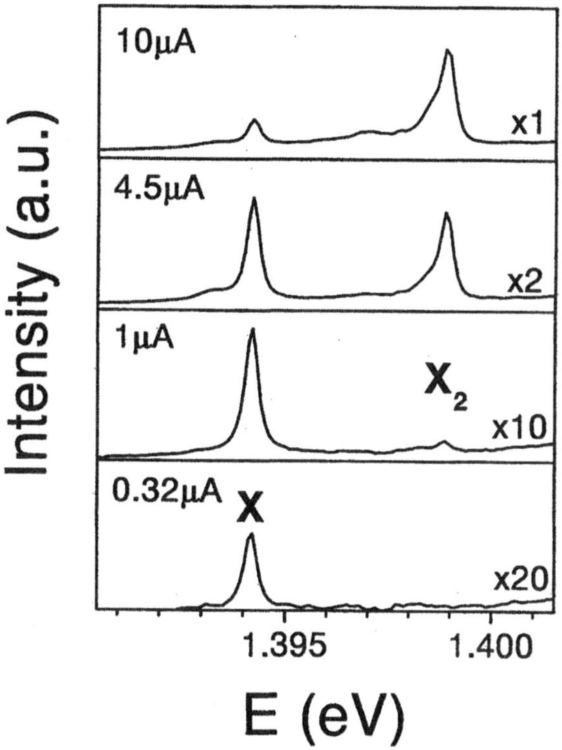

Fig. 17: Electroluminescence spectra from the device for different injection currents. Sharp emission lines (marked X and X_2) are seen, arising from a single quantum dot in the structure.

the X_2 transition also displays photon anti-bunching. After emission of a biexciton photon the dot is occupied by a single exciton and must therefore capture another electron-hole pair before a second biexciton photon can be emitted. In this case the zero delay dip is shallower, $g^{(2)}(0) = 0.75$, than for the single exciton. However, as demonstrated by the calculated curve in Fig. 19 (iv), this is largely because the recovery in $g^{(2)}(\tau)$ is faster for the biexciton, which has a shorter lifetime, than exciton, and thus is more difficult to resolve experimentally for the biexciton. These results show that at higher injection currents the diode can generate pairs of photons, one at the exciton energy of the dot and the other at the biexciton.

5.4 Single photon emission in electroluminescence

In order to regulate the emission time of the single photons, we applied a pulsed current to the diode. Figure 20(i) shows the second-order correlation function recorded by applying a bias consisting of a dc component of 1.50 V superimposed on voltage pulses with a height of 0.15 V, a width of 400 ps and a repetition rate of 80

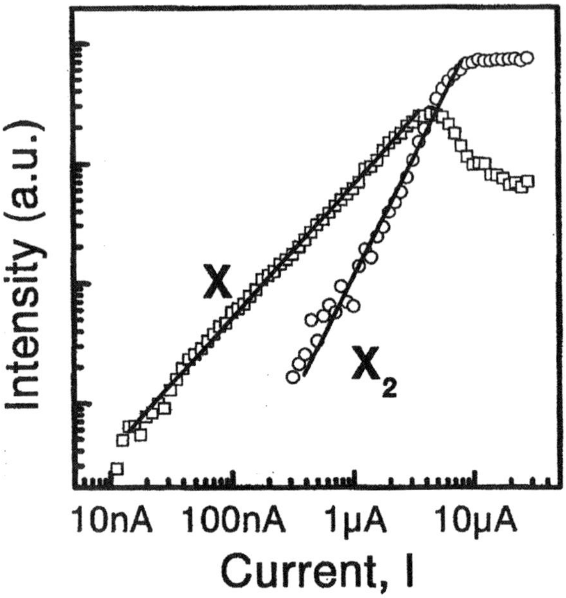

Fig. 18: Measured (symbols) dependence of the intensities of the X and X_2 lines upon the injection current. The solid lines show a linear fit to the low current data.

MHz. The dc component was chosen to be just under the turn-on voltage, so as to generate little electroluminescence. Under pulsed electrical injection the emission from the sample is also pulsed. The correlation trace therefore presents a series of peaks which are separated by the pulse repetition period. Notice that the peak at zero delay is much weaker than those at finite delay, proving a suppression of the multi-photon emission in the dot electroluminescence pulses. In comparison, for the electroluminescence measured for the wetting layer the correlation peaks are all of roughly equal height, as expected if the emission is Poissonian.

6. Analysis

We model the intensity and photon statistics of the quantum dot luminescence by regarding it, at any point in time, to contain zero, one or two excitons. This seems to be the case from experiment, as emission from tri-exciton complexes is not observed, although such an analysis ignores the intermittent periods when an excess carrier is trapped in the dots. We consider the probabilities n_0, n_1, and n_2 that the dot contains 0, 1, or 2 excitons. Scattering is neglected, and we construct rate equations by expressing the change in the probabilities due to pumping and recombination. The master equations for an effective injection rate p, which is related to the laser power in the optical injection experiments and the diode current for electrical injection, are given by,

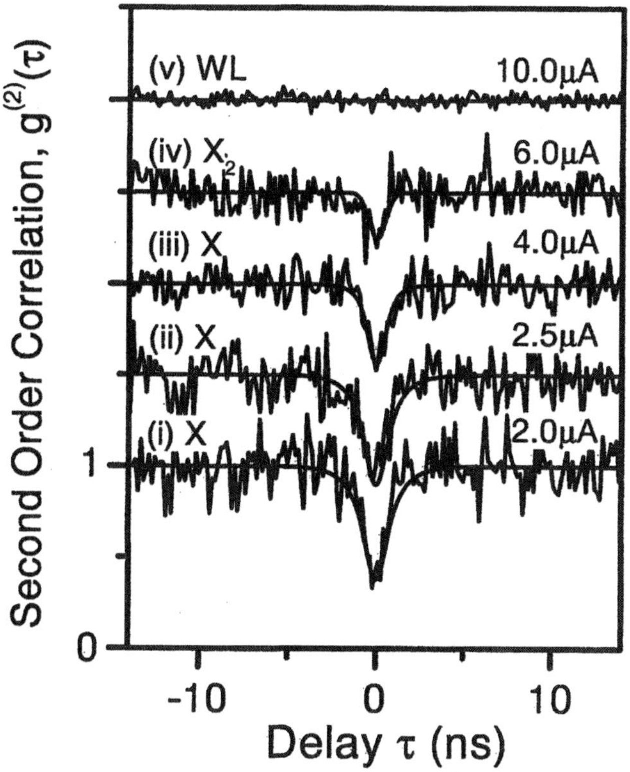

Fig. 19: Second-order correlation function, $g^{(2)}(\tau)$, of the electroluminescence of the single exciton line measured for different injection currents of (i) 2.0, (ii) 2.45 and (iii) 4.0 mA, as well as the biexciton line for 6 mA (iv). For comparison, the correlation trace of the wetting layer electroluminescence (v) is also shown. Smooth lines are calculated second-order correlation functions for electron/hole pair injection rates of (i) 0.45, (ii) 0.55, (iii) 1.31, and (iv) 2.00 pairs/ns.

$$\frac{dn_0}{dt} = -n_0 p + \frac{n_1}{\tau_1}, \tag{1}$$

$$\frac{dn_1}{dt} = n_0 p - n_1 p + \frac{n_2}{\tau_2} - \frac{n_1}{\tau_1}, \tag{2}$$

$$\frac{dn_2}{dt} = n_1 p - \frac{n_2}{\tau_2}, \tag{3}$$

where τ_1 and τ_2 are the radiative lifetimes of the exciton and biexciton. The terms on the right hand side of Eq. (2) are due to excitation of an empty dot, excitation of a dot containing one exciton, radiative decay of a dot containing two excitons, and radiative decay of a dot containing a single exciton. In addition, since the dot must contain 0, 1 or 2 excitons only, the sum of the probabilities must equal 1,

$$n_0 + n_1 + n_2 = 1. \tag{4}$$

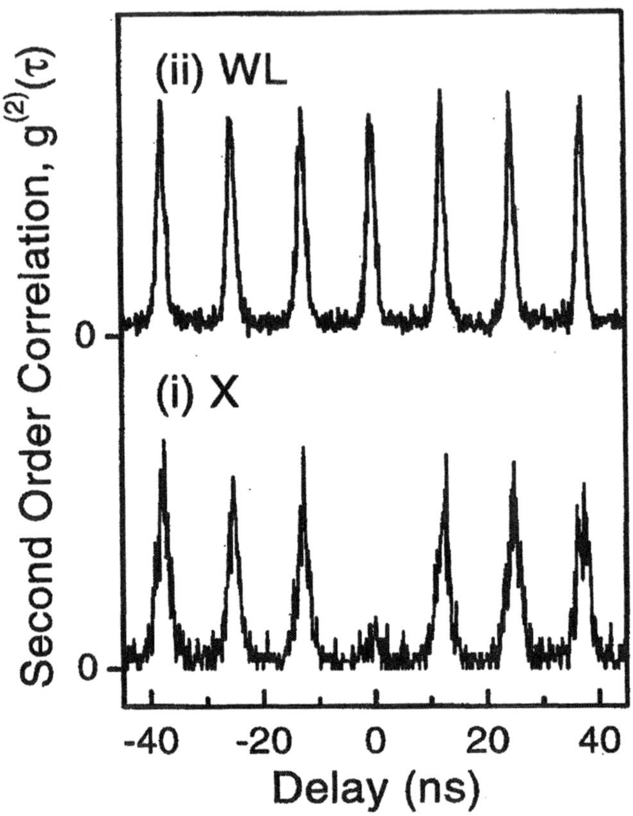

Fig. 20: Correlation measured using pulsed electrical injection for (i) quantum dot exciton and (ii) wetting layer emission.

6.1 CW solutions

6.1.1 Power dependence of luminescence intensity

To calculate the relative intensities of the exciton lines under CW laser or current excitation, we take the steady state solutions of (2) and (3), and substitute into (4) to obtain the probability at a given power of having a given number of excitons in the dot. Since the probability of photon emission is proportional to these probabilities and the radiative recombination rate, the intensities I_1 and I_2 of emission due to exciton and biexciton recombination are as follows:

$$I_1 \propto \frac{n_1}{\tau_1} = \frac{1}{\tau_1}\left(1 + \frac{1}{p\tau_1} + p\tau_2\right)^{-1}, \tag{5}$$

$$I_2 \propto \frac{n_2}{\tau_2} = \frac{1}{\tau_2}\left(1 + \frac{1}{p\tau_2} + \frac{1}{p^2\tau_1\tau_2}\right)^{-1}. \tag{6}$$

The lifetimes of the exciton (τ_1) and biexciton (τ_2) state can be determined by experiment, leaving no fitting parameters, other than scaling the I and p axes to experimental values.

Since a maximum of two excitons can be confined in these dots, the intensity of the X_2 line saturates as the power is increased. In contrast, the intensity of the X line reaches a maximum, and is then suppressed due to competition with the X_2 state. The injection rate at which maximum intensity is reached, $p_{\max}$, can be determined by differentiating (5) and solving for $dI_1/dp = 0$,

$$p_{\max} = \frac{1}{\sqrt{\tau_1 \tau_2}}. \tag{7}$$

Thus the ratio of the maximum intensity of the biexciton $I_2^{\max}$ to the maximum intensity of the exciton $I_1^{\max}$ is given by

$$\frac{I_2^{\max}}{I_1^{\max}} = \frac{\tau_1 + 2\sqrt{\tau_1 \tau_2}}{\tau_2}. \tag{8}$$

Typically, the lifetime of the biexciton is around half that of the exciton, which yields a ratio of maximum intensities of $2 + 2\sqrt{2}$. This corresponds to the maximum intensity of the exciton line to be only 21% of the maximum biexciton intensity for a dot where the biexciton lifetime is half that of the exciton. The power at which this occurs is then $\sqrt{2}/\tau_1$ photons per nanosecond, or $\sqrt{2}$ photons per exciton lifetime. These values are in agreement with the experimental data in Fig. 5.

6.1.2 Second-order correlation

In the Hanbury-Brown and Twiss experiment that we use to determine the photon emission statistics of our devices, we measure the time between two photons by starting and stopping a timer. The start event is provided by a single photon detector configured to detect a photon at either the X or the X_2 energy. The probability of generating a start event is thus proportional to the photon intensity at the selected energy. Similarly, the stop event is provided by a second single-photon detector, and the probability of generating a stop is proportional to the photon intensity at the selected stop energy. As a specific example, we now consider an experiment where both the start and the stop are tuned to the exciton energy.

As soon as the start is received, the dot has just emitted a photon by the recombination of a single exciton, and we know for certain that the dot is empty. In this case we have the initial conditions, $n_0^i = 1$, $n_1^i = 0$, and $n_2^i = 0$. The probability of detecting a stop at time t after this event can be determined from the probability that the quantum dot contains a single exciton at time t. The system of differntial equations is then solved analytically [13] to give the normalized second-order correlation function

$$g^{(2)}(\tau) = 1 + k\mathrm{e}^{-\tau/\tau_-} - (1+k)\mathrm{e}^{-\tau/\tau_+} \tag{9}$$

$$\tau_- = \frac{2}{2p + \frac{1}{\tau_1} + \frac{1}{\tau_2} - \sqrt{\frac{4p}{\tau_1} + \left(\frac{1}{\tau_1} - \frac{1}{\tau_2}\right)^2}} \tag{10}$$

$$\tau_+ = \frac{2}{2p + \frac{1}{\tau_1} + \frac{1}{\tau_2} + \sqrt{\frac{4p}{\tau_1} + \left(\frac{1}{\tau_1} - \frac{1}{\tau_2}\right)^2}} \tag{11}$$

$$k = \frac{\tau_2 - \tau_1}{\tau_- - \tau_+}. \tag{12}$$

Figure 21 shows the form of $g^{(2)}(\tau)$ calculated for different injection rates (p) and taking the exciton lifetimes determined experimentally in the photoluminescence experiments, $\tau_1 = 1.36$ ns and $\tau_2 = 0.61$ ns. The width of the dip increases with decreasing power, in line with the excitation time (average time between dot excitation events by laser). The width of the dip saturates at weak powers where the excitation time is much greater than the exciton lifetime. In this regime, the suppression of $g^{(2)}(\tau)$ is determined by the exciton lifetime. We also note that at high powers the second-order correlation becomes more complicated, showing bunching behavior at finite time delay, in addition to the anti-bunching at zero delay. This is explained by the depletion of the single exciton states by excitation into the biexciton state at higher injection rates.

6.1.2.1 Role of background luminescence

The finite value of the second-order correlation of the quantum dot emission at zero time delay derives from two factors: emission from parts of the sample other than the quantum dot, and secondly, the finite time resolution of the measuring equipment. We now consider the contribution of both.

Background counts include the dark counts in the detectors, and any stray light, probably from the substrate or wetting layer regions of the sample that enters the detection system. We can thus divide the total counts C into the signal component from the dot S, and the background component B, so that $C = S + B$. The correlation signal is proportional to C^2, yet the contribution from the dot is only S^2. The remaining terms in $C^2 = (S + B)^2$ therefore contribute a background level to the correlation measurement, and reduce the amplitude of the dip. Thus the strength of the correlation background relative to the total correlation signal is given by

$$A = \frac{2SB + B^2}{(S + B)^2} \approx 2\frac{B}{S}, \qquad \text{(for } B \ll S\text{)}. \tag{13}$$

We note that in experiment the background is typically $\sim 5\%$, which reduces the amplitude of the dip by only 10%.

6.1.2.2 Role of finite time resolution

The jitter associated with the photon counting avalanche photodiodes limits the time resolution of the Hanbury-Brown and Twiss system to 0.5–1 ns. Since the width

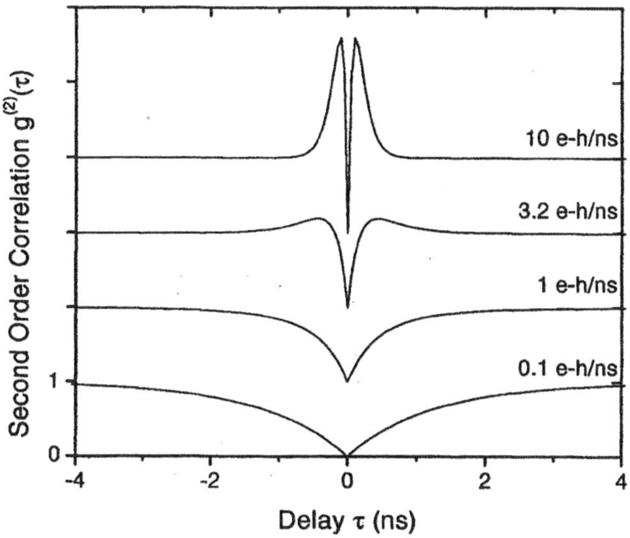

Fig. 21: Calculated second-order correlation of emission from an exciton in a quantum dot for different electron-hole pair capture rates.

of the dip in the correlation trace is comparable to this, significant broadening of the dip occurs, which is accompanied by a strong reduction of its amplitude. To account for this, we can take a correlation function calculated including background contributions, and then broaden it by convolution with a Gaussian function, which represents the measured response of the system. It is clear from Fig. 22 that the unavoidable broadening introduced by our measurement system provides a significant reduction in the amplitude of the dip, by up to a factor of 2.

6.1.2.3 Comparison with experiment

The dashed line superimposed on the experimental data in Figs. 6 and 7 shows $g^{(2)}(\tau)$ calculated using Eq. (9-12), and taking the experimentally determined lifetimes. The excitation rate, p, is determined by comparing the ratio of the X and X_2 peaks in the calculated and measured spectra. This calculation reproduces the width of the anticorrelation dip faithfully. However, it overestimates the depth of the dip, as it predicts $g^{(2)}(0) = 0$, the value for an ideal single-photon emitter.

The smooth solid lines in Figs. 6 and 7 display a more realistic simulation, which additionally include the effects of the background luminescence and the finite time resolution of the measuring system, as detailed above. The background PL level was measured directly by tuning the spectrometer to a nearby wavelength where there is no excitonic transition. Its effect is to add a constant background to the correlation trace. The time resolution of the system was determined experimentally to be 0.85 ns.

Similar considerations limit the minimum in $g^{(2)}(\tau)$ in the electroluminescence correlation measurements of Fig. 19. As in the PL experiments, this is largely due to the finite time resolution of the measurement system, rather than two-photon emission events from the dot. The smooth lines in Fig. 19 show the calculated form of $g^{(2)}(\tau)$, after taking account of the background luminescence and the effect of the finite temporal resolution of the measurement system, as described above. The excellent agreement of the measured and calculated curves, for which there are no fitting parameters, demonstrates that the finite value of $g^{(2)}(0)$ derives mostly from the limited time resolution of the measurement system, while two-photon emission from the device contributes to $g^{(2)}(0) < 0.07$ at an injection current of 2μA. This represents more than an order of magnitude reduction in multi-photon emission events compared to a classical light source displaying Poissonian statistics.

6.1.3 CW biexciton correlation with exciton

In the cross-correlation experiments, the start is provided by the detection of a photon at the biexciton energy, and the stop provided by detection of a photon at the exciton energy. To calculate the form of the second-order correlation trace, we must first divide the trace into two regimes, one for positive times, and one for negative times. Since the system is triggered by biexciton emission for $t > 0$, the initial probability of the dot containing one exciton, n_1^i, is equal to 1. For $t < 0$, we calculate the probability that an exciton is emitted just before a biexciton. Another way of looking at this is the probability that a biexciton is emitted just after an exciton. Therefore we can set the initial conditions to exciton emission, and so $n_0^i = 1$. We can then invert the t axis, and combine with the previous solution to form the CW correlation trace. Broadening is introduced in the usual way, and the calculated correlation is found to be in excellent agreement with the experimental data shown in Fig. 11.

6.2 Pulsed solutions

For pulsed excitation, it is necessary to make further assumptions to those made for CW excitation. Firstly, the pulse width is short compared to the radiative lifetimes of the exciton complexes. This is a reasonable assumption for pulsed laser excitation, since the time of our laser pulse is only a few ps, and the capture time of the excitons created in the barriers into to dot is expected to be an order of magnitude less than the lifetime of the biexciton state. This assumption is necessary to neglect re-excitation of the dot within a single period. Secondly, we need to assume that the lifetime of the excitons in the dot is less than the period of the laser. This is justified since the lifetime of the exciton state is an order of magnitude shorter than the period. This approximation is necessary so that the exciton always empties during a laser period.

6.2.1 Time integrated PL as a function of power

In a given laser pulse, the number of excitons generated in the capture region close to the dot obeys Poisson statistics, and the probability P of generating n excitons

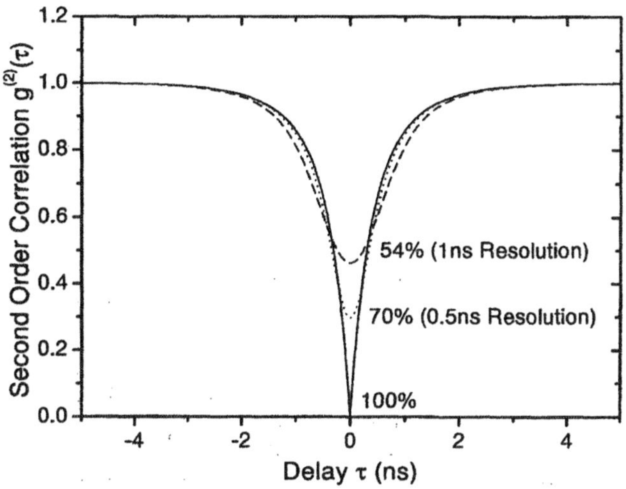

Fig. 22: Solid, dotted and dashed lines show calculated second-order correlation of emission from an exciton in a quantum dot, for a system with perfect, 0.5 ns, and 1 ns time resolution. It is clear that the finite time resolution of our experimental system causes a strong reduction in the amplitude of the anti-bunching dip.

in the dot can be expressed as

$$P(n) = \frac{p^n \mathrm{e}^{-p}}{n!}. \tag{14}$$

If one or more excitons are generated, then a photon will be emitted at the single exciton recombination energy

$$I_1 \propto n_1 = 1 - P(0) = 1 - \mathrm{e}^{-p}. \tag{15}$$

Similarly, if two or more excitons are generated, then a photon will be emitted at the biexciton recombination energy

$$I_2 \propto n_2 = 1 - P(0) - P(1) = 1 - (1+p)\mathrm{e}^{-p}. \tag{16}$$

The power dependence of pulsed excited PL intensity is compared with calculations in Fig. 3. The calculated PL intensity, shown by solid lines, is calculated using the experimentally determined lifetimes for X and X_2 of 1.36 and 0.61 ns. Again there is a close agreement between the observed and calculated dependence.

6.2.2 Second-order correlation

There are two aspects of the pulsed correlation that are of interest. They are the relative size of the zero delay peak, and the jitter of the system, characterized by the shape of the other peaks.

6.2.2.1 Suppression of zero delay peak

For an ideal single-photon emitter, the relative area of the zero delay peak will be zero, signifying the complete absence of photon-pair detection. However, in our measurements a weak peak is seen, which we expect is due to stray light. Therefore it is useful to assess the role of stray light, and to determine the significance of the area of the central peak.

In pulsed measurements, we need only consider the pulsed components of the total signal, since CW components such as dark counts only add a flat background. The major contributor to background is thus PL from the substrate and other regions of the sample. An event is registered with zero delay if a photon is detected by each detector during the same laser period. The probability of this event occurring W_0 is given by the following equation, where the factor of $\frac{1}{2}$ comes from the possibility of both photons being sent to the same detector

$$W_0 = \tfrac{1}{2}P(2). \tag{17}$$

The probability W_1 of an event being registered with one period delay is given by the probability that a single photon is generated, and directed to the start, and the probability that a single photon is generated in the subsequent period and directed towards the stop

$$W_1 = \tfrac{1}{2}P(1) \cdot \tfrac{1}{2}P(1) = \tfrac{1}{4}P(1)^2. \tag{18}$$

The ratio of the peak intensities is therefore

$$\frac{W_0}{W_1} = \frac{2P(2)}{P(1)^2}. \tag{19}$$

For a low intensity Poissonian source, $P(2) = P(1)^2/2$ and W_0/W_1 is equal to 1. Therefore for an arbitrary source, W_0/W_1 represents the suppression of multi-photon pulses relative to a Poissonian source of the same intensity. It can be shown that for an attenuated single-photon source in combination with a poissonian background, that the area of the zero delay peak is increased by the same amount as the background level in CW measurements, and is described by Eq. (13). The relative area of the zero delay peak in the experiments is consistent with the background level measured, and a perfect single-photon source.

6.2.2.2 Single photon emission jitter

The peak shape, and the jitter associated with the uncertainty in the photon emission time is calculated by the convolution of the calculated time resolved PL at the same power. Broadening can be introduced either before or after the convolution, but not in both places.

6.2.2.3 Comparison to experiments

The dashed lines in Figs. 8 and 9 display $g^{(2)}(\tau)$ calculated using the method described above and taking the exciton lifetimes and excitation rate (p) determined from experiment. As for the CW correlation traces, there is a close agreement between the measured and calculated curves. This again demonstrates that after taking account of the level of background PL, the device can be modeled as an ideal single-photon emitter.

6.2.3 Pulsed biexciton correlation with exciton

As soon as the biexciton is emitted, then the initial state of the dot is so that $n_1^i = 1$. Therefore, we can calculate the time resolved PL of the exciton from calculating $n_1(\tau)$, starting with this initial condition, and setting $p = 0$. For the other peaks, we simply convolute the time resolved PL of the exciton with that of the biexciton for the initial conditions calculated for the chosen power.

7. Discussion

The close agreement of the measured and calculated second-order correlation function demonstrates that the device can be regarded as an ideal single-photon emitter, with a Poissonian background emission from the other layers in the device. This background emission limits the rate of two-photon pulses from the device compared to a laser of the same average intensity to about 5%. This would allow quantum key distribution which is unconditionally secure from a multi-photon beamsplitting type attack [1]. A twenty-fold suppression of the multi-photon pulses compared to a laser of the same average intensity allows greater attenuation of the signal in a quantum cryptography system. This attenuation is equivalent to the extension of the optical fibre in a quantum cryptography system by 38 km at 1.3 μm and 68 km at 1.55 μm.

The two-photon emission events that we observe occur because of emission from other regions of the semiconductor. For both the optically and electrically injected device, the background emission derives mostly from the wetting layer on which the dots are formed. Although nearly all of the wetting layer luminescence is filtered out by the spectrometer, there is a weak tail at the quantum dot wavelengths, which contributes to the correlation signal. If larger quantum dots could be used, which emit at longer wavelengths away from the wetting layer peak, the background emission could be greatly reduced. However, in these experiments, the choice of quantum dot wavelength was dictated by the use of Si avalanche photodiodes.

The experiments demonstrate that each of the optically active exciton complexes formed by the dot produce a single photon per excitation pulse. There are potential advantages in designing single-photon emitting devices around biexcitonic emission from quantum dots rather than the single exciton transition. Since the radiative lifetime of the biexciton is shorter than that of the exciton, by at least a factor of two in these measurements, the maximum possible emission rate from the biexciton state can be higher. Another advantage is a reduction in the timing jitter associated with the uncertainty in the time between photons. This would allow the photon detector used in an application to be gated 'on' for a shorter time, thus reducing its dark count probability.

The cross-correlation measurements of Sect. 4.3, show that quantum dots excited with two electrons and two holes can also be used to generate pairs of photons, one at the exciton energy and another at the biexciton. Hence the emission at one photon energy could be used to herald the emission at the other. Although only polarization correlated pair emission was observed in our asymmetric dots, it has been predicted that the exciton and biexciton photon should have entangled polarization states [34]. The biexciton photon, which results from recombination of either of the two electrons with the hole of opposite spin, can be emitted in either σ^- or σ^+ circular polarization. Provided there is no scattering of the exciton state left in the dot, the exciton photon will then be emitted in the opposite polarization configuration. Hence the state of the two photons can be represented by the maximally entangled Bell state.

8. Outlook

These experiments suggest that a semiconductor technology for generating single photons, as well as photon pairs, may be within reach. The electrically driven diode structure would be a particularly attractive source, as it avoids the need for a pump laser and its costly alignment with the quantum dot. Potentially this source could be mass produced by photo-lithography on a wafer scale and thus be relatively cheap. However, a number of technological challenges remain. In particular, we find the luminescence of the structures studied to be quenched for temperatures greater than about 70K. This is due to thermal excitation of the carriers out of the shallow confinement potentials of these dots. In these experiments the choice of dot was dictated by the fact that we used Si avalanche photodiodes for detection of the single photons. Larger quantum dots, emitting at longer wavelengths, are known to be highly efficient at room temperature. They also have the advantage that they can be tailored to emit at the fibre optic wavelength of 1.3 μm. The collection efficiency from the dot must also be enhanced, which can be achieved through patterning of the device or, via its integration into a cavity structure. Quantum dot light emitting diodes may therefore provide an attractive source of either single photons, or photon pairs, for applications in quantum information technology and experiments in quantum optics.

Acknowledgements

We would like to acknowledge contributions from our colleagues, C. J. Lobo, I. Farrer, K. Cooper, N. S. Beattie, D. A. Ritchie, M. Leadbeater, and M. Pepper, at Toshiba, and the University of Cambridge Cavendish Laboratory.

References

[1] See, for instance, D. Bouwmeester, A. Ekert and A. Zeilinger, (Eds.), *The physics of quantum information* (Springer, Berlin, 2000).

[2] E. Knill, R. Laflamme, and G. J. Milburn, Nature **409**, 46 (2001).

[3] G. Brassard, N. Lutkenhaus, T. Mor, and B. C. Sanders, Phys. Rev. Lett. **85**, 1330 (2000).

[4] D. F. Walls and G. J. Milburn, *Quantum Optics* (Springer, Berlin, 1994).

[5] H. J. Kimble, M. Dagenais, and L. Mandel, Phys. Rev. Lett. **39**, 691 (1977).

[6] F. Diedrich and H. Walther, Phys. Rev. Lett. **58**, 203 (1987).

[7] Th. Basche, W. E. Moerner, M. Orrit and H. Talon, Phys. Rev. Lett. **69**, 1516 (1992).

[8] F. de Martini, G. di Giuseppe and M. Marrocco, Phys. Rev. Lett. **76**, 900 (1996).

[9] C. Brunel, B. Lounis, P. Tamarat and M. Orrit, Phys. Rev. Lett. **83**, 2722 (1999).

[10] B. Lounis and W. E. Moerner, Nature **407**, 491 (2000).

[11] L. Fleury, J.-M. Segura, G. Zumofen, B. Hecht and U. P. Wild, Phys. Rev. Lett. **84**, 1148 (2000).

[12] P. Michler, A. Imamoglu, M. D. Mason, P. J. Carson, G. F. Strouse, and S. K. Buratto, Nature **406**, 968 (2000).

[13] C. Kurtsiefer, S. Mayer, P. Zarda and H. Weinfurter, Phys. Rev. Lett. **85**, 290 (2000).

[14] R. Brouri, A. Beveratos, J. P. Poizat, P. Grangier, Opt. Letters **25**, 1294 (2000).

[15] A. Imamoglu and Y. Yamamoto, Phys. Rev. Lett. **72**, 210 (1994).

[16] J. Kim, O. Benson, H. Kan and Y. Yamamoto, Nature **397**, 500 (1999).

[17] D. Bimberg, M. Grundmann, N. N. Ledenstov, *Quantum Dot Heterostructures* (Wiley, Chichester, 1999)

[18] A. J. Shields, M. P. O'Sullivan, I. Farrer, D. A. Ritchie, K. Cooper, C. L. Foden, and M. Pepper, Appl. Phys. Lett. **74**, 735 (1999).

[19] A. J. Shields, M. P. O'Sullivan, I. Farrer, D. A. Ritchie, R. A. Hogg, M. L. Leadbeater, C. E. Norman, and M. Pepper, Appl. Phys. Lett. **76**, 3673 (2000).

[20] R. M. Thompson, R. M. Stevenson, A. J. Shields, I. Farrer, C. J. Lobo, D. A. Ritchie, M. L. Leadbeater, and M. Pepper, Phys. Rev. B **64**, 201302R (2001).

[21] R. M. Stevenson, R. M. Thompson, A. J. Shields, I. Farrer, C. J. Lobo, D. A. Ritchie, M. L. Leadbeater, and M. Pepper, Proceedings of MSS10, Linz, (2001).

[22] R. M. Thompson, R. M. Stevenson, A. J. Shields, I. Farrer, C. J. Lobo, D. A. Ritchie, M. L. Leadbeater, and M. Pepper, Proceedings of ISCS, Tokyo (2001).

[23] P. Michler, A. Kiraz, C. Becher, W. V. Scoenfeld, P. M. Petroff, Lidong Zhang, E. Hu, A. Imamoglu, Science **290**, 2282 (2000).

[24] C. Santori, M. Pelton, G. Solomon, Y. Dale, and Y. Yamamoto, Phys. Rev. Lett. **86**, 1502 (2001).

[25] V. Zwiller, H. Blom, P. Jonsson, N. Panev, S. Jeppesen, T. Tsegaye, E. Goobar, M.-E. Pistol, L. Samuelson, and G. Björk, Appl. Phys. Lett. **78**, 2476 (2001).

[26] E. Moreau, I. Robert, J. M. Grard, I. Abram, L. Manin, and V. Thierry-Mieg et al, Appl. Phys. Lett. **79**, 2865 (2001).

[27] Z. Yuan, B. E. Kardynal, R. M. Stevenson, A. J. Shields, C. J. Lobo, K. Cooper, N. S. Beattie, D. A. Ritchie, and M. Pepper, to be published in Science (2001).

[28] L. Landin, M.-E. Pistol, C. Pryor, M. Persson, L. Samuelson, and M. Miller, Phys. Rev. B **60**, 16640 (1999).

[29] K. Hinzer, P. Hawrylak, M. Korkusinski, S. Fafard, M. Bayer, O. Stern, A. Gorbunov, and A. Forchel, Phys. Rev. B **63**, 075314 (2001).

[30] F. Findeis, M. Baier, A. Zrenner, M. Bichler, and G. Abstreiter, Phys. Rev. B **63**, 121309 (2001).

[31] J. J. Finley, A. D. Ashmore, A. Lematre, D. J. Mowbray, M. S. Skolnick, I. E. Itskevich, P. A. Maksym, M. Hopkinson, and T. F. Krauss, Phys. Rev. B **63**, 073307 (2001).

[32] E. Dekel, D.V. Regelman, D. Gershoni, E. Ehrenfreund, W.V. Schoenfeld, and P.M. Petroff, Solid State Commun. **117**, 395 (2001).

[33] E. Moreau, I. Robert, L. Manin, V. Thierry-Mieg, J. M. Grard, and I. Abram, Phys. Rev. Lett. **87**, 183601 (2001).

[34] O. Benson, C. Santori, M. Pelton and Y. Yamamoto, Phys. Rev. Lett. **84**, 2513 (2000).

[35] D. S. Naik, C. G. Peterson, A. G. White, A. J. Berglund, and P. G. Kwiat, Phys. Rev. Lett. **84**, 4733 (2000).

[36] M. Bayer, A. Kuther, A. Forchel, A. Gorbunov, V. B. Timofeev, F. Schäfer, and J. P. Reithmaier. Phys. Rev. Lett. 82 1748 (1999).

[37] T. Flissikowski, A. Hundt, M. Lowisch, M. Rabe, and F. Henneberger, Phys. Rev. Lett. 86 3172 (2001).

[38] V. D. Kulakovskii, G. Bacher, R. Weigand, T. Kmmell, A. Forchel, E. Borovitskaya, K. Leonardi and D. Hommel, Phys. Rev. Lett. **82**, 1780 (1999).

[39] M. Paillard, X. Marie, P. Renucci, T. Amand, A. Jbeli, and J. M. G'erard, Phys. Rev. Lett. **86**, 1634 (2001).

Chapter 5

Spin, spin-orbit, and electron-electron interactions in mesoscopic systems

Yuval Oreg [a], P. W. Brouwer [b], X. Waintal [b] and Bertrand I. Halperin [c]

[a] *Department of Condensed Matter Physics, Weizmann Institute of Science, Rehovot 76100, Israel*
E-mail: Oreg@wisemail.weizmann.ac.il

[b] *Laboratory of Atomic and Solid State Physics, Cornell University, Ithaca, NY 14853-2501 USA*

[c] *Lyman Laboratory of Physics, Harvard University, Cambridge MA 02138 USA,*
E-mail: halperin@hall.harvard.edu

Abstract

We review recent theoretical developments about the role of spins, electron-electron interactions, and spin-orbit coupling in metal nanoparticles and semiconductor quantum dots. For a closed system, in the absence of spin-orbit coupling or of an external magnetic field, electron-electron interactions make it possible to have ground states with spin $S > 1/2$. We review here a theoretical analysis which makes predictions for the probability of finding various values of spin S for an irregular particle in the limit where the number of electron is large but finite. We also present results for the probability distribution of the spacing between successive groundstate energies in such a particle.

In a metallic particle with strong spin-orbit interactions, for odd electron number, the groundstate has a Kramers' degeneracy, which is split linearly by a weak applied magnetic field. The splitting may be characterized by an effective g-tensor whose principal axes and eigenvalues vary from one level to another. Recent calculations have addressed the joint probability distribution, including the anisotropy, of the eigenvalues. The peculiar form of the spin-orbit coupling for a two-dimensional electron system in a GaAs heterostructure or quantum well leads to a strong suppression of spin-orbit effects when the electrons are confined in a small quantum dot. Spin-effects can be enhanced, however, in the presence of an applied magnetic field parallel to the layer, which may explain recent observations on fluctuations in the conductances through such dots.

We also discuss possible explanations for the experimental observations by Davidovic and Tinkham of a multiplet splitting of the lowest resonance in the tunneling conductance through a gold nano-particle.

1. Introduction

During the last two decades the fabrication technology of small conducting islands, known as *quantum dots*, using semiconductor heterostructures, spattering of small metal grains and other methods have advanced so much that they can behave, under the right conditions, as *artificial atoms* [1]. In contrast to natural atoms, these artificial atoms do not have special symmetries, unless special experimental efforts are being performed [1].

The electron spin is responsible for a number of interesting effects in small chaotic conducting islands at low temperatures which are quite distinct from the role of spin in a bulk material. The role of spin is modified by electron-electron interactions in a way that has consequences for the distribution of the energy separations between ground states with different number of electrons, as well as for the probability of finding different spin quantum numbers at a fixed number of electrons. Other interesting effects are produced by spin-orbit coupling, which is important for the distributions of energy levels and wavefunctions in closed quantum dots and may modify their properties, and which affects the conductance distribution for an ensemble of open quantum dots.

The addition of source and drain leads, and in case of semiconductor heterostructures gates that control the charge on the dot, allows one to measure the properties of the dot-leads compound as a function of V, the difference between the potentials on the source and drain, and V_g, the potential on the gate lead.

In Sect. 2. of this chapter we discuss a few effects of electron-electron interactions in a closed dot, with no spin-orbit coupling and no significant Zeeman field. The analysis reviewed in this section has been developed by a number of research groups in the last few years. We re-derive here an effective low energy Hamiltonian that was first discussed in Ref. [2] using renormalization group (RG) arguments. Then, we extend our studies [3] for the ground state spin of a quantum dot, and analyze its influence on the Coulomb blockade peak spacing that appear in the conductance as we sweep V_g at zero bias voltage, V. Further details, concerning the parameters of the effective model in actual systems and the relation of the effective model to other models, are given in the Appendices to the paper.

In Sect. 3., we give a brief review of recent work on the effects of spin-orbit coupling in metal nanoparticles and GaAs quantum dots. The reader will be referred to the literature for a fuller account.

In Sect. 4., we discuss a problem motivated by experimental observations of Davidovic and Tinkham [4] of a multiplet splitting of the *lowest* resonance in the tunneling differential conductance through a gold nanoparticle as a function of the source-drain potential, V. We consider two possible mechanisms that may lead to effects of that type. The first is due to an almost degenerate ground state (Subsection 4.1) and the second due to a nonequilibrium phenomenon (Subsection 4.2). In Subsection 4.3 we compare these mechanisms, which both involve electron spin and/or electron-electron interaction in an essential way, and suggest experimental ways to distinguish between them.

2. Electron-electron interactions in a closed dot

When the shape of the dot is symmetric [1], a part of its single particle levels are degenerate, just like in an atom with a spherically symmetric potential. The exact electron many body configuration, which is set by the repulsive interaction between the electrons combined with the Pauli exclusion principle, is summarized in a set of rules, known as Hund's rules. The first of them assert that a partially filled set of degenerate levels will have the maximal spin that is consistent with the exclusion principle. This happens because for a system of several electrons the "most antisymmetric" coordinate wave function has the largest spin. The most antisymmetric wave function minimizes the Coulomb repulsion between the electrons because it vanishes when there are two electrons at the same point. Strong spin-orbit interactions may change Hund's rules.

To create symmetric dots, special (experimental) effort is necessary. Any generic dot, however, does not have any special symmetry and can be considered as a chaotic one, its single particle levels may assumed to be random, and are described by random matrix theory (RMT). Many theories have concentrated on the statistical properties of random levels and the way they can be used to understand chaotic dots in actual experiments, for review see Ref. [5]. The interactions between electrons in the dot and between them and the environment were commonly described by the so-called *constant interaction model*. In this model all the effects of interaction are cast into a single capacitance that describes the change in the energy of the system due to the dot's charging.

This class of models was very successful in explaining and predicting experimental results as Coulomb peak height fluctuations, conductance fluctuations through an open dot and so on [5]. However, it fails to explain experiments measuring motion of peaks in magnetic field [6], distributions of Coulomb blockade peak spacing [7–10], and multiplets that appear in a single Coulomb blockade peak [4]. Motivated by the failure of the constant interaction model, a few models that extend it were suggested [2,3,11–13].

In Subsection 2.1 we derive, by integrating out fast degrees of freedom in the RG sense, an effective low-energy-Hamiltonian, $\mathcal{H}_{\text{eff}}$. This effective Hamiltonian, first discussed in Ref. [2], describes the properties of the quantum dot at energies smaller than the Thouless energy E_{Th} of the system, which is inversely proportional to the time it takes to cross the dot along its largest dimension. We then show that a proper choice of parameters for $\mathcal{H}_{\text{eff}}$ and its analysis reproduced other models for the effects of electron–electron interaction in quantum dots. Appendix A presents some details of the relation and equivalence of different models. Throughout this article we will neglect fluctuations of the interaction parameters, we discuss the limits of this approximation in Subsection 2.1.1.

In the process of averaging over the fast motion of the electrons in the Fermi sea, three channels of interaction appear: The direct/charging channel, the exchange/spin channel and the Cooper channel. The actual values of the parameters that describe these channels in $\mathcal{H}_{\text{eff}}$ depend on the fast motion of electrons. We can have a richer situation when there are intermediate scales between the Fermi energy and the Thouless energy, *e.g.*, the inverse of the mean free time, τ, between elastic collisions with static impurities in diffusive systems. In Appendix B we find, using the Fermi-liquid-theory, the effective exchange interaction parameter for several three and two dimensional systems. Its dependance on electron density in two dimensional Si-MOSFET and AlGaAs ballistic heterostructures is estimated. Appendix C deals with the renormalization of the Cooper channel.

After a detailed discussion of the model in Subsection 2.1, we use it to analyze several physical quantities. In Subsection 2.2 we present an extended version of our study in Ref. [3] of the ground state spin configuration distribution. The following Subsection (Subsection 2.3) discusses the effects of spin fluctuations in the ground state on the Coulomb blockade peak distribution, (see also Ref. [12]). In Subsection 2.3.1 we compare some predictions of the theory with available experimental results [9,10]. We find that, although several features of the theory are observed in experiments, there is still disagreement between theory and experiment. We will discuss the possible causes for this discrepancy.

2.1 Effective Hamiltonian

To find properties, such as the ground state spin distribution of electrons in the dot in the presence of electron–electron interactions, we would like to describe them at low energy in a simple form. Ref. [2] shows that in the limit of large dots the effective Hamiltonian, at energies smaller than E_{Th}, is:

$$\mathcal{H}_{\mathrm{eff}} = \sum_{\mu,s} \varepsilon_\mu \psi^\dagger_{\mu s}\psi_{\mu s} + J_{\mathrm{s}}\boldsymbol{S}\cdot\boldsymbol{S} + J_{\mathrm{c}}T^\dagger T + U_{\mathrm{c}}N^2. \tag{1}$$

Here, $N = \sum_{\mu,s=\uparrow,\downarrow}\psi^\dagger_{\mu s}\psi_{\mu s}$, $\boldsymbol{S} = \sum_{\mu,s,s'}\frac{1}{2}\psi^\dagger_{\mu s}\sigma_{s,s'}\psi_{\mu s'}$ and $T = \sum_\mu \psi^\dagger_{\mu\uparrow}\psi^\dagger_{\mu\downarrow}$. The μ sum runs over $M \equiv g = 2\pi E_{\mathrm{Th}}/\Delta$ states. The symbol $\Delta = 1/(\nu\mathcal{V})$ denotes the average single particle level spacing in a dot of volume $\mathcal{V}$ and thermodynamic density of states ν (per spin). Notice that in our notation Δ refers to the level spacing for a single spin state in the dot. In a dirty dot of length L and diffusive constant D, $E_{\mathrm{Th}} = D/L^2$, while for a single dot we replace D by $\sim v_F L$.

The first term in (1) is universal: ε_μ is a single electron eigenenergy of a random matrix, and the operator $\psi^\dagger_{\mu,s}$ is the creation operator of an electron with spin S at an eigenstate μ of a random matrix. The set $\{\varepsilon_\mu\}$ depends on the symmetry of the problem. In the presence of time reversal symmetry the eigenenergies are taken from the Gaussian orthogonal ensemble (GOE, $\beta = 1$), and in the absence of time reversal symmetry from the Gaussian unitary ensemble (GUE, $\beta = 2$). When strong spin-orbit interactions are present they are taken from the Gaussian symplectic ensemble (GSE, $\beta = 4$) [1].

The direct Coulomb interaction constant U_{c}, the exchange constant J_{s}, and the interaction in the Cooper channel J_{c} depend on the specific system and the model one uses for the interaction. In addition, there are non universal corrections to Hamiltonian (1) that vanish, however, in the limit of $g \to \infty$. (See also in Subsection 2.1.1.) When the time reversal symmetry is broken the interaction in the Cooper channel vanishes. We will see below that even in the presence of time reversal symmetry the interaction in the Cooper channel is reduced due to a "screening" by fast electrons (see also Appendix C).

To understand what are the effective low energy interaction constants U_{c}, J_{s}, and J_{c} it is useful to describe the derivation of the effective Hamiltonian (1) in terms of a RG scheme. When the temperature decreases we integrate out progressively the fast motion of the electrons with energy far from the Fermi energy and find effective

[1] The symmetry index β counts the degrees of freedom of the matrix elements of the single particle Hamiltonian, $\beta = 1, 2$, or 4 if its elements are real complex or real quaternion numbers, respectively. A magnetic flux $\sim \phi_0/\sqrt{g}$ through the dot, where $\phi_0 = hc/e \approx 4.12 \times 10^{-13}\mathrm{Tesla}\cdot\mathrm{m}^2$, leads to a time reversal symmetry breaking.

coupling constants in the direct (U_c), exchange (J_s) and Cooper (J_c) channels [14,15]. First, the fast motion of the electrons, at energies of $O(E_F)$ away from the Fermi level, "dresses" the bare electrons and forms quasi-particles. The Landau Fermi-liquid theory describes this dressing process [14,15] and the way it renormalizes the system parameters. In Appendix B we use this theory to estimate J_s for several situations.

The Fermi-liquid "dressing" continues up to energies of order $1/\tau$. Below $1/\tau$ the motion of the electrons becomes diffusive and new diffusion singularities appear. In a situation where one of the dimensions of the system is much smaller than the others, the system may become quasi-two-dimensional at frequencies smaller than the Thouless energy that is related to the short dimension. This reduction in the dimension of the system enhances the diffusive singularities and changes the flow of the interaction parameters [16]. It appears that the RG flow in the Cooper channel is sensitive to disorder more than the RG flow in the other channels [16], (see also Appendix C). However, in certain situations, especially when disorder is strong, we also have to consider its effect on the flow in the other channels [16].

Finally, we arrive at temperatures $T < E_{\mathrm{Th}}$ and are left with an effective "zero dimensional" Hamiltonian, $\mathcal{H}_{\mathrm{eff}}$. The length scale associated with such low temperatures is larger than the system size and therefore the interaction parameters are constants that do not depend on the site or state index. We note that in ballistic samples, *i.e.*, when the electrons cross the sample before suffering substantial scattering from impurities we may use the Fermi-liquid theory, without additional complications due to the diffusive motion at intermediate scales.

2.1.1 Fluctuations in the interaction parameters

In practice, not all the samples have exactly the same shape and/or impurity configuration, we therefore expect to find sample to sample (mesoscopic) fluctuations in the RG process that will lead to fluctuations in the interaction constants of the different channel. By assumption (that is motivated and supported by numerical analysis [17,18]) the sample to sample fluctuations in the single particle levels are described by random matrix theory. The eigenenergies and eigenfunctions are those of a random matrix.

The random electron states have random charge distributions and hence we expect that the interaction parameters J_c, U_c and J_s themselves will fluctuate with the electron number. The interaction in the Cooper channel is reduced by a logarithmic factor (see Appendix C) and we will neglect its fluctuations.

When an electron is added to the system charge flows to the dot edges and leads to fluctuations in the self-consistent nonuniform potential [19]. These fluctuations scale as $\sim (r_s/\sqrt{g})\Delta$, and give the largest contributions to the fluctuations in U_c. [See Eqs. (B.3) and (B.7) for a precise definition of r_s in two and three dimensions.] Fluctuations in U_c are relevant to the Coulomb peak spacing distribution (Subsection 2.3.1) and to nonequilibrium effects (Subsection 4.2) in the conductance through the dot at finite bias voltage.

The short range part of the Coulomb interaction determines the exchange integral in the expression for the exchange interaction parameter J_s. We can therefore use a contact interaction model to find its fluctuations. The fluctuations in the interaction parameter in that case, for three dimensional samples, are [20] $\sqrt{\text{var}\, J_s} \approx J_s \max\left\{1/(2\sqrt{2N}), c_3/g\right\}$, where c_3 is a numerical number of order 1, and $N \sim E_F/\Delta$ is the total number of electrons in the dot. The first term is larger for $g^2 \gg N \Leftrightarrow L < (k_F l)l$, with L the sample size, l the mean free path, and k_F the Fermi momentum. For ballistic systems $l \approx L$, and the first term is much larger. Semiclassically [21], the first contribution arises from direct trajectories between two points in the dot, and the second is built from indirect trajectories with possible scattering on the surface or on static impurities. In two dimensional samples a similar calculation [12,13] gives, $\sqrt{\text{var}\, J_s} \approx J_s \max\left\{c_2 \log N/\sqrt{N}, c_2'/g\right\}$, where c_2 and c_2' are of order 1. For ballistic systems we find that, as in three dimensional samples, the first term is larger.

The actual contribution of the fluctuations in J_s to the fluctuations in the ground state energy is larger by a factor $\sim \sqrt{g}$, as in the calculation of the ground state energy we have to include interaction of a single spin with $\sim g$ electrons [12].

In any case, in the universal limit, when $g, N \to \infty$ we can neglect the fluctuations in the interaction parameters.

2.1.2 A toy model with contact interaction

In Ref. [3] we discussed a model with contact interaction, described by the Hamiltonian

$$\mathcal{H}_{\text{toy}} = \sum_{n,m,s} c^\dagger_{n,s} \mathcal{H}_0(n,m) c_{m,s} + uM \sum_n c^\dagger_{n,\uparrow} c^\dagger_{n,\downarrow} c_{n,\downarrow} c_{n,\uparrow}. \tag{2}$$

Now m runs over M coarse grain sites (in real space) and $\mathcal{H}_0(n,m)$ is a random matrix. The parameter u describes the strength of the interaction between the electrons.

This toy model was analyzed within the self consistent Hartree–Fock approximation in the limit of large M [3]. We will show in Appendix A that to first order in u the toy model (2) is equivalent to the Hamiltonian (1) with parameters

$$U_c = u/4, \quad J_s = -u \text{ and } J_c = u. \tag{3}$$

Higher order corrections in u preserve the symmetry and the structure of Hamiltonian (1) but give different values for its parameters. In particular, as shown in Appendix C for positive u (which corresponds to a repulsive interaction) they reduce J_c by a factor $\propto \log M$.

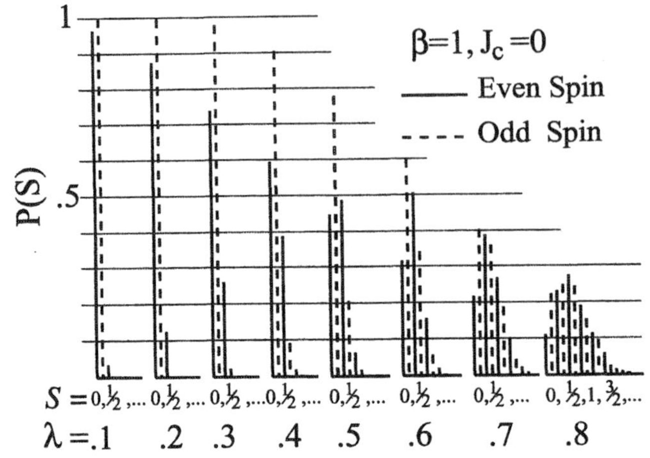

Fig. 1: The probability distribution $P(S)$ of the ground state spin of a quantum dot, computed from Eq. (4) for different values of the interaction parameter $\lambda = -J_s/\Delta$ with time reversal symmetry. Solid histograms are for integer spins, dotted ones for half-integer spins. All the graphs are starting with spin 0 increasing to the right in increments of 1/2 spin.

2.2 Ground state spin distribution

In this section we will study the ground state spin of Hamiltonian (1) assuming that the interaction constant $J_c = 0$ [as it renormalized towards zero by electrons with energy larger than the Thouless energy and is further renormalized within the toy model (see Appendix C)]. In this case, the Hamiltonian (1) becomes noninteracting within each spin sector. For a fixed number of particles N we can neglect also the charging energy U_c. The spin of the ground state is then found by minimizing the energy $E_G(S)$ of the lowest lying state with total spin S, as a function of S. Since the lowest energy state with spin S has precisely $2S$ singly occupied states, all lower lying states being doubly occupied, one has [3]

$$E_G(S) - E_G(S_0) = \sum_{\mu=1}^{S-S_0} (\varepsilon_{N+\mu+2S_0} - \varepsilon_{N+1-\mu}) + J_s\left[S(S+1) - S_0(S_0+1)\right]. \quad (4)$$

where $S_0 = 0$ (1/2) if the total number of particles N is even (odd).

Since the precise positions of the energy levels $\{\varepsilon_\mu\}$ in Eq. (4) fluctuate from grain to grain, the ground state spin S does so as well. Plots of the probability distribution $P(S)$ are shown in Fig. 1 and Fig. 2 for different values of the dimensionless parameter $\lambda = -J_s/\Delta$. The effective parameter λ, that includes renormalization from fast electrons, is calculated for small metallic dots and semiconductors dots in Appendix B. The distributions are obtained by taking the levels ε_μ from the GOE ($\beta = 1$), or, when time-reversal symmetry is broken, from the GUE, ($\beta = 2$), and

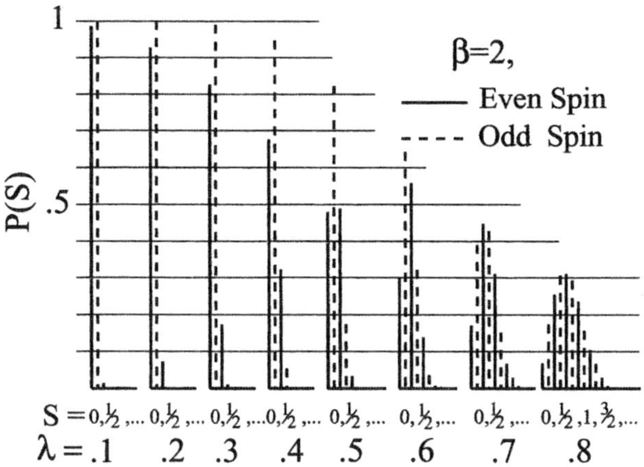

Fig. 2: The probability distribution $P(S)$ of the ground state spin of a quantum dot, computed from Eq. (4) for different values of the interaction parameter λ without time reversal symmetry. Solid histograms are for integer spins, dotted ones for half-integer spins. All the graphs are starting with spin 0 increasing to the right in increments of 1/2 spin.

minimizing Eq. (4) with respect to S. The way spin-orbit coupling reduces the effect of the interaction in the exchange channel is discussed in Sect. 3.2.

2.3 Application to Coulomb Blockade statistics

A few experimental results on the statistics of the Coulomb blockade peak spacing in a quantum dot [7–9] suggest that the predictions of the constant interaction model, with single-particle levels taken from RMT and with a constant charging interaction U_c only, fails to describe the fluctuations of Coulomb blockade peak spacing. The RMT+U_c-model predicts a Wigner surmise peak distribution, and even-odd effects. But experimentally, the peak spacing distribution is roughly Gaussian (with non Gaussian tails), it width is $\gtrsim \Delta$ [9], and even-odd effects are not observed. Numerical studies for a few electrons, with mutual Coulomb repulsion, in a disordered medium [17] deviate from the predictions of the RMT+U_c-model as well.

We will discuss now what the model (1) predicts for the Coulomb blockade peak spacing distribution. By definition, the spacing between the N'th Coulomb peak and the $N-1$'th Coulomb peak is given by

$$\Delta E = (E_{N+1} - E_N) - (E_N - E_{N-1})$$

where E_N is the energy of the system with N electrons. Fluctuations in the single particle energy levels $\{\varepsilon_\mu\}$, and in the interaction parameters U_c, J_s and J_c may lead to fluctuations in ΔE as N changes. The latter, however, vanish when $g \to \infty$. As we discussed in the previous section, the fluctuations of the single particle levels induce fluctuations in the ground state spin of the dot even when the interaction

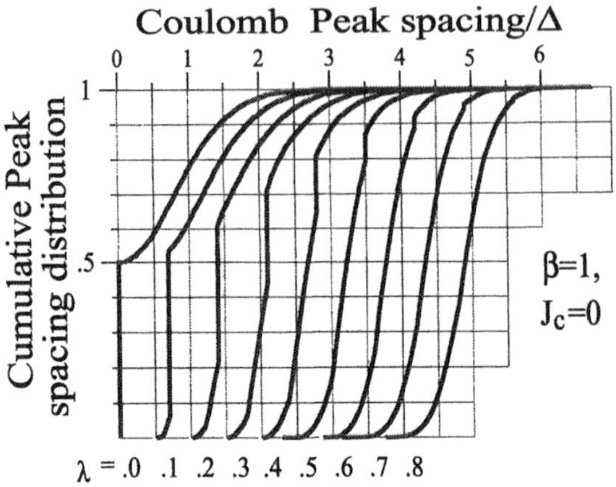

Fig. 3: Theory for cumulative Peak spacing distribution in the presence of time reversal symmetry ($\beta = 1$). We show the case without interaction in the Cooper channel ($J_c = 0$) as it is suppressed (see Appendix C). For clarity we shift the distributions with exchange parameters $\lambda = -J_s/\Delta$ in intervals of half of the average level spacing, Δ.

constants J_s, U_c and J_c do not fluctuate with N. These, in turn, cause fluctuations in the peak spacing.

In Fig. 3 we plot the cumulative Coulomb peak spacing distribution for ensembles with a GOE symmetry. As discussed in Appendix C the interaction in the Cooper channel is suppressed (by a logarithmic factor). We therefore plot in Fig. 3 the cumulative peak spacing distribution for a GOE ensemble without interaction in the Cooper channel($J_c = 0$). Figure 3 shows the distributions for exchange interaction strengths of $\lambda = -J_s/\Delta = 0, 0.1, \ldots 0.8$. We choose to plot the *cumulative* peak spacing distributions and not histograms of the peak spacing density distribution. Plotting the cumulative distribution allows us to represent delta functions in the spacing distribution and avoids the need to work with arbitrarily binning intervals, as is needed for a histogram.

Examining Fig. 3, one notices that there is a jump in the cumulative Coulomb peak spacing distribution at energy $\Delta E = 2\lambda\Delta$, corresponding to a delta function in the spacing distribution at that energy. This occurs because there is a finite probability that starting from a spin-singlet ground state, two successive electrons will enter with opposite spin into the same single-particle state, and the quantity $2\lambda\Delta$ is the exchange energy cost for adding the second electron, when $J_c = 0$. However, at large values of λ the probability that the ground state of the dot is a singlet is smaller, hence the hight of the jump decreases. There is an additional substructure of the distributions (*e.g.* a kink in the lower part of the distribution) that becomes smoother when λ increases. To understand a few details of the Coulomb peak spacing distribution curve, we have to find the ground state energies of a disordered system with $N-1$, N and $N+1$ electrons. (We assume that the system has the same disorder realization for consecutive electron entries.)

The ground state of a system with N electrons is also characterized by its spin S. Few examples of the energies of states $|N,S\rangle$ are summarized in Fig. 4. The Coulomb peak spacing is given by $\Delta E = E_{N-1,S_{N-1}} + E_{N+1,S_{N+1}} - 2E_{N,S_N}$, where $E_{N,S} = \langle N,S|\,\mathcal{H}_{\text{eff}}\,|N,S\rangle$ is the energy of the state $|N,S\rangle$. [Notice that in the presence of interaction in the Cooper channel $|N,S\rangle$ is not necessarily an eigen state of $\mathcal{H}_{\text{eff}}$, defined in Eq. (1)]. Different values of S_{N-1}, S_N, S_{N+1} give rise to different spin sequences.

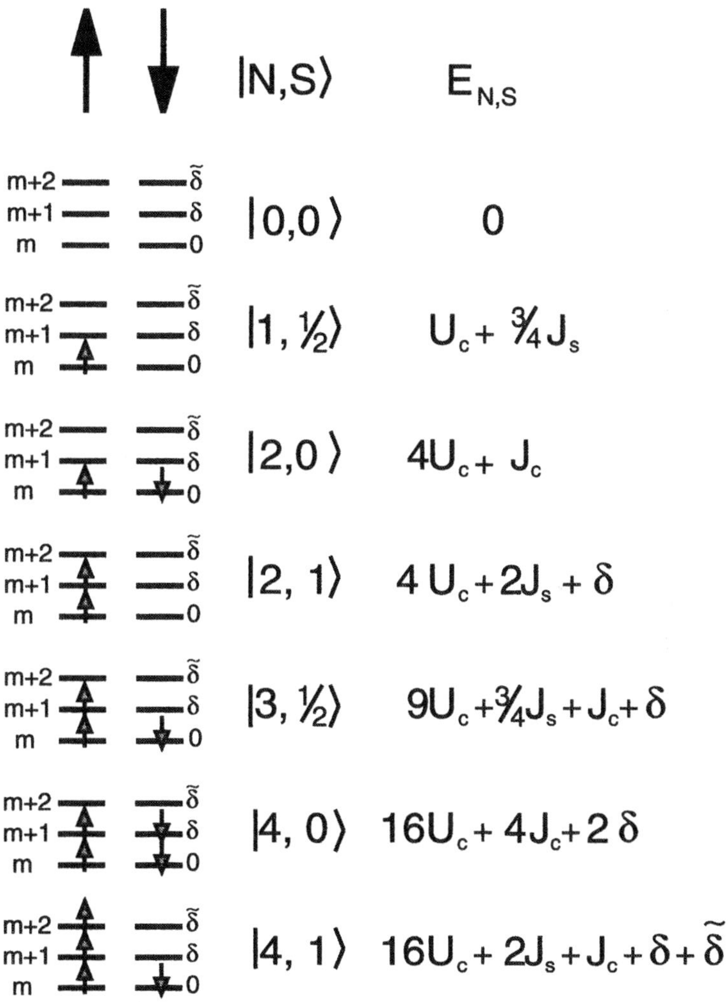

Fig. 4: Possible spin configurations of the dot and their energies according to Eq. (1). We assume that all states below level m are doubly occupied and denote this many body state $|0,0\rangle$. The single particle states are: $\epsilon_m = 0$, $\epsilon_{m+1} = \delta$, $\epsilon_{m+2} = \tilde{\delta}$.

There are few possibilities for sequences of spin entries, (see Table 1). Sequence

#1 describes a situation where initially there are $(2m-1)$ electrons in the system. The first $2(m-1)$ single particle states are doubly occupied, and the last state (state m) is singly occupied. We denote this state by $|1, {}^1\!/_2\rangle$, see Fig. 4. Then an electron is added to state m so that it is doubly occupied, we denote this state by $|2,0\rangle$. The third electron is added to state $m+1$ so that it singly occupied to form the state $|3, {}^1\!/_2\rangle$. The peak spacing of this sequence is:

$$\begin{aligned}\Delta E_1 &\equiv E_{1,{}^1\!/_2} + E_{3,{}^1\!/_2} - 2E_{2,0} \\ &= U_c + {}^3\!/_4 J_s \ + \ 9U_c + \delta + {}^3\!/_4 J_s + J_c \ - \ 2(4U_c + J_c) \\ &= 2U_c + {}^3\!/_2 J_s - J_c + \delta = \delta - u - J_c \equiv \Delta T_1.\end{aligned}$$

In the last equality we have used relation (3) for the contact interaction toy model. The peak spacings for other sequences, calculated in a similar way, are summarized in Table 1.

i	Spin Configurations	ΔE_i	ΔT_i
1	$\|1, {}^1\!/_2\rangle \Rightarrow \|2, 0\rangle \Rightarrow \|3, {}^1\!/_2\rangle$	$2U_c + {}^3\!/_2 J_s - J_c + \delta$	$-u - J_c + \delta$
2	$\|1, {}^1\!/_2\rangle \Rightarrow \|2, 1\rangle \Rightarrow \|3, {}^1\!/_2\rangle$	$2U_c - {}^5\!/_2 J_s + J_c - \delta$	$3u + J_c - \delta$
3	$\|2, 0\rangle \Rightarrow \|3, {}^1\!/_2\rangle \Rightarrow \|4, 0\rangle$	$2U_c - {}^3\!/_2 J_s + J_c$	$2u + J_c$
4	$\|2, 0\rangle \Rightarrow \|3, {}^1\!/_2\rangle \Rightarrow \|4, 1\rangle$	$2U_c + {}^1\!/_2 J_s \quad + \tilde{\delta}$	$\tilde{\delta}$
5	$\|2, 1\rangle \Rightarrow \|3, {}^1\!/_2\rangle \Rightarrow \|4, 0\rangle$	$2U_c + {}^1\!/_2 J_s + 2J_c + \delta$	$+ 2J_c + \delta$
6	$\|2, 1\rangle \Rightarrow \|3, {}^1\!/_2\rangle \Rightarrow \|4, 1\rangle$	$2U_c + {}^5\!/_2 J_s - J_c + \tilde{\delta}$	$-2u - J_c + \tilde{\delta}$

Table 1: The symbol ΔE_i denotes the spaces between the Coulomb peaks for spin sequences i. [ΔT_i is ΔE_i using the parameters of the toy model, Eq. (3).] Sequences that involve higher spins are also possible and are not included in this table. To find the actual contribution to the peak spacing we should include the probability that such a sequence occurs.

Sequence #1 occurs only if $E_{2,0} < E_{2,1} \Leftrightarrow \delta > J_c - 2J_s \Leftrightarrow \Delta T_1 > u$. In a similar way one can check that sequence #2 occurs when $\Delta T_2 > u$. Thus, both processes #1 and #2 will lead to a step in the peak spacing distribution at u. For the cumulative peak spacing distribution, this will lead to a discontinuity in the slope of the curve, which may be seen in Fig. 3 at the points $\Delta E = \lambda\Delta = -J_s$. Processes #1 and #2 are the main contributions to the approximately linear portions of the curves in the range $-J_s = \lambda\Delta < \Delta E < 2\lambda\Delta = -2J_s$, which are seen at small values of λ.

Sequence #3 will occur if $E_{2,0} < E_{2,1}$ and $E_{4,0} < E_{4,1}$. This occurs if $J_c - 2J_s < \delta < \tilde{\delta} 2J_s - 3J_c$ and leads to a step function jump in the cumulative peak distribution at ΔE_3. This is clearly seen in the theoretical curves of Fig. 3 and Fig. 5. The weight of the jump is bounded from above by $\int_{\Delta E_3}^{\infty} p(\delta')d\delta'$ where $p(\delta')$ is the Wigner distribution for consecutive levels at distance δ'.

Sequence #5 requires that $E_{2,1} < E_{2,0}$ and $E_{4,0} < E_{4,1}$, which occurs when $\delta < \min\{J_c - 2J_s, \tilde{\delta} - 3J_c + 2J_s\}$. This leads to a low energy tail in the shape of the

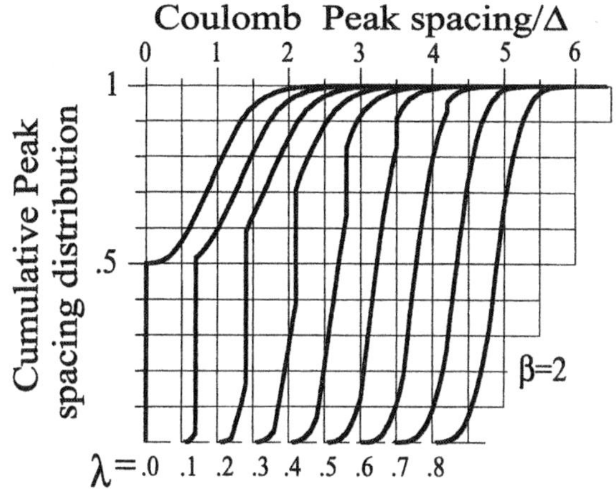

Fig. 5: Theory of cumulative peak spacing distribution in the absence of time reversal symmetry ($\beta = 2$). We assume that the external magnetic field is large enough to break time reversal symmetry, but small enough so that we can neglect Zeeman splitting. In other words, the interaction in the Cooper channel is completely suppressed and the level spacing distribution is described by a GUE ensemble. For clarity we shifted the distributions with exchange parameters $\lambda = -J_s/\Delta$ by intervals of half level spacing.

Wigner distribution curve, that extend all the way down to $\Delta E = 0$, when $J_c = 0$.

Here we have discussed some of the simplest spin-entry sequences, including situations with total spin 0, 1/2, and 1, and we have seen how these sequences appear in the cumulative peak spacing distribution. Generalizations for more complex situations are straightforward but tedious. Different sequences will lead to other singularities and smooth curve-segments in the distribution of peak spacings.

The actual numerical calculations (which includes also sequences with spin larger than 1) do not demand, however, a detailed analysis as above. We performed them in the following manner: we use 24 random levels around the center of the spectrum of a random 100×100 matrix for 1000 realizations. For each realization we find, using formula (4), the configuration of the groundstate and its energy for eight consecutive entries of electrons, this gives us 6 peak spacings for each realization, so that totally we plot a histogram of 6000 peaks.

In Fig. 5 we plot the cumulative peak spacing distribution for magnetic field large enough so that the non interacting levels may be described by the GUE ensemble and the interaction in the Cooper channel is completely suppressed. (We assume that the magnetic field is small enough so we can neglect Zeeman splitting effects.)

2.3.1 Comparison between theory and experiments

This section compares our theory for Coulomb blockade peak spacing with the available experimental results. We will see below that the agreement between theory and experiment is not very good, particularly for small peak spacings. However some

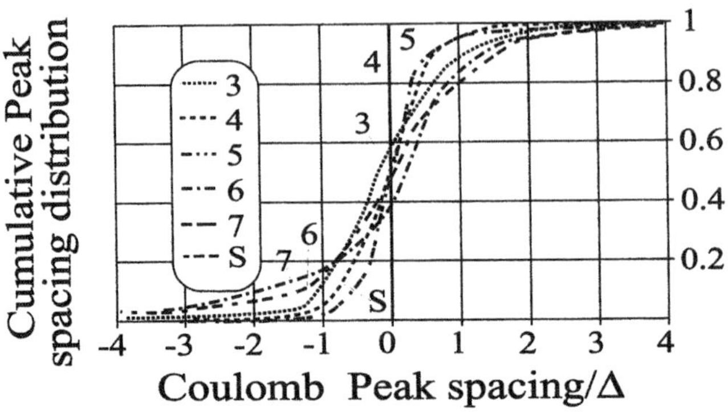

Fig. 6: Experimental results (curves "3-7" [9] and "S" [10]) of the cumulative peak spacing distributions. To compare between the dots of different sizes, we subtract from each distribution the dot average peak spacing and scaled it to the average level spacing, Δ. The dots here have time reversal symmetry.

features of the theory are found also in experiments. For example, a non-Gaussian tail at large values of peak spacing and a jump in the cumulative distribution, that we described in the preceding section, are present in few experiments.

Figure 6 depicts the cumulative peak spacing distribution of AlGaAs-dots. Dots "3-7" were studied in Ref. [9] and dot "S" in [10]. In all the experimental curves that we present here, no magnetic field is applied. We therefore assume in the theoretical analysis that time reversal symmetry is conserved. In each curve we normalize the peak spacing by the dot level spacing, which vary from dot to dot since their sizes are different. We have used here the average level spacings quoted in the experimental papers, there are, of course, some uncertainties in these values. After this normalization we would expect that the curves of dots 3-7 will be similar. This should happen because they have similar electron density and therefore similar r_s and exchange interaction parameter (see Appendix B). In addition we would expect that dot "S" will have a different curve, because it has an electron density that is larger by a factor ~ 3. However, as Fig 6 shows, the experimental results behave differently. This may be attributed to the intrinsic noise in the system (see Table 1 in [9]), but we still do not understand completely the origin for this behavior. The data of Refs. [7] and [8] show a significantly wider distribution of level spacings than found in Ref. [9], despite the apparent similarity of the systems studied by these groups. So far, there has been no clear explanation for this discrepancy.

We nevertheless plot in Fig. 7 our theory and the experimental curve of [10] (without magnetic field). The overall experimental curve fits well to a Gaussian. However, the details of the upper (right) tail of the cumulative distribution, *i.e.*, the jump and the non Gaussian tails fits better to the theory of the spin fluctuations in the Ground state. To fit best the upper tail of the experimental results in the absence of magnetic field, we choose in the theory (with GOE symmetry but without interaction in the Cooper channel) $-J_s/\Delta = \lambda = 0.4$. Notice that using the static

RPA estimate of λ (in Appendix B), with the experimental value $r_s \sim 0.72$ [10] we find, as expected, a value that is somewhat smaller than 0.4.

We expect that several effects, (not included in the theory) may be important for current experiments. The lower part of the peak spacing distribution, that is built from single particle levels that are, by chance, very close to each other, is especially sensitive to these effects. Indeed, this part appears to be far from the theoretical curves. Among them are:

(1) Non universal effects of finite g that cause fluctuations in the interaction parameter [12]. For ballistic two-dimensional dots $g = \sqrt{2\pi n A}$, where A is the dot area and n is the density of the electrons. In the dots of Ref. [7–9] $r_s \sim 1 - 3$ and $g \sim 50 - 150$ (in these of Ref. [10] $r_s \sim 0.72$ and $g \sim 35$). Thus, when we compare theory to experiment, fluctuations in the interaction constants cannot always be ignored. Ullmo and Baranger present in Ref. [12] a detailed study of the effect of fluctuations of the interaction parameters (see also the discussion here in page 152). They find indeed that when these fluctuations are included the results "are significantly more like the experimental results than the simple constant interaction model".

(2) We assume that the temperature, and the single particle levels width (due to tunneling to the leads), is smaller than the mean level spacing and therefore we ignore their effects. This assumption is not valid for the lower part of the distribution as it is built from levels whose distance from neighboring levels might be much smaller than the average level spacing.
The importance of the temperature was considered very recently by Usaj and Baranger [22]. They find that temperature effects are significant even at $T \sim 0.1\Delta$.

(3) There is experimental noise due to charge motions during the measurements time. This effect [9] might be the dominant contribution to the smearing of the distribution in the experimental curve.

3. Spin-orbit effects

Spin-orbit coupling can have major effects on the ground states or the low-energy transport properties of a mesoscopic system. In many metallic nanoparticles, spin-orbit effects arise from randomly placed heavy-ion impurities, which can simultaneously scatter electrons and flip their spin, subject to the constraints imposed by the requirement of time-reversal invariance in the absence of an applied magnetic field. In other cases, one is concerned with metal particles where the spin-orbit effects are already significant in the band-structure of the ideal host crystal, so that the "spin" variable in the Bloch states actually represents a mixture of spin and orbital degrees of freedom at the microscopic level. In this case spin-flip scattering with the requisite spin-orbit symmetry can occur whenever there is scattering: from defects, from impurities, or from the boundaries of the sample. Spin-orbit scattering in the above cases can generally be characterized by a spin-orbit scattering rate, and the importance of spin-orbit effects is determined by the ratio of this rate to other fre-

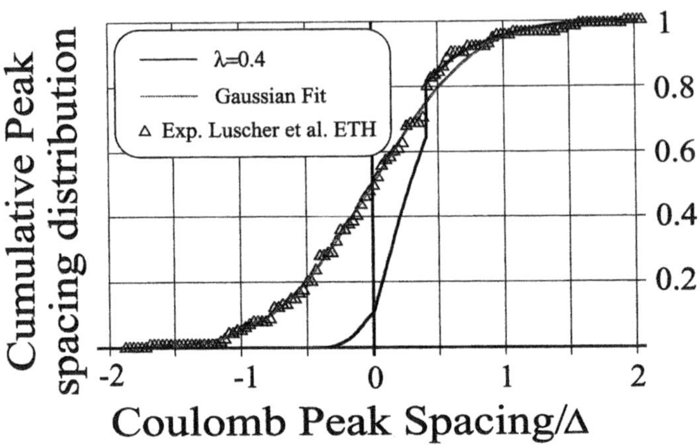

Fig. 7: Comparison between theory and experiment[10]. To fit best the upper tail of the experimental results in the absence of magnetic field, we choose in the theory (with GOE symmetry but without interaction in the Cooper channel) $-J_s/\Delta = \lambda = 0.4$.

quencies characteristic of the mesoscopic system. Effects of spin-orbit scattering on the groundstate spin-structure and on the energy splitting in an applied magnetic field will be discussed in Subsection 3.1. The effects of spin-orbit coupling on the spacing of groundstate energies, in the presence of electron-electron interactions, will be discussed in Subsection 3.2.

A peculiar situation can arise in two-dimensional electron systems in materials such as GaAs. Here the dominant spin-orbit effects arise from terms in the effective Hamiltonian in which there is a coupling to the electron spin linear in the electron velocity. (These terms arise from the asymmetry of the potential well confining the electrons to two dimensions and from the lack of inversion symmetry in the GaAs crystal structure.) The special form of this coupling leads to a large suppression of spin-orbit effects when the 2D electron system is confined in a small quantum dot. However, effects of spin-orbit coupling are again enhanced in the presence of a strong magnetic field parallel to the plane of the sample, so that they must be taken into account in such properties as the level-spacings of a closed dot or the statistics of conductance fluctuations in a dot coupled to leads through one or more open channels. These effects will be discussed in Subsection 3.3 below.

3.1 Effective g-tensor of a metal particle with spin-orbit scattering.

According to Kramers theorem, a metal particle with an odd number of electrons with no special symmetry, in zero magnetic field, must have a degenerate groundstate manifold, with pairs of states related to each other by time-reversal symmetry. In the absence of spin-orbit coupling, the total spin S is a good quantum number, and the groundstate manifold is just that expected for half-integer S. As we have seen in Sect. 2., if the electron-electron interaction is weak, we will essentially always find $S = 1/2$ for odd N and the ground state will be just two-fold degenerate. For stronger electron-electron interactions, however, there will be some probability of finding $S = 3/2$ or larger, so that four-fold or higher degeneracies are also possible. When spin-orbit interactions are turned on, the higher degeneracies will be broken into a set of doublets, so that the ground state will again be two-fold degenerate.

If we now apply a magnetic field B to the system, the degenerate ground state will be split. For sufficiently small B, one of the states will move up in energy by an amount $\delta\varepsilon$ which is linear in B, while the other state will move down by the same amount. These shifts may be measured, at least in principle, by electron-tunneling spectroscopy experiments in an applied magnetic field. We discuss here the statistical properties of the distribution of energy shifts expected under various circumstances. We concentrate on the situation where the electron-electron interaction is weak, so that the many-body ground state is well described by the picture of weakly interacting quasiparticles, as effects of electron-electron interactions will be discussed in the next subsection.

Quite generally, we may write the linear splitting of a Kramers doublet in the form

$$\delta\varepsilon = |\mu_B/2|(\boldsymbol{B} \cdot \overleftrightarrow{K} \cdot \boldsymbol{B})^{1/2} \tag{5}$$

where $\mu_B < 0$ is the electron Bohr magneton and $\overleftrightarrow{K}$ is a real, positive-definite symmetric 3×3 tensor. In the absence of spin-orbit coupling, $\overleftrightarrow{K}$ is isotropic, with $K_{ij} = 4\delta_{ij}$. When spin-orbit coupling is present, we find that $\overleftrightarrow{K}$ varies from level to level, and is in general anisotropic. We write the three eigenvalues of $\overleftrightarrow{K}$ as g_k^2, $(k = 1, 2, 3)$, with $|g_1| \leq |g_2| \leq |g_3|$, and refer to the g_k's as the three principal g-factors for the level. Although the energy-splittings in a static magnetic field only define the absolute values of the g_k, by considering the response to a time-varying magnetic field (e.g., a spin resonance experiment) one can also give an unambiguous meaning to the sign of the product of the three g-factors. Since the sign of an individual g_k has no physical meaning, we adopt the convention that g_3 and g_2 are always positive, but g_1 can be positive or negative, depending on the specific system considered.

For the case of weakly interacting electrons, which we consider here, the ground state for $2N + 1$ electrons consists of $2N$ electrons in filled Kramers doublets, plus one electron in a doublet which is singly occupied. The filled doublets give no contribution to the linear energy shift because in each case one state moves up and the other moves down by the same amount. Thus, the g-factors are determined by the properties of the singly-occupied state.

In the presence of spin-orbit coupling there are two contributions which can shift the g-values from the bare value $g = 2$. If we take into account only the interaction of B with the electron spin, then spin-orbit coupling will always reduce the g-values. For example, if the magnetic field is applied in the z-direction, the state which is shifted down in energy will be the particular linear combination of the two degenerate states which has the maximum expectation value of $-S_z$. This expectation value is $\leq 1/2$, so the spin-contribution to the g-factor will generically be reduced by spin-orbit coupling.

On the other hand, there is also an orbital contribution to the linear Zeeman effect, when spin-orbit coupling is present . (In the absence of spin-orbit coupling, the orbital states in an irregular dot will be generically non-degenerate and time-reversal-invariant, so they cannot acquire a linear energy shift in a weak magnetic field.) Both the orbital and spin contributions were considered by Matveev *et al.* [23], who discuss the expectation value and probability distribution of $\delta\varepsilon^2$ for the magnetic field in an arbitrary fixed direction.

By contrast Brouwer *et al.* [24] considered the joint probability distribution of the three g-values for a single level, so they could examine the anisotropy as well as the magnitude of the g-tensor. Their analysis concentrated on the case where orbital effects can be ignored, so that the mean-square g-factors are monotonically reduced with increasing spin-orbit coupling. The strength of the spin-orbit coupling in this case is determined by a parameter

$$\lambda_{\text{so}} = \sqrt{\frac{\pi\hbar}{\tau_{\text{so}}\Delta}} \tag{6}$$

where Δ is the mean separation between one-electron energy levels. The mean spin-orbit scattering time τ_{so} is defined so that if we prepare a state with spin up, the probability to find it in the same spin direction after time t is $\sim e^{-t/\tau_{\text{so}}}$. When $\lambda_{\text{so}} \gg 1$, one finds that the g-factors are greatly reduced from their bare values, and one can obtain an analytic form for the joint probability distribution:

$$P(g_1, g_2, g_3) \propto \prod_{i<j} \left|g_i^2 - g_j^2\right| \prod_i e^{-3g_i^2/2\langle g^2\rangle} \tag{7}$$

For intermediate values of the coupling parameter λ_{so}, one can perform numerical simulations to study the distribution, using random-matrix theory. In Fig. 8 we show the λ_{so}-dependence of the mean values of g_k^2, as well as the values of g_k for a particular realization of the random matrices.

Very recently, Petta and Ralph [Ref. [25]] have measured effective g-values for a number of levels in each of several different nanoparticles, of Cu, Ag and Au, with diameters in the range 3-5 nm. They did not vary the direction of the applied magnetic field, so they could not study the anisotropy of the g-tensor. However, the statistical distributions of the g-factors (normalized to their means) for different levels in a given particle were found to be in good agreement with the theories of Refs. [23] and [24]. For the mean g-factor, an agreement with Refs. [23] and [24] is found if the spin contribution is taken into account only; the mean g-values observed in the Au particles (ranging from 0.12 to 0.45) were significantly smaller than what one might expect from the orbital contribution, according to the theory of Ref. [23], unless one assumes a very short mean-free-path for the electrons. (Using the formulas in Ref. [23], one would need a mean-free-path of order 0.1 nm to get g-values this

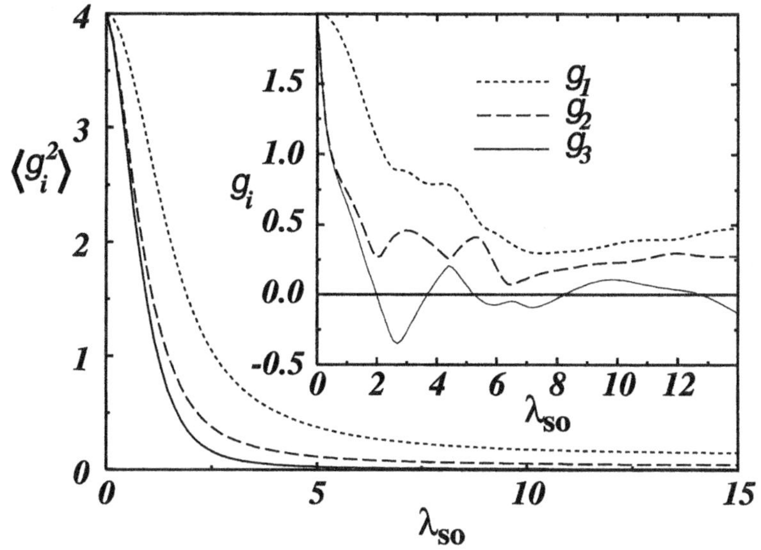

Fig. 8: Average of the squares of principal g-factors versus spin-orbit scattering strength λ_{so}, obtained from numerical simulation of a random matrix model. Inset: g_1, g_2, and g_3 for a specific realization. We have included the sign of g_1.

small.) Small g-values (below 0.5) for Au nanoparticles were also observed previously by Davidovic and Tinkham [4].

3.2 Effects of spin-orbit coupling on interaction-corrections to ground-state configurations and energy spacings.

So far, we have ignored the effects of electron-electron interactions. This should generally be valid if the exchange interaction is small compared to the threshold for the Stoner instability, so that the probability of finding $S > 1/2$ is small in the absence of spin-orbit coupling. When the spin-orbit coupling parameter λ_{so} is large, the effective exchange interaction between two electrons in states close to the Fermi-energy of the particle will be even further reduced, as the mean spin in any state becomes small compared to $1/2$, and the local spin orientations have different spatial distributions for one-electron states belonging to different Kramers doublets.

In the limit of very large spin-orbit coupling, where the mean spin tends to zero, the exchange interaction should also tend to zero. This means that the parameter J_s in the effective Hamiltonian (1) of Sect. 2.1 should be set to zero. This is consistent with the fact that spin is no longer a good quantum number of the system, and the term proportional to J_s is no longer invariant under the set of allowed unitary transformations of the random matrices.

A consequence of this analysis is that if a spin-orbit scatterers are added to a system with fixed electron-electron interaction (fixed r_s) the probability distribution for the separation of successive groundstate energies, measured by the Coulomb-blockade peak separations, should approach that of a non-interacting electron system in the symplectic ensemble. This means that there should be a bimodal distribution with an even-odd alternation. The chemical potential to add a second electron to a Kramers doublet is the same as the energy to add the first electron, after the coulomb blockade energy [U_c in Eq. (1)] is subtracted, whereas the chemical potential for the next electron will be larger by an amount approximately given by the mean level spacing Δ.

3.3 Spin-orbit effects in a GaAs quantum dot in a parallel magnetic field.

The most important spin-orbit terms in the effective Hamiltonian for a 2D electron gas (2DEG) in a GaAs heterostructure or quantum well may be written in the form

$$\mathcal{H}_{so} = \gamma_1 v_x \sigma_y - \gamma_2 v_y \sigma_x \tag{8}$$

where $\boldsymbol{v}$ is the electron velocity operator. We have assumed that the 2DEG is grown on a [001] GaAs plane, and we have chosen the x and y axes to lie in the [110] and [$1\bar{1}0$] directions. For an open 2DEG this leads to a spin-orbit scattering rate of order $\gamma^2 D$, where D is the diffusion constant and γ is the geometric mean of the two coupling constants in Eq. (8). For a confined dot of radius R, in zero magnetic field, however, the effects of spin-orbit coupling are suppressed if the typical angle of spin precession for an electron crossing the dot, given by $\theta = \gamma R$, is small compared to unity. One finds in this case that the matrix elements of $\mathcal{H}_{so}$ are greatly reduced for energy states whose energy separation is small.

Halperin *et al.* [26] have argued that the effects of spin-orbit coupling can be enhanced, however, in the presence of an applied magnetic field in the plane. The enhancement is maximum when the Zeeman energy becomes comparable to the Thouless energy (*i.e.*, the inverse of the transit time for an electron in the dot), in which case there is an effective spin-mixing rate comparable to the spin-orbit scattering rate for an open system with an electron mean free-path equal to the mean free path in the dot. For a closed dot, the spin-mixing would be manifest in the repulsion of energy levels for different spin, and the appearance of anti-crossings of the levels as when the Zeeman field is varied.

Motivated by experiments of Folk *et al.* [27], Halperin *et al.* [26] considered the "universal conductance fluctuations" of a dot connected to a pair of leads with one or more channels open in each lead. They considered explicitly the case where there is a weak magnetic field perpendicular to the dot, so that time reversal symmetry is broken, and the system is in the class of the unitary ensemble, even in the absence of spin-orbit coupling. It was shown that effects of spin-orbit coupling in large Zeeman field could then lead to a factor of two reduction in the variance in the conductance, which is in addition to the factor of two reduction caused by breaking of the spin degeneracy. Calculations of the cross-over, as a function of the in-plane magnetic field, were in at least qualitative agreement with the experimental observations.

Very recently, Aleiner and Fal'ko [28] have considered the case without a perpendicular magnetic field, so that the system without spin-orbit coupling would be in the orthogonal ensemble. They have shown that the application of a parallel magnetic field in this case turns on a spin-orbit perturbation with a special symmetry, so that the system retains an effective time-reversal symmetry even in the presence of the large Zeeman field. The spin-orbit coupling leads to a reduction in the size of conductance fluctuations, but not as much as one would obtain if the time-reversal symmetry was also broken. The spin-orbit coupling also leads to a reduction in the "weak localization" correction to the average conductance, but does not lead to complete suppression as one would find for a broken time reversal symmetry. (However, as noted by Meyer *et al.* [29] and by Fal'ko and Jungwirth [30], for an asymmetric quantum well of finite thickness, application of a strong magnetic field parallel to the sample can lead to broken time-reversal symmetry due to orbital effects, even in the absence of spin-orbit coupling.)

4. Origin of multiplets in the differential conductance

In a recent experiment Davidovic and Tinkham [4] studied tunneling into individual Au nanoparticles of estimated diameters 25 nm, at dilution refrigerator temperatures. The differential conductance dI/dV, as a function of the source-drain voltage V, indicate resonant tunneling via discrete energy levels of the particle. Unlike previously studied normal metal particles of Au and Al, in these samples they find that the *lowest* energy tunneling resonances are split into clusters of 2-10 sub-resonances. The distance between resonances within one cluster is much smaller than the mean level spacing of the Au grain.

This situation is illustrated schematically in Fig. 9. The differential conductance dI/dV shows resonances, where each resonance in dI/dV is actually a multiplet, the splitting between the peaks of the multiplets being a factor ~ 30 smaller than distance between the resonances (which is of the order single-particle level spacing in the grain).

In this section we outline two-different mechanisms which can lead to a fine structure of the first conductance peak. We first show how such a fine structure can occur if the ground state has a finite spin with small energy splittings between states of different magnetic quantum number. In this model it is necessary to have a relatively large total spin in order to split the conductance peak into many sub-peaks. This mechanism would also be suppressed by large spin-orbit coupling.

In the second mechanism, following Agam *et al.* [31], we show how such a fine structure can arise from nonequilibrium processes induced by the large bias voltage V used in the experiment. This mechanism seems to us to be the more likely one for explaining the observations of Ref. [4]. We also indicate how, experimentally, one might distinguish between the two proposed explanations.

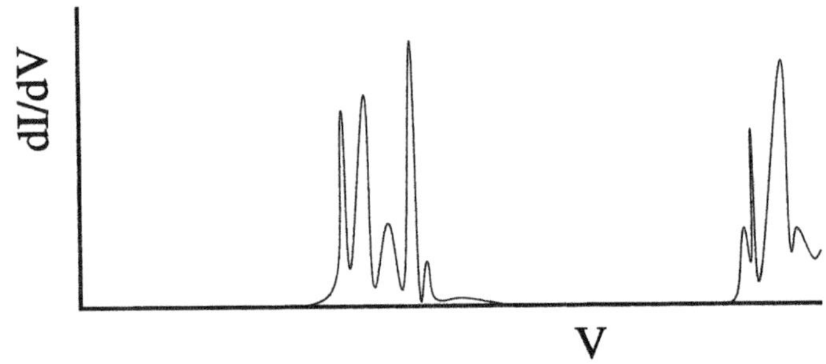

Fig. 9: Schematic illustration of the experimental results of Ref. [4]: The peaks in the differential conductance are split. The distance between the multiplets is of the order of the single particle level spacing Δ; the distance between peaks in the same multiplet is much smaller.

4.1 Multiplets from an almost degenerate groundstate

In general, a peak in the differential conductance as a function of the bias voltage V may occur if an additional channel for tunneling onto or from the metal grain is opened at that V. The relation between V at the peaks and the ground state energies is complicated; it depends on the capacitive division between the left and right contact and on the conductances of the two tunneling contacts. A detailed account of the possible scenarios can be found in the review by von Delft and Ralph [32].

Here, we make the simplifying assumption that the left point contact has the bigger resistance and the smaller capacitance, so that the electrostatic potential of the dot equals that of the right reservoir, and the contact to the left reservoir can be seen as the "bottleneck" for current flow. Then, if the grain has N electrons at zero bias, a conductance peak occurs when

$$eV = E_{N+1} - E_N,$$

i.e., when the bias voltage V is precisely equal to the difference of the energies of any two many-body states of the grain with N and $N+1$ electrons (for $V > 0$), provided the initial N-particle state is populated at the corresponding bias. Below we focus on the first peak in the differential conductance and discuss when and how a fine structure of that first peak can arise. We assume that the temperature is small compared to any splittings in the energy levels, and we assume (for the moment) that there is no spin-orbit coupling.

If the N and $N+1$-particle ground states have perfect degeneracies, the difference $E_N - E_{N+1}$ can only take a single value, and a single peak will be observed, no matter what the spin of the ground state is. Hence to observe a fine structure, the degeneracy of the ground state has to be lifted. This can be done by application of a uniform magnetic field, as is illustrated in Fig. 11 for $S_N = 0$ or $S_N = 1$ and $S_{N+1} = 1/2$. In the case where $S_{N+1} = S_N + 1/2$, the difference $E_{N+1} - E_N$ between the energies of

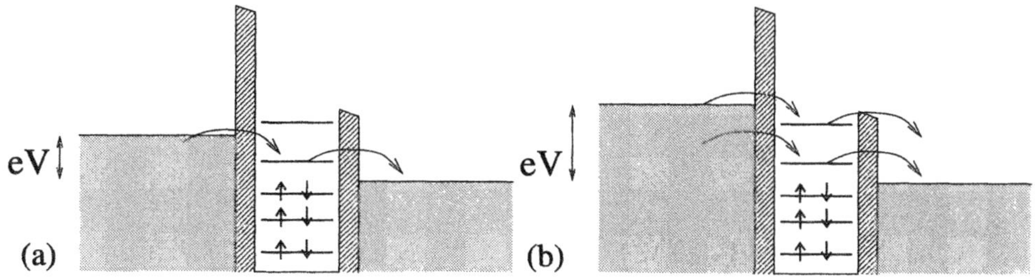

Fig. 10: Schematic drawing of the tunneling process considered here. The left point contact has the smaller capacitance and the smaller conductance. When the bias voltage is increased, peaks in the differential conductance occur, whenever a new channel for tunneling onto or from the grain is opened. Compare the bias voltages in (a) and (b).

the many-body states for N and $N+1$ particles can take two values,

$$E_{N+1} - E_N = E^0_{N+1} - E^0_N \pm (1/2) g\mu_B B,$$

where E^0_N and E^0_{N+1} are the N and $(N+1)$-particle energies in the absence of the magnetic field. The differential conductance shows a double peak at voltages

$$eV_\pm = E^0_{N+1} - E^0_N \pm (1/2) g\mu_B B,$$

as is seen in Fig. 11a. On the other hand, only a single peak at bias voltage $V_+ = E_{N+1} - E_N + (1/2) g\mu_B B$ is found if $S_{N+1} = S_N - 1/2$. Although the bias voltage V_- corresponds as well to a transition energy between many-body states with N and $N+1$ particles for $S_N > S_{N+1}$, no peak in the differential conductance is found at that bias voltage, because the initial state of that transition is an excited state, which is not populated at $V = V_-$. Population of an excited N-particle state is only possible at higher bias voltages $V \geq V_+$ via inelastic processes that use the $N+1$-particle state as an intermediate step. (A small nonequilibrium population of the excited N-particle state, and hence a small peak at $V = V_-$, may, however, occur as a result of inelastic cotunneling, as is explained in Ref. [33].) If the difference in the total spin quantum numbers for N and $N+1$ is greater than $1/2$, then there can be no conduction peak at all in the absence of spin-orbit coupling or inelastic cotunneling processes.

The situation changes in the presence of weak static magnetic impurities. "Weak" here means that they can be seen as a small perturbation on top of the picture sketched in the previous sections. "Static" means that the impurity ion has a large intrinsic angular-momentum and large crystal-anisotropy, so that we can neglect the matrix elements for transitions between different impurity spin-states. The impurity spins could be in the grain itself or could be located close to the grain in the surrounding insulator. Then if the many-body state of the grain has non-zero spin, the spin degeneracy will be lifted by the coupling to the impurity spin, which can give rise to a splitting of the lowest conductance peak even in the absence of an applied magnetic field. A significant difference between this case and the splitting

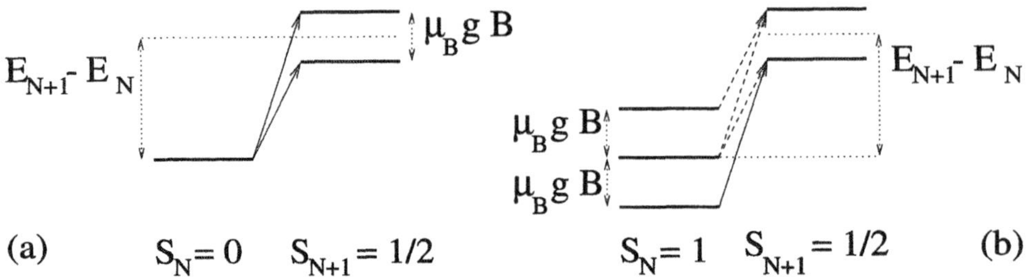

Fig. 11: Possible transitions between Zeeman split states with N and $N+1$ particles, for $S_N = 0$, $S_{N+1} = 1/2$ (a), and $S_N = 1$ and $S_{N+1} = 1/2$ (b). Note that the transitions starting out of the excited states of the triplet in (b), denoted by the dashed arrows, do not give rise to peaks in the differential conductance, (assuming that equilibrium is reached between successive tunneling events), because the excited states are not populated at $eV = E_{N+1} - E_N - \mu_B g B/2$.

due to an external field is that the effective coupling now depends on the microscopic details of the electron wavefunctions close to the impurity, and the level splitting will generally be different for the N and $N+1$ electron states. As we shall see, this makes it possible for the lowest conductance resonance to split into more than two sub-peaks.

We first consider the case of a single impurity spin. According to the Wigner-Eckart theorem, if the coupling to the impurity spin is weak, an electronic many-body state with total spin S will be split into $(2S+1)$ equally-spaced levels, characterized by the quantum number of the magnetic moment in the direction parallel to that of the frozen impurity spin. The size of the splitting depends on the concentration and microscopic details of the impurities. Since a peak in dI/dV can occur whenever $eV = E_{N+1} - E_N$, many close peaks appear when the degeneracy of the ground state is lifted. The total number of possible transitions is $(2S_N+1)(2S_{N+1}+1)$, since now E_N and E_{N+1} can take $2S_N+1$ and $2S_{N+1}+1$ values, respectively. However, for the same reasons as discussed above, not all possible transitions give rise to peaks in the differential conductance: Only transitions at energy differences $\Delta E = E_{N+1} - E_N$ where the initial N-particle state is already populated at a bias voltage $eV \leq \Delta E$ are reflected as peaks in the differential conductance, and the spin component parallel to the frozen impurity spin can only change by $\pm 1/2$. Some examples are shown in Fig. 12 for $S_N = 1/2$, $S_{N+1} = 1$ and $S_N = 1$, $S_{N+1} = 1/2$. In the figure, the transitions that correspond to true peaks in dI/dV are shown as solid arrows, the other ones are shown with dashed arrows. In practice, since eV is typically much bigger than the fine structure of the N and $N+1$-particle levels, all transitions appearing at energy differences ΔE bigger than the difference $eV_{\rm th} = E^g_{N+1} - E^g_N$ between the ground state energies for N and $N+1$ particles will show up as true peaks at $eV = \Delta E$, while no peaks appear for $eV < eV_{\rm th}$ ($V_{\rm th}$ is the threshold voltage for current flow).

If there are several frozen impurity spins coupling to the electrons, we can again use the Wigner-Eckart theorem, in the case of weak coupling, to show that a groundstate

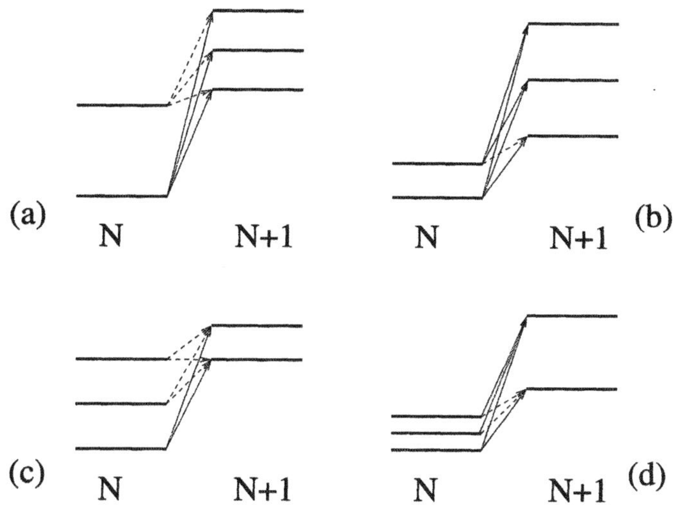

Fig. 12: Possible transitions between states with $S_N = 1/2$, $S_{N+1} = 1$ (top) and $S_N = 1$, $S_{N+1} = 1/2$ (bottom), when the spin degeneracy is broken by the presence of (several) static magnetic impurities. The transitions indicated by solid arrows give rise to peaks in the differential conductance; the transitions indicated by dashed arrows do not, because the state they are starting from is not populated at the corresponding bias voltage. How many peaks are visible depends on the actual splitting of the energies.

with spin S is split into $(2S+1)$ equally spaced levels classified by the spin component along some direction which is a weighted vector sum of the several frozen impurity spins. The weights in this sum will be different in the states with N particles and with $N+1$ particles, so that the quantization axes will generally be different in the two states, as well as the level splittings. This will lift the selection rule that the quantum number can only change by $\pm 1/2$ on the addition of a single electron, so the number of lines in the multiplet may increase accordingly.

The model of coupling to one or more frozen spins can also be generalized to the case of dynamical spins. For example, in the case of coupling to a single dynamical localized spin S_i in the insulating material close to the grain, the localized spin and the spin of the electrons in the metal grain S together will form states with total angular momentum ranging from $|S - S_i|$ to $|S + S_i|$, and will split up in the corresponding multiplet. In this case the number of possible transitions depends on the detailed selection rules governing transitions of the localized spin.

Finally, we consider the situation where spin-orbit coupling is present. In the case of weak spin-orbit coupling, a groundstate with spin greater than $1/2$ can be split into several different energy levels even in the absence of an applied magnetic field and in the absence of magnetic impurities (although the splitting by spin-orbit coupling only arises in second order perturbation theory, whereas the splitting caused by magnetic impurities already appears in first order perturbation theory). In general, the various states will be split by different amounts, and so multiple subpeaks can be observed, for large S, just as we found for the case with a frozen magnetic impurity. In the present case, however, states with odd N remain twofold degenerate by Kramers' degeneracy, which reduces the number of possible transitions roughly by a factor of two, compared to the case of static magnetic impurities.

If the spin-orbit coupling is too strong, however, spin-orbit splittings will become comparable to or larger than the single-particle level spacings. In this case, it is no longer possible for splittings between spin states to give rise to fine structure of the conductance peak on a scale small compared to Δ. (Also, as we have seen previously in Subsection 3.2, exchange-splittings tend to be reduced in this case, so that the one electron picture should be valid for the ground states.) The gold particles studied in Ref. [4] appear to be in the strong spin-orbit coupling regime.

4.2 Multiplets from nonequilibrium processes

A second mechanism to observe multiple peak structures in the differential conductance is via nonequilibrium population of highly excited states of the metal grain, as was first proposed by Agam *et al.* [31]. This mechanism does not need a degeneracy, or near-degeneracy, of the ground state. The idea of Ref. [31] is as follows: Since the bias voltage is typically much larger than the spacing Δ between single particle levels, after an electron has tunnelled on and off the grain, the grain may be left in an excited N-particle state, with an occupation of the single-particle levels that differs from the ground state, see Fig. 13. The fact that there is a different occupation of the single-particle levels will slightly shift the addition energies $E_{N+1} - E_N$, thus giving rise to peaks in dI/dV at different values of the bias voltage V. In Ref. [31] spinless particles were considered. In that case, nonequilibrium processes cause a fine structure of the second and higher resonances, but not the first one [31]. For spin 1/2 particles and even N, the scenario of Ref. [31] can also lead to a fine structure for the first resonance, as we will now describe.

We denote the highest occupied (self-consistent) single-particle level in the N-particle ground state by $\varepsilon_{N/2}$ and assume that the N-particle ground state has zero total spin. When the bias voltage exceeds the threshold $eV_{\text{th}} = E^g_{N+1} - E^g_N$, current flow can leave the grain in an excited N-electron state, when, after an electron has tunnelled into the level $\varepsilon_{N/2+1}$, another electron tunnels out of a lower-lying level ε_ν, see Fig. 13. Note that the excited state can have a total spin $S = 0$ or $S = 1$. Since the grain is now in an N-particle state that is different from the ground state (compare Figs. 13a and c), the energy cost $E_{N+1} - E_N$ for addition of an electron, and hence the position of a peak in dI/dV, is, in general, different from $E^g_{N+1} - E^g_N$. *A priori*, this difference can have three contributions: (1) An electron can tunnel into different single-particle levels than in the ground state. (2) The transition energy $E_{N+1} - E_N$ depends on the spin S of the states involved, which can be different from the ground state spin. (3) All transition energies depend uniquely (but weakly) on the populations of the initial and final states through mesoscopic fluctuations of the interaction contribution to the energy [31]. The characteristic energy scales for the first two of these contributions are Δ and J_s, while the third is of order $\Delta/\sqrt{g}$, as we shall see below, where $g = 2\pi E_{\mathrm{Th}}/\Delta$ is the dimensionless conductance of the grain. For an even N, however, there exist excited N-particle states for which the level $\varepsilon_{N/2+1}$ is only singly occupied and $S = 0$, so that the first two contributions vanish. Then only the contribution from mesoscopic fluctuations remains. If g is large, the energy scale for the fluctuations is small, and one finds multiple conductance peaks close to $eV = E^g_{N+1} - E^g_N$, where the total width of the multiplet is of order $\Delta/\sqrt{g}$.

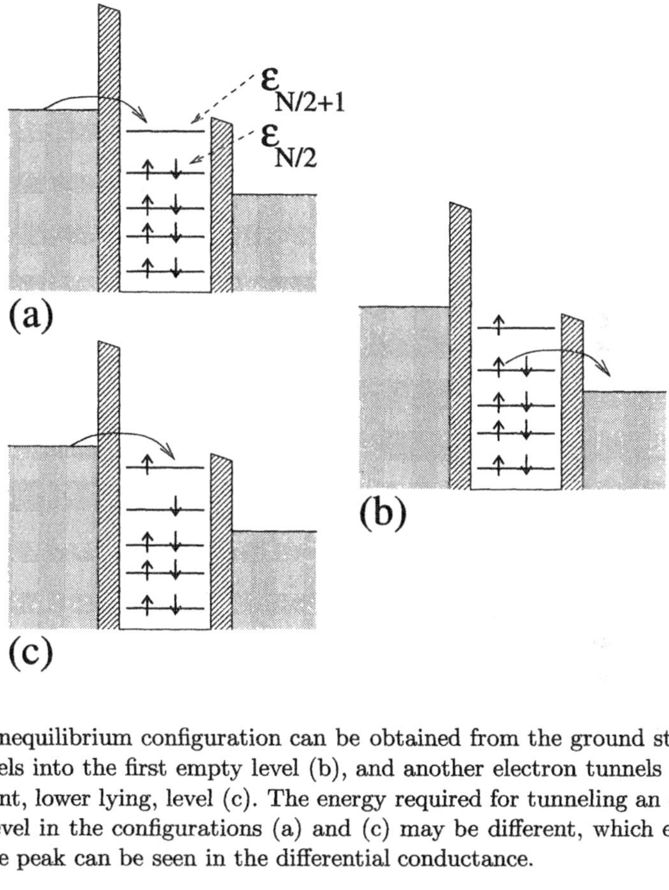

Fig. 13: A nonequilibrium configuration can be obtained from the ground state (a) if an electron tunnels into the first empty level (b), and another electron tunnels off the grain from a different, lower lying, level (c). The energy required for tunneling an electron into the highest level in the configurations (a) and (c) may be different, which explains that more than one peak can be seen in the differential conductance.

Although all nonequilibrium configurations have their own characteristic transition energy $E_{N+1} - E_N$, not all of them need to correspond to a peak in the conductance; no peak is observed if the corresponding voltage is below the threshold V_{th}, which was needed to populate the corresponding excited state. We estimate that the number of sub-peaks in the first peak mutiplet, $N_{\text{first-peak}}$ due to the nonequlibrium effect is of order $eV/2\Delta$, which is roughly the ratio of the Coulomb blockade energy to the single-particle level spacing.

To understand the reasons for this estimate let us first consider the case where we can neglect the exchange coupling J_s in the effective Hamiltonian (1). (This is correct in the case where spin-orbit coupling is strong). We also neglect the Cooper-channel interaction J_c, for the reasons discussed in Appendix C. However, we take into account fluctuations in the size of U_c between different pairs of levels. In this case, the many-body states can be labelled by the occupancies of the single-particle levels, and the ground states with energy E_N^g and energy E_{N+1}^g are described by Fig. 13a and b respectively. Let us assume that V is close to, but slightly above, the threshold voltage $V_{\rm th} = E_{N+1}^g - E_N^g$. We further assume that when an electron enters or leaves the system it does not excite other electrons by multi-electron processes; in particular we ignore Auger-like processes. Similarly we assume that the only important contributions to the conductance are via real, energy conserving, transitions.

Under these assumptions, one finds that the excited states that contribute to $N_{\rm first-peak}$ have precisely one hole below $\varepsilon_{N/2}$. The energy distance between these single-hole states is roughly Δ. Since only excited states within an energy eV from the ground state can be populated, the number of possible hole states is $\sim eV/\Delta$. Of these, roughly half will lead to positive energy shifts, which are necessary for contributions to the multiplet structure, so we find $N_{\rm first-peak} \sim eV/2\Delta$. The width of the peak is proportional to the size of the fluctuations in $U_{\rm c}$ that we have discussed in Subsection 2.1.1; in case of long range Coulomb interaction it is $\sim \Delta/\sqrt{g}$.

Many-body states with two or more holes below $\varepsilon_{N/2}$ cannot contribute to the conductance for voltages close to $V_{\rm th}$, because if two holes are present when there are N electrons on the particle, the level $\varepsilon_{N/2+1}$ will necessarily be doubly occupied. Then, the next electron would have to enter through the level $\varepsilon_{N/2+2}$, which would require an additional energy, of order Δ, that is not available for V slightly above $V_{\rm th}$. If, by chance, the distance of level $\varepsilon_{N/2+2}$ from level $\varepsilon_{N/2+1}$ is smaller than $J_{\rm s}$ then transition through nonequilibrium states with two holes may occur. This situation will increase $N_{\rm first-peak}$ significantly, it should be proportional now to $\sim (U/\Delta)^2/2$.

Let us now consider the case where the exchange parameter J_s is not negligible. Then, the N electron states having one electron in the level $\varepsilon_{N/2+1}$ and one hole below the Fermi energy will have different energies depending on whether they have $S = 0$ or $S = 1$. For $S = 1$, the energy is reduced by $|J_s|$, so that the energy to add the next electron to the level $\varepsilon_{N/2+1}$ is increased by $|J_s|$, and the peaks arising from the triplet configuration will be shifted up by this amount relative to the singlet contributions. If this shift is comparable to the level spacing Δ, then only the singlet peaks will appear close to the threshold, and the number of peaks in the lowest multiplet will be the same as before, $N_{\rm first-peak} \sim eV/2\Delta$. If $|J_s|$ is sufficiently small, but not negligible, however, the singlet and triplet peaks may appear to form a single multiplet, with twice as many peaks as before.

4.3 A comparison between the mechanisms of Subsections 4.1 and 4.2

The main difference between the two explanations offered here is that the former entails many different transitions between states very close to the N and $N+1$-particle ground states, whereas the latter makes use of, in principle, the same transition between different pairs of states that are highly excited above the ground state. Thus, one may distinguish between the two scenarios, when the electrostatic potential of the grain can be changed by the voltage V_g on a nearby gate: Fine tuning of V_g affects the threshold bias voltage V_{th}, and hence the number of possible nonequilibrium configurations. Hence, if the fine structure of the first resonance is due to nonequilibrium processes, the peaks will disappear one by one when V_g is tuned to the charge degeneracy point. This is illustrated in Fig. 14. In this figure, we have simulated the differential conductance from the rate equations of Ref. [5,34], for the case where all relaxation of excited states inside the grain occurs due to the coupling to the leads. The four panels show the differential conductance for four different values of the gate voltage, where the multiplet consists of 5, 4, 3, and 2 peaks. The closer the gate voltage is to a charge degeneracy point, the fewer nonequilibrium peaks can be observed. On the other hand, if the fine structure is due to a degenerate ground state, no highly excited states are involved, and the number of peaks in the multiplet is insensitive to V_g. Of course, both explanations (nonequilibrium processes and an almost degenerate ground state) can apply at the same time. In that case, multiple peaks will disappear at the same time, when V_g is tuned to the charge degeneracy point. An alternative way to distinguish between the two scenarios is if the parity of the number of electrons N can be changed by a gate voltage. Nonequilibrium processes cannot explain a fine structure of the first resonance if N is odd.

5. Conclusions

In this article, we have reviewed several effects related to the electrons' spin in the presence of spin-orbit coupling and/or electron–electron interaction, in small quantum dots with relatively large number of electrons.

At low temperatures, in the absence of spin-orbit coupling, and in the absence of an applied Zeeman field, our system can be described by an effective Hamiltonian of the form (1). In the limit where the number of electrons in a chaotic dot is large, the effective Hamiltonian contains three interaction parameters, in the direct, exchange and Cooper channels. In Appendices B and C we estimate these parameters for realistic systems. We have used use them in Sect. 2. to calculate (i) the probability for the dot to have a non-zero total spin in its ground state and (ii) the distribution of the Coulomb blockade peak spacings. The theory for the latter describes many features of the experimental observations, and is qualitatively much better than what one would have obtained if one ignored the exchange interaction. However, the simple effective model does not do well in describing the low energy tail of the distribution, and it does not account for the large differences in the data obtained by different experimental groups.

In Sect. 3., we reviewed some recent theoretical studies on the effect of spin-orbit coupling in a quantum dot or metal particle. In Subsection 3.1 we discussed the relation between the splitting of the ground state Kramers degeneracy (in the case of a strong spin-orbit coupling and a weak magnetic field) and an effective g-tensor. The joint probability distribution of the eigenvalues of the tensor was presented.

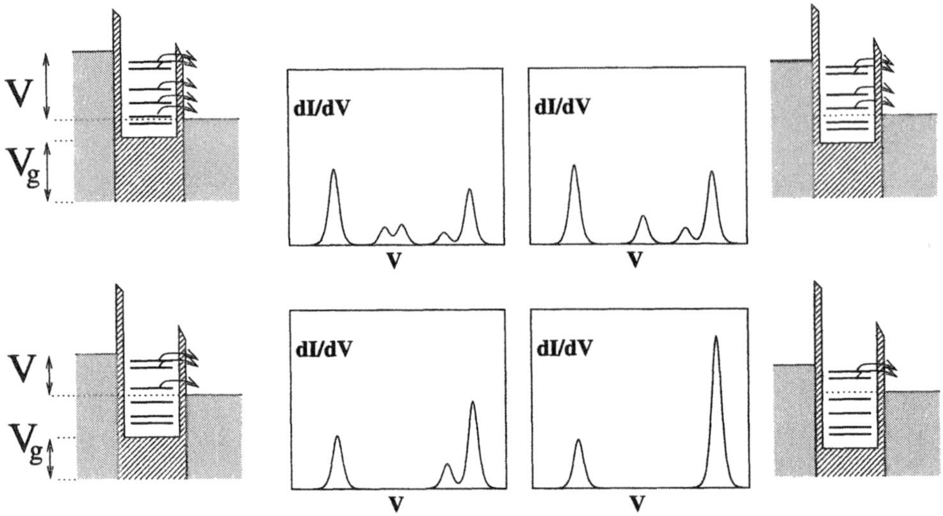

Fig. 14: When the voltage V_g of a nearby gate is varied, the number of excited N-particle states that can be populated by current flow is also changed. The four panels show how the peaks in the differential conductance disappear one by one when V_g is tuned in the direction of the charge-degeneracy point (at which current flow happens at zero bias voltage). As seen in the figure, the (minimal) bias voltage, V, for nonzero current flow decreases as V_g approaches the charge-degeneracy point. In the four panels shown, the number of peaks decreases from five (upper left) to two (lower right panel). The single-particle levels are for the $N+1$-particle system; the arrows indicate from which levels electrons can escape to the right reservoir. The dI/dV graphs were calculated using the rate equations of Refs. [5,34], with randomly chosen values for the interaction matrix elements that determine the dependence of transition energies on the precise population of the single-particle levels in the grain, see Subsection 2.1.1. The typical distance between the peaks is of order $\Delta/\sqrt{g}$, where g is the dimensionless conductance of the metal grain.

Recent experiments [25] that measured the distribution of the g-tensor in particles of Cu, Ag, and Au, found good agreement with many aspects of the theoretical predictions.

The combined effects of spin-orbit coupling and electron-electron interactions were discussed in Subsection 3.2. It was argued that strong spin-orbit coupling will tend to inhibit the appearance of effects due to electron-electron interactions.

In Subsection 3.3, we reviewed how the peculiar form of the spin-orbit coupling for a two-dimensional electron system in a GaAs heterostructure of quantum well leads to a strong suppression of spin-orbit effects when the electrons are confined in a small quantum dot. We explained how a magnetic field, parallel to a quantum dot in a AlGaAs 2DEG, enhances the weak spin-orbit effects in these dots. This observation may be used to tune the strength of spin-orbit coupling in quantum dots and may explain recent observations on fluctuations in the conductance through such dots.

We also discussed in Subsection 4, possible explanations, based on non equilibrium phenomena, and an almost degenerate ground states due to spin-orbit coupling and electron–electron interaction, for the observations [4] of a multiplet splitting of the lowest resonance in the tunneling conductance through a gold nanoparticle.

A few recent developments in the young field of spin and interaction effects in small quantum systems have been examined in this article. At present, there is no quantitative theory that can explain all the experimental observations in this area. Current theories describe well several aspects of small dots (for example the g-tensor eigenvalue distribution), but in others aspects, such as Coulomb blockade peak spacing, the agreement between theory and experiment is far from being satisfactory.

Acknowledgments

It is our pleasure to thank our collaborators J. N. H. J. Cremers, Ady Stern, J. A. Folk and C. M Marcus. Special thank are due to S. R. Patel and C. M. Marcus for allowing us using their data in this publication and for enlightening conversations. We also thank M. Tinkham and D. Davidovic for helpful discussions, and S. Lüscher, T. Heinzel and K. Ensslin for sending us their data. This work was supported at Harvard by NSF grants DMR-9809363 and DMR-9981283, at Cornell by NSF grant DMR-0086509 and by the Sloan Foundation, and at Weizmann by the German-Israeli Project Corporation DIP-c7.1.

6. Appendices

A Derivation of the effective Hamiltonian [Eq. (1)] from the toy model with contact interaction [Eq. (2)]

In order to analyze the ground state energy for the toy model Hamiltonian (2), the electron-electron interaction is separated into mean and fluctuations,

$$\begin{aligned} uM\sum_m n_{m\uparrow}n_{m\downarrow}, = uM\sum_m (\langle n_{m\uparrow}\rangle n_{m\downarrow} + n_{m\uparrow}\langle n_{m\downarrow}\rangle) \\ + uM\sum_m \delta n_{m\uparrow}\delta n_{m\downarrow} \\ - uM\sum_m \langle n_{m\uparrow}\rangle\langle n_{m\downarrow}\rangle, \end{aligned} \tag{A.1}$$

where $\delta n_{ms} = n_{ms} - \langle n_{ms}\rangle$ and the average occupation $\langle n_{m,\uparrow}\rangle = \langle n_{m,\downarrow}\rangle$ is calculated using the self-consistent (Hartree-Fock) Hamiltonian

$$\mathcal{H}^{\mathrm{HF}} = \sum_{n,m,s} c^\dagger_{n,s}\mathcal{H}_0(n,m)c_{m,s} + uM\sum_m (\langle n_{m\uparrow}\rangle n_{m\downarrow} + n_{m\uparrow}\langle n_{m\downarrow}\rangle)\,. \tag{A.2}$$

For technical convenience, we define the occupancies $\langle n_{ms}\rangle$ in a reference state with $2N-2K$ electrons and zero spin, where K is a number of order unity, chosen such that all relevant particle-hole excitations have energy less than $K\Delta$. For a disordered metal grain, we may assume that the eigenvectors and eigenvalues $\varepsilon^{\mathrm{HF}}_\mu$ of $\mathcal{H}^{\mathrm{HF}}$ are distributed like those of a random matrix (with the possible exception of the spacing of two eigenvalues closest to the Fermi level, see Refs. [3,18]; a reference state with $2N-2K$ electrons, rather than with $2N$ electrons is chosen, to ensure that only eigenvalues of $\mathcal{H}^{\mathrm{HF}}$ above the Fermi level are needed). The last term on the r.h.s. of Eq. (A.1) is a constant shift of the energy and can be omitted.

We then construct a state with $2N$ (or $2N+1$) electrons by adding $2K$ (or $2K+1$) electrons to the reference state, and find an effective Hamiltonian for low-lying particle-hole excitations using the remaining third term on the r.h.s. of Eq. (A.1) as a perturbation. The derivation of this effective Hamiltonian proceeds along the lines sketched in Sect. 2.2 and Ref. [2]. In the limit $M\to\infty$, the effective Hamiltonian has the form (1), where to lowest nontrivial order in u one has $\varepsilon_\mu = \varepsilon^{\mathrm{HF}}_\mu$ and $-J_{\mathrm{s}} = J_{\mathrm{c}} = u$.

When studying the effective interaction amplitudes perturbatively in u, one finds that virtual particle-hole excitations involving states with energies ε far away from E_{F} ($\Delta < \varepsilon - E_{\mathrm{F}} < E_{\mathrm{Th}}$) contribute to the $O(u^2)$ term and to higher order terms. This leads to an effective renormaliziation of the interaction constants J_{s} and J_{c}, and of the spacing between the levels ε_μ. For example,

$$J_{\mathrm{s}}(u) = u + u^2\frac{1}{M}\left[\int^{E_{\mathrm{F}}} d\varepsilon d\varepsilon'\frac{\nu(\varepsilon)\nu(\varepsilon')}{\varepsilon+\varepsilon'} - \int_{E_{\mathrm{F}}} d\varepsilon d\varepsilon'\frac{\nu(\varepsilon)\nu(\varepsilon')}{\varepsilon+\varepsilon'}\right] + \mathcal{O}(u^3), \tag{A.3}$$

$$\varepsilon_\mu(u) = \varepsilon_\mu^{\mathrm{HF}} - u^2 \varepsilon_\mu^{\mathrm{HF}} \frac{1}{M} \left[\int^{E_\mathrm{F}} d\varepsilon d\varepsilon' \int_{E_\mathrm{F}} d\varepsilon'' \frac{\nu(\varepsilon)\nu(\varepsilon')\nu(\varepsilon'')}{(\varepsilon + \varepsilon' - \varepsilon'')^2} \right.$$
$$\left. + \int^{E_\mathrm{F}} d\varepsilon \int_{E_\mathrm{F}} d\varepsilon' d\varepsilon'' \frac{\nu(\varepsilon)\nu(\varepsilon')\nu(\varepsilon'')}{(\varepsilon - \varepsilon' - \varepsilon'')^2} \right] + \mathcal{O}(u^3), \tag{A.4}$$

where $\nu(\varepsilon)$ is the density of states for the Hamiltonian $\mathcal{H}^{\mathrm{HF}}$, and J_c is renormalized to zero (see appendix C).

Beyond the first order in the interaction strength u, the symmetry of the effective Hamiltonian (1), and of the result (4) that was derived from it, differs from that of the equivalent expression in Ref. [3], which was obtained from the toy model (2) using the selfconsistent Hartree-Fock approximation. The reason of this difference is that, in higher orders of perturbation theory, the selfconsistent Hartree-Fock approximation neglects certain contributions to the ground state energy. (For example, the first correction to J_s is of second order in u [second term in Eq. (A.3)], and not, as in the Hartree-Fock approximation, of third order [3].)

When all contributions are taken into account, the symmetry of Eq. (4) and the form of the effective Hamiltonian (1) is preserved to all orders in u.

B The relation between the parameter $\lambda = -J_\mathrm{s}/\Delta$ and r_s

Landau-Fermi-liquid theory expresses various properties of the system in terms of the coefficients $f_{p\sigma p\sigma'}$. Using the notations of Ref. [35] we find:

$$\frac{C}{C_b} = \frac{m^\star}{m_b} = 1 + \frac{F_1^s}{d}; \quad \frac{\chi_P}{\chi_{Pb}} = \frac{m^\star}{m_b} \frac{1}{1 + F_0^a}, \tag{B.1}$$

where C is the specific heat, χ_P the Pauli susceptibility. The quantities with the suffix $-_b$ include band effects, but do not include electron–electron interaction corrections. The latter are encompassed in the F-coefficients of the Landau Fermi liquid theory. The letter d denotes the dimension of the system. For ballistic systems there are no anomalous renormalizations of the Fermi-Liquid coefficients and we have:

$$\lambda = -J_\mathrm{s}/\Delta = -F_0^\mathrm{a} = 1 - \frac{\chi_P}{\chi_{Pb}} \frac{m^\star}{m_b} = 1 - \frac{\chi_{Pb}}{\chi_P} \frac{C}{C_b}. \tag{B.2}$$

Thus, in principle the ratio of the specific heat and the susceptibility gives the desired interaction parameter J_s.

B1 Three dimensions

There are various ways to calculate theoretically the Landau F parameters, using different approximations for the electron–electron interaction. They varied from a simple static RPA approximation to more complicated approaches like density functional theory. Ref. [36] reviews the subject.

The relevant parameter to describe the strength of the interaction effects is r_s, the ratio of the typical potential energy to the kinetic energy of electrons. In metals it is defined by:

$$\frac{4\pi}{3} r_s^3 a_B^3 = \frac{1}{n}, \quad a_B = \frac{\hbar^2 \epsilon 4\pi\epsilon_0}{e^2 m_b} \tag{B.3}$$

where n is the density of electrons and a_B is the Bohr radius in the metal. Some values for r_s in metals are giving in Ref. [37] (page 74) and in Ref. [38].

For $0 < r_s < 5$ the effective mass [37] ranges between $0.96 < m^\star/m_b < 1.06$ where for small r_s (< 3), $m^\star < m_b$, and for larger r_s (> 3), $m^\star > m_b$. (See table VII on page 103 in [37].)

Using the approximation of Rice [39] for the effective mass and for the susceptibility one can roughly approximate

$$\lambda_{\text{Rice}}(r_s) \sim (3 + r_s)/25, \text{ for } 1 < r_s < 5. \tag{B.4}$$

Another approximation for the susceptibility is given in [36] (see page 256). Assuming that the effective mass is renormalized as in the Rice approach (*i.e.*, not very significant renormalization) we find that:

$$\lambda_{\text{Singwi}}(r_s) \sim (2 + r_s)/16, \text{ for } 1 < r_s < 5. \tag{B.5}$$

For small $r_s \sim 1$ the difference between the estimates is only 15 % while for $r_s \sim 5$ it close to 30 %. The second estimate reproduces quite well experimental measurements of λ in a wide range of metals. The parameter λ is determined by various experimental methods such as electron spin resonance, spin wave, Knight shift and total susceptibility. [See p. 256 of Ref. [36] and reference therein for further details.]

Using Eqs. (B.4) and (B.5) we can estimate the parameter J_s in different materials; however the estimates are rough and should be taken with a grain of salt. Typical values for metallic elements are given in Table B.1.

	Li	Na	K	Rb	Cs	Cu	Ag	Au
r_s	3.25	3.93	4.86	5.20	5.62	2.67	3.02	3.01
$\lambda_{\text{Rice}}(r_s)$	0.25	0.28	0.31	0.33	0.34	0.23	0.24	0.24
$\lambda_{\text{Singwi}}(r_s)$	0.33	0.37	0.43	0.45	0.48	0.29	0.31	0.31
	Be	**Mg**	**Ca**	**Sr**	**Ba**	**Nb**	**Fe**	**Pb**
r_s	1.87	2.66	3.27	3.57	3.71	3.07	2.12	2.30
$\lambda_{\text{Rice}}(r_s)$	0.19	0.23	0.25	0.26	0.27	0.24	0.20	0.21
$\lambda_{\text{Singwi}}(r_s)$	0.24	0.29	0.33	0.35	0.36	0.32	0.26	0.27

Table B.1: Estimates for $\lambda = -J_s/\Delta$ in selected metals.

B2 Two dimensions

Most of the calculations for the Landau-Fermi-liquid parameters in two dimensional systems were performed for Silicon MOSFET. For a review see Ref. [36] (especially page 257) and Ref. [40] (pages 454 and 468).

We note that as we sweep an external magnetic field, perpendicular to the sample area, the spin susceptibility oscillates because the difference in the occupations of Landau levels with spins up and down. This effect makes the comparison between theory and experiment complicated. We will not be interested in such anomalously large exchange enhancement.

In case of silicon MOSFET we should include also the valley degeneracy, and the difference in the dielectric constants of Si and SiO that causes the dielectric function be space dependent. The screening from the metallic electrodes may influence the results as well.

For GaAs/AlGaAs heterostructures the first two complications are absent. It appears that due to the absence of valley degeneracy in GaAs/AlGaAs the parameter J_s should be larger than in the case of the Si MOSFET. Therefore, GA/AlGaAs might be more appropriate to the study of spin configurations and there dependance on interaction constants.

A static random phase approximation for GaAs gives [41]

$$\lambda = \frac{m^*}{m_b} G(\frac{r_s}{\sqrt{2}}); \quad G(x) = \begin{cases} \frac{x}{\pi}\frac{\operatorname{arccosh}(1/x)}{\sqrt{1-x^2}} & \text{for } x \leq 1 \\ \frac{x}{\pi}\frac{\arccos(1/x)}{\sqrt{x^2-1}} & \text{for } x > 1 \end{cases} \tag{B.6}$$

where in two dimensions

$$r_s = \frac{e^2 m_b}{\hbar^2 \sqrt{\pi n}} \frac{1}{\epsilon 4\pi\epsilon_0} = \frac{5.45 * 10^5}{\sqrt{n(cm^2)}}. \tag{B.7}$$

In the last equality we take $\epsilon = 12.9$, $m_b = 0.067 m_e$ and m_e is the free electron mass.

It can be verified that

$$G(x) \stackrel{x\to\infty}{\longrightarrow} 1/2 \quad \text{and} \quad G(x) \stackrel{x\to 0}{\longrightarrow} (x/\pi)\log(2/x).$$

The factor 1/2 for large x is due to the spin degeneracy and appears because both spins participate in the screening in the RPA approximation.(In case of a MOSFET the factor 1/2 is substituted by 1/4 due to the valley degeneracy.)

The same static RPA approximation gives for the effective mass:

$$m_b/m^* = 1 - (\sqrt{2}/\pi) r_s + r_s^2/2 + (1 - r_s^2) G(r_s/\sqrt{2}). \tag{B.8}$$

Numerically, in this approximation $0.95 < m^*/m_b < 1$. In other words within the static RPA approximation the mass renormalization is not very significant.

Using this approximation we plot $\lambda = -J_s/\Delta$ as a function of r_s and n in Fig. B.1. For example, $\lambda(n = 0.7 * 10^{11} cm^{-2}) \cong \lambda(r_s \cong 2) \sim 0.34$,

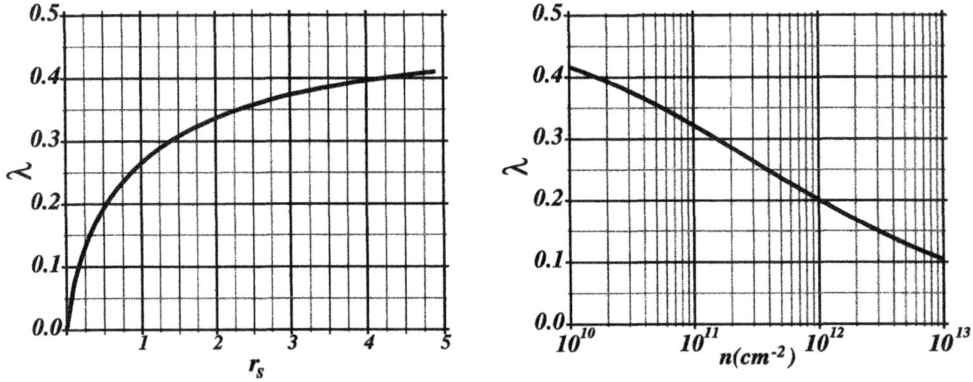

Fig. B.1: λ as a function of the ratio of the typical potential energy to the kinetic energy of electrons r_s, and as a function of the electron density n, for GaAs/AlGaAs heterostructures in the static RPA approximation.

C Renormalization of the interaction in the Cooper channel

In Sect. 2.1 we described how to integrate out the interaction between electrons at high frequency in the RG sense. The interaction in the Cooper channel deserves a special consideration, as we will see below it reduces substantially when the temperature decreases.

To see how it works in practice we look at the Dyson equation for the vertex part of the interaction in the Cooper channel (for a precise definition of the vertex part see Ref. [42] Sect. 33.3). Since the divergencies in the Cooper channel are logarithmic we can write the Dyson equation, for T larger than the inverse of the elastic mean free time τ, in a RG form [16]:

$$dJ_c(T)/dl = -J_c^2(l)/\Delta, \;\; l = \log(E_F/T), \;\; T > 1/\tau. \tag{C.1}$$

Integration of this equation, from E_F to T gives

$$J_c(T) = \frac{J_c(E_F)}{1 + (J_c(E_F)/\Delta)\log(E_F/T)}. \tag{C.2}$$

This logarithmic suppression of the interaction in the Cooper channel was first discussed in Ref. [43] and is known as the Tolmachev-Anderson-Morel log or pseudo-electron-potential log.

In quasi-two-dimensional samples, for $T < 1/\tau$ the RG equation (C.1) is modified to [16]:

$$dJ_c(T)/dl = \Delta/(g\pi) - J_c^2(T)/\Delta, \;\; 1/\tau > T > E_{Th}. \tag{C.3}$$

We have neglected here the effects of the diffusive motion on the other channels. The presence of the term $1/(\pi g)$ slows down the logarithmic reduction in the Cooper channel, and physically describes the enhancement of the interaction between the electrons in the Cooper channel due to their diffusive motion. In case of quasi-one-dimensional systems the full Dyson equation should be solved [44].

Finally, the process of integration of the motion at high frequencies reaches the Thouless energy, and we find the effective Hamiltonian (1). We will analyze the Cooper channel for energies below the Thouless energy using the contact model (2). In principle, the behavior of the Cooper channel can be solved exactly by the method of Richardson [45]. But, to understand qualitatively the reduction of the interaction in the Cooper channel, at energy below E_{Th} it is sufficient to solve the Dyson equation for the interaction matrix element $\langle\alpha\bar{\alpha}|\,\hat{u}\,|\bar{\alpha}\alpha\rangle$ in the Cooper channel. Formally this equation is:

$$\langle\alpha\bar{\alpha}|\,\hat{u}\,|\alpha\bar{\alpha}\rangle = \langle\alpha\bar{\alpha}|\,\hat{u}^0\,|\alpha\bar{\alpha}\rangle - \sum_{\nu\neq\alpha} \frac{\langle\alpha\bar{\alpha}|\,\hat{u}^0\,|\nu\bar{\nu}\rangle\,\langle\nu\bar{\nu}|\,\hat{u}\,|\alpha\bar{\alpha}\rangle}{|\varepsilon_a - \varepsilon_\nu|}, \tag{C.4}$$

with $\hat{u}^0 = uMc^\dagger_{n,\uparrow}c^\dagger_{n,\downarrow}c_{n,\downarrow}c_{n,\uparrow}\delta_{nm}$, $\hat{u} = c^\dagger_{n,\uparrow}c^\dagger_{n,\downarrow}u(n,m)c_{m,\downarrow}c_{m,\uparrow}$, the operator $\psi^\dagger_{\alpha\uparrow(\downarrow)} = \sum_{n=1}\phi_{\alpha\uparrow(\downarrow)}(n)c^\dagger_{n,\uparrow(\downarrow)}$ where the functions $\phi_\mu(n)$ are real eigenfunctions of a random matrix with real elements, n runs on the sites of the random systems, M is the total number of sites, and $|\alpha\bar{\alpha}\rangle = \psi^\dagger_{\alpha\downarrow}\psi^\dagger_{\alpha\uparrow}\,|0\rangle$. Using the anti-commutation relations of $c_{n,s}$ operators we find an equation for the unknown amplitudes $u(n,m)$

$$\sum_{nm}\phi_\alpha^2(n)u(n,m)\phi_\alpha^2(m) =$$

$$\sum_{nm} \phi_\alpha^2(n)(uM\delta_{nm})\phi_\alpha^2(m) + \sum_{n,m,l,k,\nu\neq\alpha} \frac{\phi_\alpha^2(n)(uM\delta_{nl})\phi_\nu^2(l)\phi_\nu^2(k)u(k,m)\psi_\alpha^2(m)}{\varepsilon_\alpha - \varepsilon_\nu}. \tag{C.5}$$

Now with the relation $\left\langle\phi_\mu^2(l)\phi_\mu^2(k)\right\rangle = 1/M^2(1+2\delta_{lk})$ (that is valid in the limit $M \to \infty$) we find, comparing the elements in the series that sum over n and m

$$u(n,m) = uM\delta_{nm} - \lambda\frac{\log M}{M}\left(\sum_l u(l,m) + 2u(n,m)\right), \tag{C.6}$$

where the logarithmic factor appears from the summation over the energies ν. The term $\propto 2u(n,m)$ on the left hand side can be neglected as it is small [by a factor $(\log M)/M$] compared to the one on the right hand side. A summation over n gives now

$$\sum_l u(l,m) = \frac{uM}{1+\lambda\log M}$$

substituting in (C.6) and taking the limit $M \gg 1$ we find:

$$u(n,m) = u\frac{M\delta_{nm} + \lambda\log M(M\delta_{nm} - 1)}{1+\lambda\log M}, \tag{C.7}$$

where $\lambda = u/\Delta$ and Δ is the average level spacing in the dot. Hence:

$$\langle\alpha\bar{\alpha}|\,\hat{u}\,|\alpha\bar{\alpha}\rangle = \frac{1}{M^2}\sum_{m,n}(1+2\delta_{nm})u(n,m) = 2u + \frac{u}{1+\lambda\log M}. \tag{C.8}$$

The logarithmic factor reduces the term that is $\propto \frac{1}{M^2}\sum u(n,m)$ and does not involve the δ_{nm} factor. This term correspond to the Cooper channel as it involves contraction of two wave functions, associated with two creation (or two annihilation) operators. Thus, we find that from the Thouless energy up to energies of the order of the level spacing, similarly to the pseudo-electron-potential log [see after Eq. (C.2)], there is an additional logarithmic suppression of the interaction in the Cooper channel. For that reason we took $J_c = 0$ in the analysis of the ground state spin distributions in Sect. 2.2 and the Coulomb blockade peak spacing in Sect. 2.3.

References

[1] L. L. Sohn, L. P. Kouwenhoven, and G. Schon, *Mesoscopic Electron Transport* (Kluvers, Dordrecht, 1997), nato ASI Series E 345.

[2] I. L. Kurland, I. L. Aleiner, and B. L. Altshuler, Phys. Rev. B **62**, 14886 (2000).

[3] P. W. Brouwer, Y. Oreg, and B. I. Halperin, Phys. Rev. B **60**, R13977 (1999).

[4] D. Davidovic and M. Tinkham, Phys. Rev. B **61**, R16359 (2000).

[5] C. W. J. Beenakker, Phys. Rev. B **44**, 1646 (1991). Y. Alhassid, Rev. Mod. Phys. 72, 895 (2000).

[6] Y. Oreg, K. Byczuk, and B. I. Halperin, Phys. Rev. Lett. **85**, 365 (2000).

[7] U. Sivan *et al.*, Phys. Rev. Lett. **77**, 1123 (1996).

[8] F. Simmel, T. Heinzel, and D. A. Wharam, Euro. Lett. **38**, 123 (1997).

[9] S. R. Patel *et al.*, Phys. Rev. Lett. **80**, 4522 (1998).

[10] S. Lüscher *et al.*, Phys. Rev. Lett. **86**, 2118 (2001).

[11] H. U. Baranger, D. Ullmo, and L. I. Glazman, Phys. Rev. B **61**, R2425 (2000).

[12] D. Ullmo and H. U. Baranger, cond-mat/0103098, (2001).

[13] I. L. Aleiner, P. W. Brouwer, and L. I. Glazman, cond-mat/0103008, (2001).

[14] R. Shankar, Rev. Mod. Phys. **66**, 129 (1994).

[15] J. Polchinski, Effective Field Theory and the Fermi Surface, Proceedings of 1992 Theoretical Advanced Studies Institute in Elementary Particle Physics, edited by J. Harvey and J. Polchinski (World Scientific, Singapore, 1993), 1993, hep-th/9210046.

[16] A. M. Finkel'stein, in *Soviet Scientific Review*, edited by I. M. Khalatnikov (Harwood Academic Publisher GmbH, Moscow, 1990), Vol. 14.

[17] R. Berkovits, Phys. Rev. Lett. **81**, 2128 (1998).

[18] S. Levit and D. Orgad, Phys. Rev. B **60**, 5549 (1999).

[19] Y. M. Blanter, A. D. Mirlin, and B. A. Muzykantskii, Phys. Rev. Lett. **78**, 2449 (1997).

[20] Y. M. Blanter, Phys. Rev. B **54**, 12807 (1996), (See formula 42).

[21] Y. M. Blanter, A. D. Mirlin, and B. A. Muzykantskii, Phys. Rev. Lett. **80**, 4161 (1998).

[22] G. Usaj and H. U. Baranger, cond-mat/0108027, (2001).

[23] K. A. Matveev, L. I. Glazman, and A. I. Larkin, Phys. Rev. Lett. **85**, 2789 (2000).

[24] P. W. Brouwer, X. Waintal, and B. I. Halperin, Phys. Rev. Lett. **85**, 369 (2000).

[25] J. R. Petta and D. C. Ralph, cond-mat/0106452, (2001).

[26] B. I. Halperin *et al.*, Phys. Rev. Lett. **86**, 2106 (2001).

[27] J. A. Folk *et al.*, Phys. Rev. Lett. **86**, 2102 (2001).

[28] I. L. Aleiner and V. I. Fal'ko, cond-mat/0107385, (2001).

[29] J. S. Meyer, A. Altland, and B. L. Alt'shuler, cond-mat/0105623, (2001).

[30] V. I. Fal'ko and T. Jungwirth, cond-mat/0106019, (2001).

[31] O. Agam *et al.*, Phys. Rev. Lett. **78**, 1956 (1997).

[32] J. von Delft and D. C. Ralph, Phys. Rep., to be published.

[33] O. Agam and I. L. Aleiner, Phys. Rev. B **56**, R5759 (1997).

[34] D. V. Averin, A. N. Korotkov, and K. K. Likharev, Phys. Rev. B **44**, 6199 (1991).

[35] D. Pines and P. Nozières, *The theory of quantum liquids* (W .A. Benjamin Imc, New York, 1966).

[36] K. S. Singwi and M. P. Tosi, Sol. Stat. Phys. **36**, 177 (1981).

[37] L. Hedin and S. Lundqvist, Sol. Stat. Phys. **23**, 1 (1969).

[38] N. W. Ashcroft, *Solid State Physics* (CBS publishing Asia Ltd., Philadelphia, 1987).

[39] T. M. Rice, Ann. Phys. **31**, 100 (1965).

[40] T. Ando, A. B. Fowler, and F. Stern, Rev. Mod. Phys. **54**, 437 (1982).

[41] J. F. Janak, Phys. Rev. **178**, 1416 (1969). Notice that there is a mistake in the way that the valley degeneracy is introduced.

[42] A. A. Abrikosov, L. P. Gorkov, and I. E. Dzyaloshinski, in *Methods of Quantum Field Theory in Statistical Physics*, edited by R. A. Silverman (Prentice-Hall, Inc. Englewood Cliffs, New-Jersey, 1963).

[43] P. Morel and P. W. Anderson, Phys. Rev. **125**, 1263 (1962).

[44] Y. Oreg and A. M. Finkel'stein, Phys. Rev. Lett. **83**, 191 (1999).

[45] J. von Delft, Annalen der Physik (Leipzig) **10**, 219 (2001).

Chapter 6

Kondo effect in quantum dots with an even number of electrons

Mikio Eto

Faculty of Science and Technology, Keio University,
3-14-1 Hiyoshi, Kohoku-ku, Yokohama 223-8522, Japan,
E-mail: eto@rk.phys.keio.ac.jp

Abstract

We theoretically investigate the Kondo effect in semiconductor quantum dots when the number of electrons in the dots, N, is even. First we overview the Kondo effect with odd N which strongly influences various transport properties. Then we construct a theory for the Kondo effect with even N when the spin-singlet and -triplet states are almost degenerate. We evaluate the Kondo temperature as a function of the energy difference between the states, Δ, by the scaling method, and show that the Kondo effect is significantly enhanced by the competition between the spin states. This is in agreement with the experimental results using "vertical" quantum dots in which Δ can be tuned by applying a magnetic field. We also examine the Kondo effect by the mean-field theory. The enhancement of the Kondo effect can be understood in terms of the overlap between the Kondo resonant states created around the Fermi level. These resonant states provide the unitary limit of the conductance; $G \sim 2e^2/h$.

---

1. Introduction

Recent microfabrication techniques on semiconductors have enabled us to fabricate zero-dimensional systems on the submicron scale, the quantum dots. In such systems the charging energy can be larger than the thermal energy, which significantly influences the current through the dots. The current is suppressed unless the "resonant" condition of $E(N) = E(N-1) + \mu$ is fulfilled, where $E(N)$ is the energy of N-electron state and μ is the Fermi energy of the external leads connected to the dots by the tunnel junctions. This is called Coulomb blockade. When the gate voltage is applied to change the electrostatic potential in the dots, the current flows at the resonance. In consequence, the conductance shows a quasi-periodic peak structure as a function of the gate voltage (Coulomb oscillation) [1]. Between the current peaks is the Coulomb blockade region where the number of electrons in the dot, N, is almost fixed to an integer value. N is changed one by one with increasing the gate voltage [1]. In quantum dots, the one-electron energy is quantized to discrete levels. These levels are filled consecutively with increasing N in the usual case.

The conductance is usually very small in the Coulomb blockade region at low temperatures. Then the transport properties are dominated by the higher-order tunneling processes, so-called cotunneling [2]. The Kondo effect surprisingly enhances the cotunneling conductance to a value of the order of $2e^2/h$ below the Kondo temperature, T_K [3–7].

Generally the Kondo effect takes place when a localized spin is brought in contact with electron Fermi sea. It gives rise to a new many-body ground state that has a lesser spin, which influences the transport properties of conduction electrons. The Kondo effect was originally discovered in metals with dilute magnetic impurities [8–10]. Recently the Kondo effect has been observed in semiconductor quantum dots coupled to external leads through tunneling barriers [11–14]. In this case, the localized spin is formed by the electrons confined in the dots. The effect is usually observed for an odd number of electrons in a quantum dot with spin $S = 1/2$, whereas it is not relevant for an even number of electrons with spin-singlet ($S = 0$).

More recently the Kondo effect has been reported with an even N, in "vertical" quantum dots [15], in "lateral" dots [16,17] and in carbon nanotubes [18]. In the case of vertical dots, the energy difference between the spin-singlet and -triplet states, Δ, can be tuned by applying a magnetic field [19,20]. Sasaki *et al.* has found a large

[1] In the capacitance model, $E(N)$ is given by $(eN)^2/(2C)$ and thus the resonant condition is that $E(N) - E(N-1) = (2N-1)e^2/(2C) = \mu$. As a result, the Coulomb oscillation has a equi-distance spacing of e^2/C in electrostatic energy.

Kondo effect when the spin states are nearly degenerate ($\Delta \approx 0$) [15]. Tuning of the energy difference between the spin states is hardly possible in traditional Kondo systems of dilute magnetic impurities in metal and hence this situation is quite unique to the quantum dot systems.

In this article, we theoretically examine the Kondo effect in quantum dots, with odd N and even N [21,22]. We begin with the usual case of the Kondo effect in quantum dots with $S = 1/2$. We overview the Kondo physics and various transport phenomena induced by the Kondo effect in quantum dots (Sect. 2). In Sect. 3, we study the Kondo effect with an even number of electrons to explain the experimental results by Sasaki *et al.* We evaluate the Kondo temperature, T_{K}, as a function of Δ using the scaling method. We show that the competition between the spin-singlet and triplet states significantly enhances the Kondo effect, in good agreement with the experimental finding. To qualitatively understand the enhancement of the Kondo effect, the mean-field theory is useful. Section 4 is devoted to the mean-field theory of the Kondo effect. In the last section, we present the conclusions and mention another mechanism of the Kondo effect with an even number of electrons, which is relevant in the carbon nanotubes [18].

2. Kondo effect in quantum dots with $S = 1/2$

The Kondo effect in quantum dots with $S = 1/2$ was proposed theoretically [3–7] before its observation in the experiments. In this section, we consider a simple situation in which an electron is localized in a quantum dot by the Coulomb blockade. The electron occupies a single level with spin either up or down. We show that the Kondo effect increases the cotunneling conductance significantly and results in various transport phenomena.

2.1 Cotunneling current

Let us consider a level (ε_0) in a quantum dot, which is connected to two leads, L and R, by the tunneling coupling. The Hamiltonian reads

$$H = H_{\mathrm{leads}} + H_{\mathrm{dot}} + H_{\mathrm{T}}, \tag{1}$$

$$H_{\mathrm{leads}} = \sum_{\alpha=L,R} \sum_{k\sigma} \varepsilon_k c^{\dagger}_{\alpha,k\sigma} c_{\alpha,k\sigma}, \tag{2}$$

$$H_{\mathrm{dot}} = \sum_{\sigma} \varepsilon_0 d^{\dagger}_{\sigma} d_{\sigma} + U d^{\dagger}_{\uparrow} d_{\uparrow} d^{\dagger}_{\downarrow} d_{\downarrow}, \tag{3}$$

$$H_{\mathrm{T}} = \sum_{\alpha=L,R} \sum_{k\sigma} (V_{\alpha} c^{\dagger}_{\alpha,k\sigma} d_{\sigma} + \mathrm{H.c.}), \tag{4}$$

where $c^{\dagger}_{\alpha,k\sigma}$ ($c_{\alpha,k\sigma}$) is the creation (annihilation) operator of free electrons in lead α with momentum k and spin σ ($=\uparrow,\downarrow$), whereas $d^{\dagger}_{\sigma}$ (d_{σ}) creates (annihilates) an electron with spin σ in the quantum dot. The "charging energy" U is taken into account by the second term in H_{dot}.

The level ε_0 in the quantum dot can be tuned by the gate voltage. The ground state energy with N electrons, $E(N)$, is $E(0) = 0$, $E(1) = \varepsilon_0$, and $E(2) = 2\varepsilon_0 + U$. The current flows at the resonance of $E(1)-E(0)=\varepsilon_0=\mu$ and $E(2)-E(1)=\varepsilon_0+U=\mu$. Thus the Coulomb blockade with $N=1$ is realized when $\varepsilon_0<\mu<\varepsilon_0+U$.

It is convenient to avoid the complication due to the fact that there are two leads $\alpha=L,R$ [3]. We perform a unitary transformation for electron modes in the leads;

$$c_{k\sigma}=(V_L^* c_{L,k\sigma}+V_R^* c_{R,k\sigma})/V,\quad \bar{c}_{k\sigma}=(-V_{R,i}c_{L,k\sigma}+V_{L,i}c_{R,k\sigma})/V,$$

with $V=\sqrt{|V_L|^2+|V_R|^2}$. The modes $c_{k\sigma}$ are coupled to the dot with the tunneling amplitude V, whereas the modes $\bar{c}_{k\sigma}^{(i)}$ are not coupled to the quantum dot and shall be disregarded hereafter.

Owing to the tunneling coupling between the dot and leads, the energy level ε_0 has a finite line-broadening

$$\Gamma=\pi\nu V^2,$$

where ν is the density of states in the leads. In the Coulomb blockade region with $N=1$, the addition and extraction energies are given by $E^+=\varepsilon_0+U-\mu$ and $E^-=\mu-\varepsilon_0$, respectively. We assume that the energies $E^\pm$ are positive and much larger than Γ and thermal energy $k_\mathrm{B}T$, and thus the state with one electron in the dot is stable. In this region, higher-order tunneling processes make a significant contribution to the transport of electrons: (i) A process involves the virtual state with $N=0$: Just after an electron goes out of the dot, another electron enters the dot. (ii) It is also possible to involve the virtual state with $N=2$: A conduction electron enters the dot from the leads first, and then an electron goes out of the dot. The tunneling amplitudes are given by $-V^2/E^-$ and V^2/E^+, respectively, by the second-order perturbation with respect to H_T. These processes are called cotunneling since they include the simultaneous tunneling of more than one electron [2,23].

The cotunneling processes can flip the spin in the dot. For example, just after an electron with spin-up goes out of the dot, an electron with spin-down enters the dot. Then the dot spin is changed from up to down. These spin-flip processes are essentially important for the Kondo effect. To describe the processes, we make an effective low-energy Hamiltonian by integrating out the high energy states with $N=0$ or 2 in the dot. Using the second-order perturbation with respect to H_T (or, by the Schrieffer-Wolff transformation) [9,10], we obtain

$$H=\sum_{k\sigma}\varepsilon_k c_{k\sigma}^\dagger c_{k\sigma}+J\sum_{kk'}\left[\hat{S}_+c_{k'\downarrow}^\dagger c_{k\uparrow}+\hat{S}_-c_{k'\uparrow}^\dagger c_{k\downarrow}+\hat{S}_z(c_{k'\uparrow}^\dagger c_{k\uparrow}-c_{k'\downarrow}^\dagger c_{k\downarrow})\right], \quad (5)$$

where $\hat{S}$ is the operator of the dot spin; $\hat{S}_-=d_\downarrow^\dagger d_\uparrow$, $\hat{S}_+=d_\uparrow^\dagger d_\downarrow$, $\hat{S}_z=(d_\uparrow^\dagger d_\uparrow-d_\downarrow^\dagger d_\downarrow)/2$. The exchange coupling is given by $J=V^2/E_\mathrm{c}$ where $1/E_\mathrm{c}=1/E^++1/E^-$. This is the sd Hamiltonian which has been used in the study of the traditional Kondo effect [8–10]. It should be noted that there are cotunneling processes which are not accompanied by the spin-flips (potential scattering terms). They are omitted in the Hamiltonian (5) since they are not relevant in the Kondo effect, as seen in the next

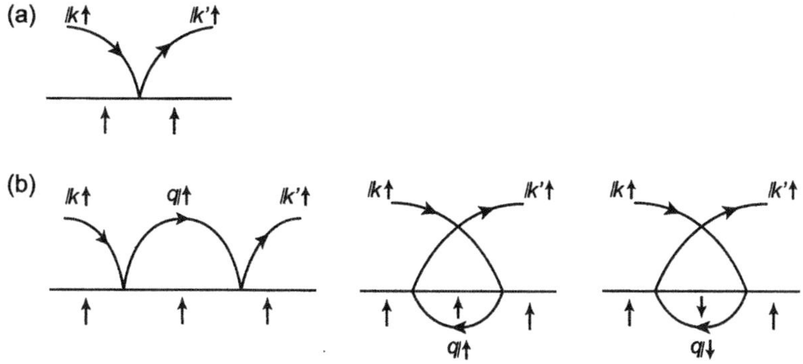

Fig. 1: The diagrams of the T-matrix, $\langle\uparrow; k' \uparrow |\hat{T}| \uparrow; k \uparrow\rangle$, (a) in the first-order perturbation and (b) in the second-order perturbation. The horizontal straight line represents the spin state in the quantum dot.

subsection.[2]

2.2 Kondo effect in quantum dots

The transport through the dot can be calculated as a "scattering problem" of the conduction electrons by the J term in Hamiltonian (5). If the term is denoted by H_I and the first term is by H_0 ($H = H_0 + H_I$), the T-matrix is written as

$$\hat{T} = H_I + H_I \frac{1}{\varepsilon - H_0 + \mathrm{i}\delta} H_I + \cdots, \tag{6}$$

for an incident conduction-electron with energy ε. To the first order of H_I in Eq. (6),

$$\langle\uparrow; k' \uparrow |\hat{T}^{(1)}| \uparrow; k \uparrow\rangle = J/2.$$

This is the scattering amplitude of an electron from state $k \uparrow$ to state $k' \uparrow$, with keeping spin up in the dot. This process can be represented by a diagram in Fig. 1(a).

To the second order of H_I, three processes contribute to $\langle\uparrow; k' \uparrow |\hat{T}^{(2)}| \uparrow; k \uparrow\rangle$. In the first and second diagrams in Fig. 1(b), there is no spin flip: In the first diagram, an electron propagates in the virtual state. In the second diagram, an electron-hole pair is created, and then the hole disappears with an incident electron. They yield

$$\sum_q (J/2)^2 \frac{1}{\varepsilon - \varepsilon_q + \mathrm{i}\delta}[1 - f(\varepsilon_q)] + \sum_q (J/2)^2 \frac{-1}{\varepsilon - (2\varepsilon - \varepsilon_q) - \mathrm{i}\delta} f(\varepsilon_q)$$
$$= \sum_q (J/2)^2 \frac{1}{\varepsilon - \varepsilon_q + \mathrm{i}\delta}.$$

[2] In the Coulomb blockade region with $N = 0$ and $N = 2$, there is no spin in a dot. In the effective Hamiltonian to the second order of H_{T}, there are only potential scattering terms. Hence the Kondo effect does not take place in this case.

The Fermi distribution functions $f(\varepsilon_q)$ are cancelled out by each other. This results in a small value of the order of $J^2\nu$. In general, we do not have any anomalous results in the absence of the spin-flip processes.

The third diagram in Fig. 1(b) includes spin-flips; an electron-hole pair is created with spin-up and -down, changing the dot spin. It should be noted that there is no counterpart of the diagram in which an electron with spin-down propagates in the virtual state. Consequently the Fermi distribution function remains in the final result, as

$$\sum_q J^2 \frac{1}{\varepsilon - \varepsilon_q + \mathrm{i}\delta} f(\varepsilon_q) = J^2\nu \int_{-D}^{D} \frac{1}{\varepsilon - \varepsilon' + \mathrm{i}\delta} f(\varepsilon')$$
$$\approx \begin{cases} -J^2\nu \ln|\varepsilon|/D & \text{for} \quad |\varepsilon| \gg k_\mathrm{B}T \\ -J^2\nu \ln k_\mathrm{B}T/D & \text{for} \quad |\varepsilon| \ll k_\mathrm{B}T. \end{cases}$$

We have assumed that the density of states is constant ν in the energy band of $[-D, D]$. This logarithmic divergence was first found by J. Kondo [8]. It stems from the Fermi edge singularity of the Fermi distribution function of the conduction electrons in the leads.

By summing the leading order logarithmic terms, we obtain

$$\langle \uparrow; k' \uparrow |\hat{T}| \uparrow; k \uparrow \rangle = \frac{J/2}{1 + 2J\nu \ln k_\mathrm{B}T/D},$$

for $|\varepsilon| \ll k_\mathrm{B}T$. Hence the result of the perturbation calculations diverges at $T = De^{-1/2J\nu}$. This temperature is called the Kondo temperature T_K [9,10].

Indeed the conductance through the quantum dot, G, shows a logarithmic T dependence in a range of temperature, in accordance with the perturbation calculations. However, G does not diverge but goes to a finite value of the order of $2e^2/h$ at $T \ll T_\mathrm{K}$ (see Fig. 2(b)). This is called unitary limit. The perturbation calculations, Eq. (6), are applicable only at $T \gg T_\mathrm{K}$. At $T < T_\mathrm{K}$, a many-body ground state is formed between the spin in the dot and conduction electrons. The many-body state is a spin-singlet, or the dot spin is completely screened out. The perturbation in which conduction electrons are "scattered" by a localized spin does not treat this situation. The many-body state allows the conduction electrons at μ to be transmitted resonantly through the dot, which results in the unitary limit. This is the scenario of the Kondo physics [9,10].

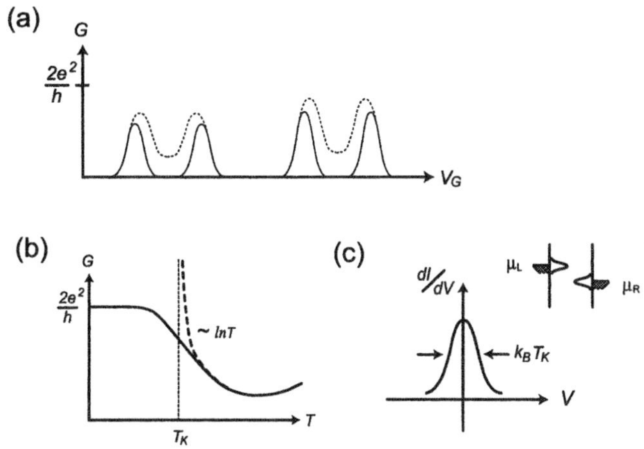

Fig. 2: Observation of the Kondo effect in semiconductor quantum dots. (a) Conductance through the dot, G, as a function of the gate voltage, V_G (Coulomb oscillation) at $T > T_K$ (solid line) and at $T < T_K$ (dotted line). G is enhanced at $T < T_K$ only in the valleys with an odd number of electrons. (b) The conductance in the Kondo valley, as a function of the temperature. The result by the perturbation calculations is indicated by the broken line. (c) The differential conductance dI/dV as a function of the bias voltage V. The inset schematically shows that two Kondo resonant levels are separated from each other under a finite bias voltage.

2.3 Observation of the Kondo effect in quantum dots

The Kondo effect gives rise to a "resistance minimum" in the traditional Kondo system of the magnetic impurities in metals. In this case, the conduction electrons are scattered by the impurities. At $T < T_K$, the Kondo state is created around an impurity spin, which is a "local singlet state." The scattering probability of the conduction electrons is enhanced by the resonant scattering through the Kondo state. In the case of quantum dots, we observe an opposite phenomenon, "conductance minimum." At $T > T_K$, the conductance through a quantum dot is suppressed by the Coulomb blockade. At $T < T_K$, the electrons can be transported through the Kondo resonant level at μ, which increases the conductance.

There are several ways to observe the Kondo effect in quantum dots. First, the conductance G increases with decreasing temperature in the Coulomb blockade regions with an odd number of electrons, whereas the Kondo effect is usually not relevant in the blockade region with an even number of electrons (Fig. 2(a)). Second, in the Kondo valley with an odd number of electrons, G shows a logarithmic T dependence, say, between $0.1T_K$ and $10T_K$ [11]. G becomes $\sim 2e^2/h$ at $T \ll T_K$ (Fig. 2(b)). Third, under finite bias voltages, V, the Kondo resonant level on one lead is off that on the other lead. As a result, the differential conductance dI/dV has a sharp peak at $V = 0$. The width of the peak is of the order of T_K (Fig. 2(c)) since the Kondo resonances have a width $\sim T_K$. Fourth, in a quite strong magnetic field, B, the Kondo effect is weakened by the Zeeman splitting $\Delta E_Z = g\mu_B B$ ($g \approx 0.4$

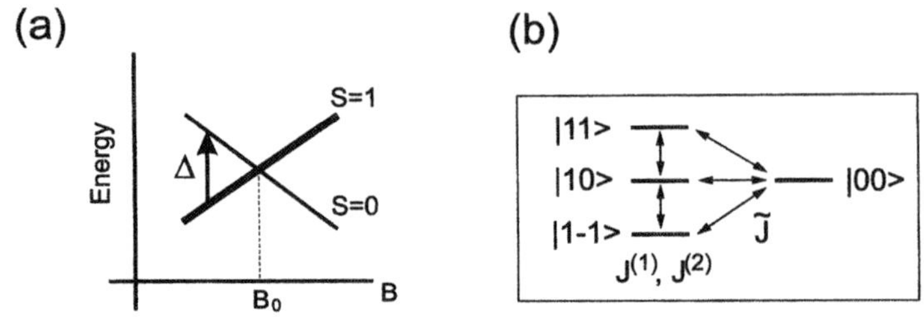

Fig. 3: (a) The energies of spin-singlet and -triplet states in a quantum dot, as functions of magnetic field B. The energy difference, $\Delta = E_{S=0} - E_{S=1}$, can be controlled by the magnetic field. $\Delta = 0$ at $B = B_0$, where the transition of the ground state occurs. (b) Spin-flip processes between the spin states. The exchange couplings $J^{(i)}$ involving the spin-triplet state only are accompanied by the scattering of conduction electrons of channel i. Those involving both the spin-triplet and -singlet states, $\tilde{J}$, are accompanied by the interchannel scattering of conduction electrons.

in GaAs, μ_B is the Bohr magneton). However, the Kondo effect is recovered by the bias voltage if the resonant condition of $eV = \pm\Delta E_Z$ is satisfied. Hence the zero-bias peak of dI/dV is split in two by the Zeeman splitting [24].

3. Kondo effect with an even number of electrons

The Kondo effect is not relevant in quantum dots with spin-singlet state. In usual case, the discrete spin-degenerate levels in a quantum dot are consecutively occupied, and hence the total spin is zero or 1/2 for an even and odd number of electrons, respectively. Therefore the conductance is enhanced by the Kondo effect with odd N, but not with even N, as shown in Fig. 2(a).

Sasaki *et al.* has found a large Kondo effect in so-called "vertical" quantum dots with an even N [15]. The spacing of discrete levels in such dots is comparable with the strength of electron-electron Coulomb interaction. Hence the electronic states deviate from the simple picture mentioned above [19,20]. If two electrons are put into nearly degenerate levels, the exchange interaction favors a spin triplet (Hund's rule) [19]. This state is changed to a spin singlet by applying a magnetic field perpendicularly to the dots, which increases the level spacing. Hence the energy difference between the singlet and triplet states,

$$\Delta = E_{S=0} - E_{S=1}, \tag{7}$$

can be controlled experimentally by the magnetic field (Fig. 3(a)). The Kondo effect is significantly enhanced around the degeneracy point between the triplet and singlet states, $\Delta \approx 0$. In this section, we theoretically study the Kondo effect when both spin-singlet and triplet states are involved [21,22].

3.1 Model

We consider two extra electrons in a quantum dot at the background of a singlet state of all other $N-2$ electrons, which we will regard as the vacuum $|0\rangle$. These two extra electrons occupy two levels of different orbital symmetry. The energies of the levels are ε_1 and ε_2. Possible two-electron states are (i) the spin-triplet state, (ii) the spin-singlet state of the same orbital symmetry as the triplet state, $1/\sqrt{2}(d_{1\uparrow}^\dagger d_{2\downarrow}^\dagger - d_{1\downarrow}^\dagger d_{2\uparrow}^\dagger)|0\rangle$, and (iii) two singlet states of different orbital symmetry, $d_{1\uparrow}^\dagger d_{1\downarrow}^\dagger|0\rangle$, $d_{2\uparrow}^\dagger d_{2\downarrow}^\dagger|0\rangle$. Among the singlet states, we only consider a state of the lowest energy, which belongs to group (iii). Thus we restrict our attention to four states, $|SM\rangle$:

$$|11\rangle = d_{1\uparrow}^\dagger d_{2\uparrow}^\dagger|0\rangle, \tag{8}$$

$$|10\rangle = \frac{1}{\sqrt{2}}(d_{1\uparrow}^\dagger d_{2\downarrow}^\dagger + d_{1\downarrow}^\dagger d_{2\uparrow}^\dagger)|0\rangle, \tag{9}$$

$$|1-1\rangle = d_{1\downarrow}^\dagger d_{2\downarrow}^\dagger|0\rangle, \tag{10}$$

$$|00\rangle = \frac{1}{\sqrt{2}}(C_1 d_{1\uparrow}^\dagger d_{1\downarrow}^\dagger - C_2 d_{2\uparrow}^\dagger d_{2\downarrow}^\dagger)|0\rangle, \tag{11}$$

where $d_{i\sigma}^\dagger$ creates an electron with spin σ in level i. The coefficients in the singlet state, C_1 and C_2 ($|C_1|^2 + |C_2|^2 = 2$), are determined by the electron-electron interaction and one-electron level spacing $\delta = \varepsilon_2 - \varepsilon_1$. We assume that $C_1 = C_2 = 1$ in this section. This is the case for $\delta = 0$.[3] The general case of $C_1 \neq C_2$ will be discussed in section 3.3. The energy difference between the states is given by Δ, Eq. (7). Although Δ is tuned experimentally by a magnetic field ($B \sim 0.2$T), we can neglect the Zeeman splitting of the triplet state. The splitting is $E_Z = 40$mK which is much smaller than the Kondo temperature, $T_K \approx 350$mK [15].

The dot is connected to two external leads L, R with free electrons being described by

$$H_{\text{leads}} = \sum_{\alpha=L,R} \sum_{k\sigma i} \varepsilon_k^{(i)} c_{\alpha,k\sigma}^{(i)\dagger} c_{\alpha,k\sigma}^{(i)},$$

where $c_{\alpha,k\sigma}^{(i)\dagger}$ ($c_{\alpha,k\sigma}^{(i)}$) is the creation (annihilation) operator of an electron in lead α with momentum k, spin σ, and orbital symmetry i ($=1,2$). The tunneling between

[3] One finds that $C_1 = C_2 = 1$ in the case of $\delta = 0$, considering the matrix element of the Coulomb interaction between $d_{1\uparrow}^\dagger d_{1\downarrow}^\dagger|0\rangle$ and $d_{2\uparrow}^\dagger d_{2\downarrow}^\dagger|0\rangle$, $\langle 22|e^2/r|11\rangle \equiv K$. In rectangular dots [15], the matrix element K is not zero and of the same order as the exchange interaction, $\langle 12|e^2/r|21\rangle \equiv J$, and smaller than the Hartree term, $\langle 11|e^2/r|11\rangle \sim$ (charging energy), typically by one order. In a case of $\langle 11|e^2/r|11\rangle = \langle 22|e^2/r|22\rangle$, $C_1/C_2 = \left(\delta + \sqrt{\delta^2 + K^2}\right)/K$. When the singlet and triplet states are degenerate, $\delta \sim J$ and thus $C_1 \approx C_2$.

the dot and the leads is written as

$$H_T = \sum_{\alpha=L,R} \sum_{k\sigma i} (V_{\alpha,i} c^{(i)\dagger}_{\alpha,k\sigma} d_{i\sigma} + \text{H.c.}).$$

We assume that the orbital symmetry is conserved in the tunneling processes (two channels in the leads).[4] As in the previous section, we perform a unitary transformation for electron modes in the leads; $c^{(i)}_{k\sigma} = (V^*_{L,i} c^{(i)}_{L,k\sigma} + V^*_{R,i} c^{(i)}_{R,k\sigma})/V_i$, $\bar{c}^{(i)}_{k\sigma} = (-V_{R,i} c^{(i)}_{L,k\sigma} + V_{L,i} c^{(i)}_{R,k\sigma})/V_i$, with $V_i = \sqrt{|V_{L,i}|^2 + |V_{R,i}|^2}$. The modes $\bar{c}^{(i)}_{k\sigma}$ are not coupled to the quantum dot and shall be disregarded. Then H_{leads} and H_T are rewritten as

$$H_{\text{leads}} = \sum_{k\sigma i} \varepsilon^{(i)}_k c^{(i)\dagger}_{k\sigma} c^{(i)}_{k\sigma}, \tag{12}$$

$$H_T = \sum_{k\sigma i} V_i (c^{(i)\dagger}_{k\sigma} d_{i\sigma} + \text{H.c.}). \tag{13}$$

We consider the situation in which the N-electron state is stable, so that the addition/extraction energies, $E^{\pm} \equiv E(N\pm1) - E(N) \mp \mu$, exceed the level broadening $\Gamma^i = \pi\nu V_i^2$ ($i = 1, 2$) and $k_B T$. We also assume that $E^{\pm} \gg |\Delta|$, δ. The cotunneling processes give rise to spin-flips among four states, $|SM\rangle$. For example, starting from the state $|11\rangle$, an electron with spin-up in level 2 goes out of the dot and then another electron with spin-down enters level 2 in the dot from a lead. Then the dot state changes to $|10\rangle$. If an electron with spin-up in level 2 goes out of the dot and then an electron with spin-down enters level 1 from a lead, the dot state $|11\rangle$ changes to $|00\rangle$. There are several spin-flip processes in our model.

To investigate the spin-flip processes, we make an effective low-energy Hamiltonian. By integrating out the states with one or three extra electrons, we obtain an extended sd Hamiltonian;

$$H_{\text{eff}} = H_{\text{leads}} + H_{\text{dot}} + H^{S=1} + H^{S=1\leftrightarrow 0} + H'_{\text{eff}}. \tag{14}$$

The Hamiltonian of the dot H_{dot} reads

$$H_{\text{dot}} = \sum_{S,M} E_{SM} f^\dagger_{SM} f_{SM}, \tag{15}$$

using pseudofermion operators $f^\dagger_{SM}$ (f_{SM}) which create (annihilate) the state $|SM\rangle$. The condition of

$$\sum_{SM} f^\dagger_{SM} f_{SM} = 1 \tag{16}$$

[4] The different symmetry means different orbital quantum number for the one-electron states in a quantum dot. This is the case of vertical quantum dots in which the tunneling processes conserve the symmetry for the motion of electrons in the transverse direction.

should be fulfilled. The third term $H^{S=1}$ represents the spin-flip processes among three components of the spin-triplet state. This resembles the sd Hamiltonian for $S=1$ in terms of the spin operator $\hat{S}$ for the dot spin

$$\begin{aligned} H^{S=1} =& \sum_{kk'}\sum_{i=1,2} J^{(i)} \left[\hat{S}_+ c^{(i)\dagger}_{k'\downarrow} c^{(i)}_{k\uparrow} + \hat{S}_- c^{(i)\dagger}_{k'\uparrow} c^{(i)}_{k\downarrow} + \hat{S}_z (c^{(i)\dagger}_{k'\uparrow} c^{(i)}_{k\uparrow} - c^{(i)\dagger}_{k'\downarrow} c^{(i)}_{k\downarrow})\right] \\ =& \sum_{kk'}\sum_{i=1,2} J^{(i)} \Big[\sqrt{2}(f^\dagger_{11} f_{10} + f^\dagger_{10} f_{1-1}) c^{(i)\dagger}_{k'\downarrow} c^{(i)}_{k\uparrow} + \sqrt{2}(f^\dagger_{10} f_{11} + f^\dagger_{1-1} f_{10}) c^{(i)\dagger}_{k'\uparrow} c^{(i)}_{k\downarrow} \\ &+ (f^\dagger_{11} f_{11} - f^\dagger_{1-1} f_{1-1})(c^{(i)\dagger}_{k'\uparrow} c^{(i)}_{k\uparrow} - c^{(i)\dagger}_{k'\downarrow} c^{(i)}_{k\downarrow})\Big]. \end{aligned} \quad (17)$$

The exchange coupling, $J^{(i)} = V_i^2/(2E_c)$ with $1/E_c = 1/E^+ + 1/E^-$, is accompanied by the scattering of conduction electrons of channel i. The fourth term $H^{S=1\leftrightarrow 0}$ in H_{eff} describes the conversion between the spin-triplet and -singlet states accompanied by the interchannel scattering of conduction electrons

$$\begin{aligned} H^{S=1\leftrightarrow 0} =& \sum_{kk'} \tilde{J} \Big[\sqrt{2}(f^\dagger_{11} f_{00} - f^\dagger_{00} f_{1-1}) c^{(1)\dagger}_{k'\downarrow} c^{(2)}_{k\uparrow} + \sqrt{2}(f^\dagger_{00} f_{11} - f^\dagger_{1-1} f_{00}) c^{(1)\dagger}_{k'\uparrow} c^{(2)}_{k\downarrow} \\ &- (f^\dagger_{10} f_{00} + f^\dagger_{00} f_{10})(c^{(1)\dagger}_{k'\uparrow} c^{(2)}_{k\uparrow} - c^{(1)\dagger}_{k'\downarrow} c^{(2)}_{k\downarrow}) + (1 \leftrightarrow 2)\Big], \end{aligned} \quad (18)$$

where $\tilde{J} = V_1 V_2/(2E_c)$. The spin-flip processes included in our model are shown in Fig. 3(b).

The last term of H_{eff} represents the scattering processes without spin-flip in the dot (potential scattering),

$$H'_{\mathrm{eff}} = \sum_{kk'\sigma}\sum_{i=1,2} \left[J'^{(i)} c^{(i)\dagger}_{k'\sigma} c^{(i)}_{k\sigma} \sum_M f^\dagger_{1M} f_{1M} + J''^{(i)} c^{(i)\dagger}_{k'\sigma} c^{(i)}_{k\sigma} f^\dagger_{00} f_{00}\right]. \quad (19)$$

The potential scattering is not relevant for the Kondo effect in the case of $C_1 = C_2$ in Eq. (11), and can be neglected for a while.

3.2 Scaling calculation of the Kondo temperature (1)

The perturbation calculations with respect to $H^{S=1} + H^{S=1\leftrightarrow 0}$ in H_{eff} yield the logarithmic terms. Figure 4 shows two diagrams in the second-order perturbation, which give us

$$-2J^{(1)2}\nu \ln|\varepsilon|/D, \quad -2\tilde{J}^2 \nu \ln|\varepsilon + \Delta|/D,$$

respectively. The former stems from the spin-flip among the triplet state and is familiar with the usual Kondo effect. On the other hand, the latter involves the mixing between the triplet and singlet states, which is original to our model.

Based on these perturbation consideration, we evaluate the Kondo temperature T_{K} using the scaling method. We adopt the "poor man's" scaling technique which was developed by Anderson [25–27]. We assume that the density of states in the leads is constant ν in the energy band $[-D, D]$. By changing the energy scale (bandwidth D) from D to $D - |dD|$, we renormalize the coupling constants, $J^{(1)}$, $J^{(2)}$ and $\tilde{J}$,

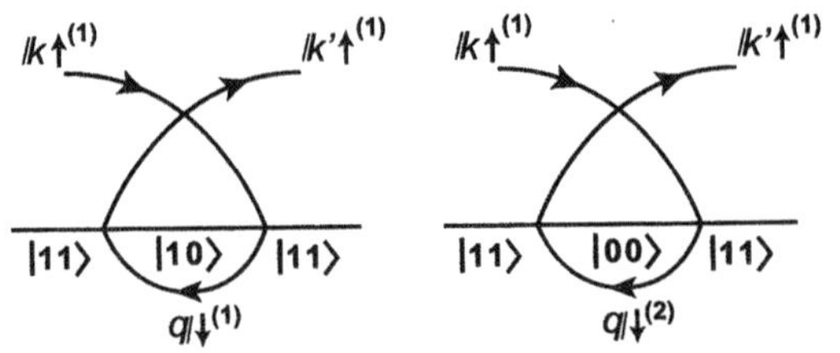

Fig. 4: Two second-order diagrams of the T-matrix, $\langle 11; k' \uparrow |\hat{T}| 11; k \uparrow \rangle$, which yield logarithmically divergent terms, in our model involving spin-singlet and -triplet states in a quantum dot. The horizontal straight line represents the spin state $|SM\rangle$ in the dot.

so as not to change the low-energy physics, within the second-order perturbation calculations. Then we obtain the scaling equations. According to the equations, the exchange couplings grow with decreasing D, and finally become so large that the perturbation does not work. The Kondo temperature is determined as the energy scale D at which the perturbation breaks down.

We obtain a closed form of the scaling equations in two limits. (i) When the energy scale D is much larger than the energy difference $|\Delta|$, H_{dot} can be safely disregarded in H_{eff}. The scaling equations can be written in a matrix form;

$$\frac{d}{d\ln D}\begin{pmatrix} J^{(1)} & \tilde{J} \\ \tilde{J} & J^{(2)} \end{pmatrix} = -2\nu \begin{pmatrix} J^{(1)} & \tilde{J} \\ \tilde{J} & J^{(2)} \end{pmatrix}^2 . \tag{20}$$

(ii) For $D \ll \Delta$, the ground state of the dot is a spin triplet and the singlet state can be disregarded. Then $J^{(1)}$ and $J^{(2)}$ evolve independently,

$$\frac{d}{d\ln D} J^{(i)} = -2\nu J^{(i)2}, \tag{21}$$

whereas $\tilde{J}$ does not change.

In the case of $|\Delta| \ll T_{\text{K}}$, the scaling equations (20) remain valid till the scaling ends. The matrix in Eq. (20) has eigenvalues of

$$\begin{aligned} J_{\pm} &= (J^{(1)} + J^{(2)})/2 \pm \sqrt{(J^{(1)} - J^{(2)})^2/4 + \tilde{J}^2} \\ &= J^{(1)} + J^{(2)}, 0. \end{aligned} \tag{22}$$

The larger one, J_+, diverges upon decreasing the bandwidth D and determines T_{K}:

$$\begin{aligned} T_{\text{K}}(0) &= D_0 \exp[-1/2\nu J_+] \\ &= D_0 \exp[-1/2\nu (J^{(1)} + J^{(2)})]. \end{aligned} \tag{23}$$

Here D_0 is the initial bandwidth, which is given by $\sqrt{E^+ E^-}$ [28].

When $\Delta > D_0$, the scaling equations (21) work in the whole scaling region. This yields

$$T_{\text{K}}(\infty) = D_0 \exp[-1/2\nu J^{(1)}] \tag{24}$$

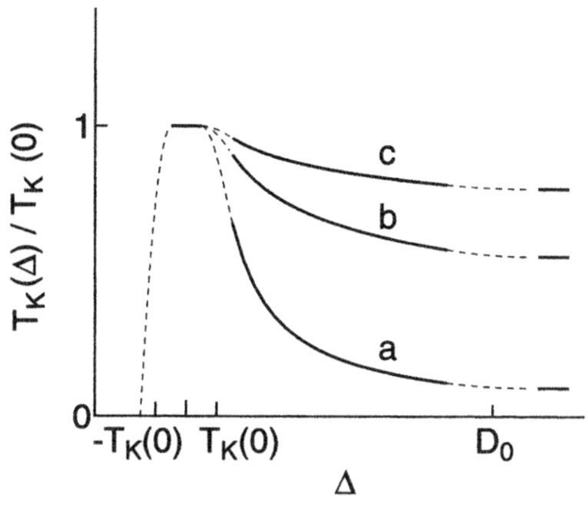

Fig. 5: The scaling calculations of the Kondo temperature T_K as a function of $\Delta = E_{S=0} - E_{S=1}$, in the model involving spin-singlet and -triplet states in a quantum dot. $J^{(2)}/J^{(1)} = \tan^2\theta$ with *a*, $\theta/\pi = 0.25$; *b*, 0.15; and *c*, 0.10. D_0 is the bandwidth in the leads.

for $J^{(1)} \geq J^{(2)}$. This is the Kondo temperature for spin-triplet localized spins [29].

In the intermediate region of $T_K(0) \ll \Delta \ll D_0$, the exchange couplings develop by Eq. (20) for $D \gg \Delta$. Around $D = \Delta$, $\tilde{J}$ saturates while $J^{(1)}$ and $J^{(2)}$ continue to grow with decreasing D, following Eq. (21) for $D \ll \Delta$. We match the solutions of these scaling equations at $D \simeq \Delta$ and obtain a power law of $T_K(\Delta)$

$$T_K(\Delta) = T_K(0) \cdot (T_K(0)/\Delta)^{\tan^2\theta}, \tag{25}$$

with

$$\begin{aligned}\tan\theta &= \tilde{J}/[\sqrt{(J^{(1)} - J^{(2)})^2/4 + \tilde{J}^2} + (J^{(1)} - J^{(2)})/2] \\ &= \sqrt{J^{(2)}/J^{(1)}}\end{aligned} \tag{26}$$

for $J^{(1)} \geq J^{(2)}$. Here $(\cos\theta, \sin\theta)^T$ is the eigenfunction of the matrix in Eq. (20) corresponding to J_+. $\theta \sim 0$ for $J^{(1)} \gg J^{(2)}$ and $\theta = \pi/4$ for $J^{(1)} = J^{(2)}$. In general, $0 < \theta \leq \pi/4$ and thus $0 < \tan^2\theta \leq 1$.

Finally, for $\Delta < 0$, all the coupling constants saturate and no Kondo effect is expected, provided $|\Delta| \gg T_K(0)$. Thus T_K quickly decreases to zero at $\Delta \sim -T_K(0)$.

The Kondo temperature as a function of Δ is schematically shown in Fig. 5. Our results clearly indicate that the Kondo effect is enhanced by the competition between singlet and triplet states. This is in accordance with the experimental findings by Sasaki *et al.* [15]: The Kondo effect has been observed around $\Delta = 0$. The Kondo temperature elsewhere is probably low in comparison with the actual temperature, so that no Kondo effect has been seen.

3.3 Scaling calculation of the Kondo temperature (2)

We have considered the case of $C_1 = C_2$ in the singlet state, Eq. (11), in the previous section and obtained $T_K(\Delta)$ in an analytical form. After our work [21], Pustilnik and Glazman examined a different model for the "triplet-singlet Kondo effect" [30]. In their model, $C_2 = 0$ in the singlet state. Their results are qualitatively the same as ours but quantitatively different: They have derived a power law of $T_K(\Delta)$ with a universal exponent, $\gamma = 2 + \sqrt{5}$, considering a fixed point of renormalization flow, whereas our results yield nonuniversal exponent, $\gamma = J^{(2)}/J^{(1)}$ (Eqs. (25) and (26)).

To elucidate the discrepancy between the models, we examine a general situation with $C_1 \neq C_2$ [31]. In this case, the effective Hamiltonian to describe the singlet-triplet conversion, Eq. (18), is generalized to

$$\begin{aligned} H^{S=1\leftrightarrow 0} = \sum_{kk'} \Big[& \sqrt{2}(\tilde{J}_1 f_{11}^{\dagger} f_{00} - \tilde{J}_2 f_{00}^{\dagger} f_{1-1}) c_{k'\downarrow}^{(1)\dagger} c_{k\uparrow}^{(2)} \\ & + \sqrt{2}(\tilde{J}_2 f_{00}^{\dagger} f_{11} - \tilde{J}_1 f_{1-1}^{\dagger} f_{00}) c_{k'\uparrow}^{(1)\dagger} c_{k\downarrow}^{(2)} \\ -(\tilde{J}_1 f_{10}^{\dagger} f_{00} + \tilde{J}_2 f_{00}^{\dagger} f_{10}) & (c_{k'\uparrow}^{(1)\dagger} c_{k\uparrow}^{(2)} - c_{k'\downarrow}^{(1)\dagger} c_{k\downarrow}^{(2)}) + (1 \leftrightarrow 2) \Big], \end{aligned} \tag{27}$$

where $\tilde{J}_1 = C_1 V_1 V_2/(2E_c)$ and $\tilde{J}_2 = C_2 V_1 V_2/(2E_c)$. The coefficients of the potential scattering terms in H'_{eff}, $J'^{(i)}$, $J''^{(i)}$, are also involved in the renormalization.

The poor man's scaling method yields the following equations. (i) For $D \gg |\Delta|$,

$$\begin{aligned} dJ^{(i)}/d\ln D &= -2\nu \left[J^{(i)2} + (\tilde{J}_1^2 + \tilde{J}_2^2)/2 \right], \\ d\tilde{J}_1/d\ln D &= -2\nu (J^{(1)} + J^{(2)}) \tilde{J}_1 + \nu J' \tilde{J}_1, \\ d\tilde{J}_2/d\ln D &= -2\nu (J^{(1)} + J^{(2)}) \tilde{J}_2 - \nu J' \tilde{J}_2, \\ dJ'/d\ln D &= 8\nu (\tilde{J}_1^2 - \tilde{J}_2^2), \end{aligned} \tag{28}$$

where $J' = J'^{(1)} - J'^{(2)} - J''^{(1)} + J''^{(2)}$. (ii) For $D \ll \Delta$, the scaling equations are the same as before and given by Eqs. (21). (iii) At $D \simeq \Delta$, we match the solutions of Eqs. (28) and (21).

In the case of $C_1 = C_2$, $\tilde{J}_1 = \tilde{J}_2$ ($= \tilde{J}$ in the previous section). In the general case of $C_1 \neq C_2$, however, we find that a line of $\tilde{J}_1 = \tilde{J}_2$ is not stable in the scaling by Eqs. (28). The renormalization flow goes to a fixed point of $J^{(1)} = J^{(2)} = \infty$, $(\tilde{J}_1/J^{(1)})^2 = 2(2 + \sqrt{5})$, $\tilde{J}_2 = 0$, and $J'/J^{(1)} = -2(1 + \sqrt{5})$, as the energy scale D decreases to T_K. This is the same fixed point as discussed in Ref. [30] with $\tilde{J}_2 = 0$ and $J^{(1)} = J^{(2)}$. If the exchange couplings are expanded around the fixed point to the first order of $1/\ln(D/T_K)$, Δ dependence of T_K shows a power law with the universal exponent; $\gamma = 2 + \sqrt{5}$ [30]. However, Eqs. (28) have been derived by the perturbation, which is valid for νJ's $\lesssim 1$ and $D \gg |\Delta|$, and hence the range of energy scale should be limited where the physical properties are determined by the fixed point.

We examine a realistic situation by solving the scaling equations, (28) and (21), numerically and evaluate T_K as a function of Δ. As the initial conditions, we choose

$\nu(J^{(1)} + J^{(2)}) = 0.1.$ [5] This corresponds to the experimental situation in which $E_c \approx$ 10K and line width $\Gamma \approx 1.5$K [15]. The Kondo temperature is determined as the energy scale D at which $\nu(J^{(1)} + J^{(2)}) = 1$. We denote the Kondo temperature obtained in the previous section with $C_1 = C_2$ in Eq. (11) by $T_{K0}(\Delta)$.

Figure 6(a) shows the Kondo temperature T_K, normalized by $T_{K0}(0)$, as a function of Δ. The case with $C_1 = C_2$ ($T_{K0}(\Delta)$) is presented by solid lines, $C_2 = 0$ by broken lines, and $C_2/C_1 = 0.6$ by dotted lines. We also change $J^{(2)}/J^{(1)}$ ($= \tan^2\theta$): *a*, 1 ($\theta/\pi = 0.25$); *b*, 0.5 ($\theta/\pi = 0.2$); and *c*, 0.3 ($\theta/\pi = 0.15$). In all the cases, the Kondo temperature increases with decreasing Δ. For large Δ, Δ dependence of T_K is almost the same as $T_{K0}(\Delta)$, irrespectively of C_2/C_1. With decreasing Δ, $T_K(\Delta)$ deviates gradually from $T_{K0}(\Delta)$ when $C_1 \neq C_2$. Figure 6(b) shows its logarithmic-logarithmic plot. The exponent is a nonuniversal value of $J^{(2)}/J^{(1)}$ at large Δ and increases slowly to the universal value of $2 + \sqrt{5}$ (≈ 4.2) with decreasing Δ. The universal exponent can be seen in quite limited situations.

We find that $T_K(\Delta)$ is minimum with $C_1 = C_2$ ($T_{K0}(\Delta)$) and maximum with $C_2 = 0$. In a case of $C_2 = 0$, the development of the exchange couplings is restricted in a subspace with $\tilde{J}_2(D) = 0$, and in consequence they reach the fixed point fastest. As the initial values are deviated from this condition, the renormalization flow to the fixed point needs more "time" in a larger space. Since our previous model of $C_1 = C_2$ is the farthest from the condition of $\tilde{J}_2(D) = 0$, $T_{K0}(\Delta)$ yields a lower limit of $T_K(\Delta)$.

4. Mean-field theory of the Kondo effect

In this section, we present the mean-field theory of the Kondo effect. It is useful to qualitatively understand the enhancement of the Kondo effect due to the competition between the spin-triplet and singlet states.

The mean-field theory of the Kondo effect was pioneered by Yoshimori and Sakurai [32] and is commonly used for the Kondo lattice model [33]. It enables us to capture main qualitative features of the Kondo effect; renormalizability at the scale of T_K, resonance at the Fermi level, resonant transmission, etc. The Kondo resonant state which appears at $T < T_K$ consists of the dot states $|SM\rangle = f^\dagger_{SM}|0\rangle$ and the Fermi sea of conduction electrons $\Pi c^\dagger_{k\sigma}|0\rangle$. We take into account the spin couplings between them by the mean field, $\langle f^\dagger_{SM} c_{k\sigma}\rangle$, neglecting their fluctuations. [6] These spin couplings give rise to resonant states around the Fermi level μ with the width of the order of T_K. The conduction electrons can be transported through the resonant levels, which yields the unitary limit of the conductance $G \sim 2e^2/h$.

---

[5] Initially, $\tilde{J}_i/(J^{(1)} + J^{(2)}) = C_i\lambda/(1+\lambda^2)$, where $\lambda = \sqrt{J^{(2)}/J^{(1)}}$, and $J'/(J^{(1)} + J^{(2)}) = (-C_1^2 + C_2^2)/2$, assuming that $E^+ = E^-$.

[6] This mean-field theory is equivalent to the mean-field theory using the slave bosons for $U = \infty$ Anderson model in the Kondo region [9]. This method is exact in the large-N limit when the dot state is N-fold degenerate.

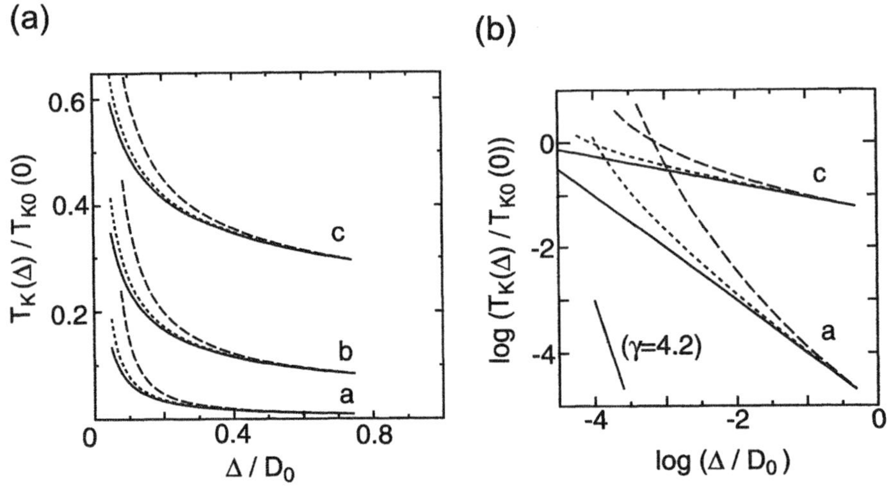

Fig. 6: (a) Numerical results of the Kondo temperature T_K as a function of Δ, in an extended model for the "triplet-singlet Kondo effect" (section 3.3), and (b) its logarithmic-logarithmic plot. T_K is normalized by $T_{K0}(0) = D_0 \exp[-1/2\nu(J^{(1)} + J^{(2)})]$. The case with $C_1 = C_2$ is drawn by solid lines, $C_2 = 0$ by broken lines, and $C_2/C_1 = 0.6$ by dotted lines. $J^{(2)}/J^{(1)} = \tan^2\theta$ with a, $\theta/\pi = 0.25$; b, 0.2; and c, 0.15. D_0 is the bandwidth in the leads. In (b), a slope of $\gamma = 2 + \sqrt{5} \approx 4.2$ is indicated.

4.1 Kondo resonance for spin $S = 1/2$

To illustrate the mean-field theory for the Kondo effect in quantum dots, we begin with the usual case of $S = 1/2$. We rewrite the Hamiltonian, Eq. (5), using the pseudofermion operators, $f_\sigma^\dagger$ (f_σ) which create (annihilate) the state $|\sigma\rangle$ ($\sigma = \uparrow, \downarrow$). By replacing the spin operator $\hat{S}$ by $\hat{S}_+ = f_\uparrow^\dagger f_\downarrow$, $\hat{S}_- = f_\downarrow^\dagger f_\uparrow$, $\hat{S}_z = (f_\uparrow^\dagger f_\uparrow - f_\downarrow^\dagger f_\downarrow)/2$, Eq. (5) becomes

$$H = \sum_{k\sigma} \varepsilon_k c_{k\sigma}^\dagger c_{k\sigma} + \sum_\sigma E_\sigma f_\sigma^\dagger f_\sigma + J \sum_{kk'} \sum_{\sigma,\sigma'} f_\sigma^\dagger f_{\sigma'} c_{k'\sigma'}^\dagger c_{k\sigma}. \tag{29}$$

The constraint

$$f_\uparrow^\dagger f_\uparrow + f_\downarrow^\dagger f_\downarrow = 1 \tag{30}$$

is required. In the second term in Eq. (29), we have included the Zeeman splitting, $E_{\uparrow,\downarrow} = \varepsilon_0 \pm E_Z$.

In the mean-field theory, we introduce the order parameter

$$\langle \Xi \rangle = \frac{1}{\sqrt{2}} \sum_k (\langle f_\uparrow^\dagger c_{k\uparrow} \rangle + \langle f_\downarrow^\dagger c_{k\downarrow} \rangle) \tag{31}$$

to describe the spin couplings between the dot states and conduction electrons. The mean-field Hamiltonian reads

$$H_{\rm MF} = \sum_{k\sigma} \varepsilon_k c^\dagger_{k\sigma} c_{k\sigma} + \sum_\sigma E_\sigma f^\dagger_\sigma f_\sigma - \sum_{k,\sigma} (\sqrt{2} J \langle \Xi \rangle c^\dagger_{k\sigma} f_\sigma + \text{H.c.}) + 2J|\langle \Xi \rangle|^2 + \lambda \left(\sum_\sigma f^\dagger_\sigma f_\sigma - 1 \right). \tag{32}$$

The constraint, Eq. (30), is taken into account by the last term with a Lagrange multiplier λ. By minimizing the expectation value of $H_{\rm MF}$, $\langle \Xi \rangle$ is determined self-consistently (Appendix A).

In the absence of the Zeeman effect, $E_\uparrow = E_\downarrow = \varepsilon_0$. The mean-field Hamiltonian, $H_{\rm MF}$, represents a resonant tunneling through an "energy level," $\tilde{E}_0 = \varepsilon_0 + \lambda$, with "tunneling coupling," $\tilde{V} = -\sqrt{2} J \langle \Xi \rangle$. $\tilde{V}$ provides a finite width of the resonance, $\tilde{\Delta}_0 = \pi\nu|\tilde{V}|^2$, with ν being the density of states in the leads. The constraint, Eq. (30), requires that the states for the pseudofermions are half-filled, that is, $\tilde{E}_0 = \mu$. Hence the Kondo resonant state appears just at the Fermi level μ, as indicated in the inset (A) in Fig. 7(a). The self-consistent calculations give us the resonant width

$$\tilde{\Delta}_0 = \pi\nu \left| \sqrt{2} J \langle \Xi \rangle \right|^2 = D_0 \exp[-1/2\nu J]. \tag{33}$$

This is identical to the Kondo temperature $T_{\rm K}$.

In the presence of the Zeeman splitting, $E_\uparrow = \varepsilon_0 - E_{\rm Z}$ and $E_\downarrow = \varepsilon_0 + E_{\rm Z}$. Hence the resonant level is split for spin-up and -down electrons, $\tilde{E}_{\uparrow/\downarrow} = E_{\uparrow/\downarrow} + \lambda$. The constraint, Eq. (30), yields $\varepsilon_0 + \lambda = \mu$ (see inset (B) in Fig. 7(a)). The resonant width $\tilde{\Delta}$ is determined as

$$\tilde{\Delta}^2 + E_{\rm Z}^2 = \tilde{\Delta}_0^2, \tag{34}$$

where $\tilde{\Delta}_0$ is given by Eq. (33). The Kondo temperature is evaluated by this width, $T_{\rm K}(E_{\rm Z}) = \tilde{\Delta}$. $T_{\rm K}$ decreases with increasing $E_{\rm Z}$ and disappears at $E_{\rm Z} = T_{\rm K}(0)$, as shown in Fig. 7(a).

The conductance G through the dot is expressed, using $\Gamma_\alpha = \pi\nu|V_\alpha|^2$, as

$$G = \frac{2e^2}{h} \frac{4\Gamma_L \Gamma_R}{(\Gamma_L + \Gamma_R)^2} \left[1 - \left(\frac{E_{\rm Z}}{T_{\rm K}(0)} \right)^2 \right] \tag{35}$$

(Appendix A). This is the conductance in the unitary limit for $E_{\rm Z} = 0$. Figure 7(b) presents the $E_{\rm Z}$ dependence of the conductance. With increasing $E_{\rm Z}$, the splitting between the resonant levels for spin-up and -down becomes larger. In consequence the amplitude of the Kondo resonance decreases at μ, which reduces the conductance.

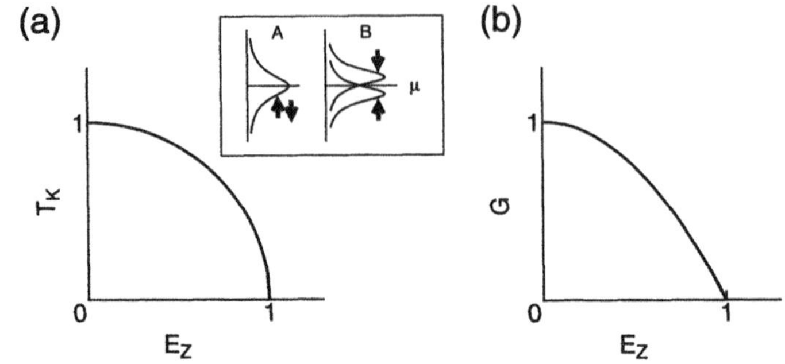

Fig. 7: The mean-field calculations for the Kondo effect in a quantum dot with $S = 1/2$. (a) The Kondo temperature $T_{\rm K}$ and (b) conductance through the dot, G, as functions of the Zeeman splitting $E_{\rm Z}$. $T_{\rm K}$ and $E_{\rm Z}$ are in units of $D_0 \exp(-1/2\nu J)$ and G is in units of $(2e^2/h)\cdot 4\Gamma_L\Gamma_R/(\Gamma_L+\Gamma_R)^2$. Inset in (a): The Kondo resonant states created around the Fermi level μ in the leads, (A) in the absence and (B) presence of the Zeeman splitting. The resonant width is given by $T_{\rm K}$.

4.2 Kondo resonance in the present model

Now we apply the mean-field theory to our model which has spin-triplet and -singlet states in a quantum dot. The discussion is restricted to the case of $C_1 = C_2$ in Eq. (11). The spin states of the coupling to a conduction electron are $(S = 1) \otimes (S = 1/2) = (S = 3/2) \oplus (S = 1/2)$ for the triplet state in the dot, and $(S = 0) \otimes (S = 1/2) = (S = 1/2)$ for the singlet state in the dot (Appendix B). To represent the competition between the triplet and single states, therefore, the order parameter should be a spinor of $S = 1/2$. It is $\langle \Xi \rangle$ where

$$\Xi = \sum_k \begin{pmatrix} \cos\varphi \left(\sqrt{2} f^\dagger_{11} c^{(1)}_{k\uparrow} + f^\dagger_{10} c^{(1)}_{k\downarrow}\right)/\sqrt{3} + \sin\varphi f^\dagger_{00} c^{(2)}_{k\downarrow} \\ \cos\varphi \left(\sqrt{2} f^\dagger_{1-1} c^{(1)}_{k\downarrow} + f^\dagger_{10} c^{(1)}_{k\uparrow}\right)/\sqrt{3} - \sin\varphi f^\dagger_{00} c^{(2)}_{k\uparrow} \end{pmatrix} \tag{36}$$

for $J^{(1)} > J^{(2)}$. A mode of the largest coupling is taken into account in this approximation. The Hamiltonian reads

$$H_{\rm MF} = H_{\rm lead} + H_{\rm dot} - J_{\rm MF}\left[\langle \Xi^\dagger \rangle \Xi + \Xi^\dagger \langle \Xi \rangle - |\langle \Xi \rangle|^2\right] + \lambda \left(\sum_{SM} f^\dagger_{SM} f_{SM} - 1\right), \tag{37}$$

where

$$J_{\rm MF} = J^{(1)} + \sqrt{J^{(1)2} + 3\tilde{J}^2}, \tag{38}$$

and

$$\tan\varphi = \sqrt{3}\tilde{J}/J_{\rm MF}. \tag{39}$$

The last term in $H_{\rm MF}$ considers the restriction of Eq. (16). The expectation value of $H_{\rm MF}$ is minimized with respect to $|\Xi|^2$. The Kondo temperature can be estimated by

$$T_{\rm K} = \pi\nu|J_{\rm MF}\langle\Xi\rangle|^2, \tag{40}$$

using $\langle\Xi\rangle$ determined by the self-consistent calculations (Appendix B).

The resonant level for the triplet state is threefold degenerate at $\tilde{E}_{S=1} = E_{S=1}+\lambda$, whereas the resonant level for the singlet state is at $\tilde{E}_{S=0} = E_{S=0}+\lambda$. These levels are separated by the energy Δ. The Lagrange multiplier λ is determined to fulfill Eq. (16). Figure 8(a) shows the calculated results of $T_{\rm K}$ as a function of Δ. Both of $T_{\rm K}$ and Δ are in units of $D_0\exp(-1/\nu J_{\rm MF})$. We find that (i) $T_{\rm K}(\Delta)$ reaches its maximum at $\Delta = 0$, (ii) for $\Delta \gg T_{\rm K}(0)$, $T_{\rm K}(\Delta)$ obeys a power law

$$T_{\rm K}(\Delta)\cdot\Delta^{\tan^2\varphi} = \text{const.}, \tag{41}$$

and (iii) for $\Delta < 0$, $T_{\rm K}$ decreases rapidly with increasing $|\Delta|$ and disappears at $\Delta = \Delta_{\rm c} \sim -T_{\rm K}(0)$:

$$\Delta_{\rm c} = -D_0\exp(-1/\nu J_{\rm MF})(1+\tan^2\varphi)(\tan^2\varphi)^{-\sin^2\varphi}. \tag{42}$$

These features are in agreement with the results of the scaling calculations.

The behaviors of $T_{\rm K}(\Delta)$ can be understood as follows. The inset of Fig. 8(a) schematically shows the Kondo resonant states. The resonance of the triplet state is denoted by solid lines, whereas that of the singlet state is by dotted lines. (A) When $\Delta \gg T_{\rm K}(0)$, the triplet resonance appears around μ, whereas the singlet resonance is far above μ. (B) With a decrease in Δ, the two resonant states are more overlapped at μ, which raises $T_{\rm K}$ gradually. This results in a power law of $T_{\rm K}(\Delta)$, Eq. (41). The largest overlap yields the maximum of $T_{\rm K}$ at $\Delta = 0$. (C) When $\Delta < 0$, the singlet and triplet resonances are located below and above μ, respectively, being sharper and farther from each other with increasing $|\Delta|$. Finally the Kondo resonance disappears at $\Delta = \Delta_{\rm c}$.

The conductance through the dot is given by

$$\begin{aligned} G/(e^2/h) = {} & \frac{4\Gamma_L^1\Gamma_R^1}{(\Gamma_L^1+\Gamma_R^1)^2}\left(\frac{\tilde{\Delta}_{11}^2}{(\varepsilon-\tilde{E}_{S=1})^2+\tilde{\Delta}_{11}^2}+\frac{\tilde{\Delta}_{10}^2}{(\varepsilon-\tilde{E}_{S=1})^2+\tilde{\Delta}_{10}^2}\right) \\ & + \frac{4\Gamma_L^2\Gamma_R^2}{(\Gamma_L^2+\Gamma_R^2)^2}\frac{\tilde{\Delta}_{00}^2}{(\varepsilon-\tilde{E}_{S=0})^2+\tilde{\Delta}_{00}^2}\Bigg|_{\varepsilon=\mu}, \end{aligned} \tag{43}$$

where $\Gamma_\alpha^i = \pi\nu|V_{\alpha,i}|^2$. The resonant widths are $\tilde{\Delta}_{11}/\tilde{\Delta}_0 = 2\cos^2\varphi/3$, $\tilde{\Delta}_{10}/\tilde{\Delta}_0 = \cos^2\varphi/3$, and $\tilde{\Delta}_{00}/\tilde{\Delta}_0 = \sin^2\varphi$ with $\tilde{\Delta}_0 = \pi\nu|J_{\rm MF}\langle\Xi\rangle|^2$. The conductance G as a function of Δ is shown in Fig. 8(b), in a symmetric case of $\Gamma_L^i = \Gamma_R^i$ $(i = 1,2)$. $G = 2e^2/h$ for $\Delta > 0$, whereas G goes to zero suddenly for $\Delta < 0$. Around $\Delta = 0$, G is larger than the value in the unitary limit, $2e^2/h$, which should be attributable to nonuniversal contribution from the multichannel nature of our model [21].

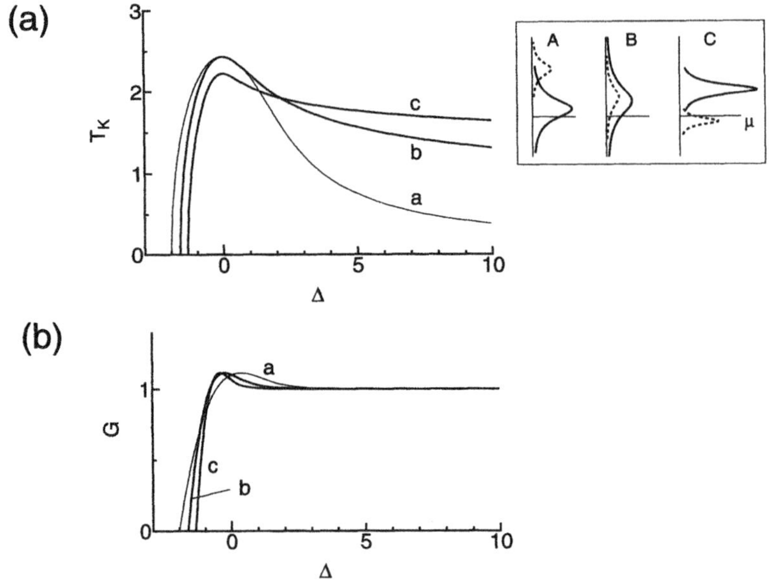

Fig. 8: The mean-field calculations for the Kondo effect in the model involving spin-singlet and -triplet states in a quantum dot. (a) The Kondo temperature T_K and (b) conductance through the dot, G, as functions of $\Delta = E_{S=0} - E_{S=1}$. T_K and Δ are in units of $D_0 \exp(-1/\nu J_{MF})$. G, in units of $2e^2/h$, is evaluated in a symmetric case of $\Gamma_L^i = \Gamma_R^i$ ($i = 1, 2$). $\tan\varphi = \sqrt{3}\tilde{J}/J_{MF}$ where *a*, $\varphi/\pi = 0.25$; *b*, 0.15; and *c*, 0.10. Note that $\varphi/\pi \leq 1/6$ in this approximation (case *a* is only for reference). Inset in (a): The Kondo resonant states for $S = 1$ (solid line) and for $S = 0$ (dotted line) when (A) $\Delta \gg T_K(0)$, (B) $\Delta \sim T_K(0)$, and (C) $\Delta < 0$.

We note that the mean-field theory is not quantitatively accurate for the evaluation of T_K. (In the case of $S = 1/2$, the exact value of T_K is obtained accidentally.) In our model, the scaling calculations indicate that all the exchange couplings, $J^{(1)}$, $J^{(2)}$, and $\tilde{J}$, are renormalized altogether following Eq. (20) for $D \gg |\Delta|$. In consequence two channels in the leads are coupled effectively for an increase in T_K. In the mean-field calculations, the interchannel couplings are taken into account in Eq. (38) only partly. In fact, conduction electrons of channel 1 and 2 independently take part in the conductance, Eq. (43). By the perturbation calculations with respect to the exchange couplings, we find that mixing terms between the channels appear in the logarithmic corrections to the conductance [21].

5. Conclusions and Discussion

The Kondo effect in quantum dots with an even number of electrons has been investigated theoretically. The Kondo temperature T_K has been calculated as a function of the energy difference $\Delta = E_{S=0} - E_{S=1}$, using the poor man's scaling method. We have found that the competition between the spin-triplet and -singlet states significantly enhances the Kondo effect: T_K is maximal around $\Delta = 0$ and decreases with increasing Δ. For $\Delta < 0$, T_K drops to zero suddenly at $\Delta \sim T_K(0)$. For $\Delta \gg T_K(0)$, $T_K(\Delta)$ shows a crossover behavior between power laws with a nonuniversal exponent ($\gamma = J^{(2)}/J^{(1)}$) and with a universal exponent ($\gamma = 2 + \sqrt{5}$). Our previous calculations [21,22] with $C_1 = C_2$ in the singlet state, Eq. (11), yield a lower limit of $T_K(\Delta)$ in an analytical form (Eqs. (23), (24) and (25)).

The mean-field theory yields a clear-cut view for the Kondo effect in quantum dots. Considering the spin couplings between the dot states and conduction electrons as a mean field, $\langle f^{\dagger}_{SM} c^{(i)}_{k,\sigma} \rangle$, we find that the resonant states are created around the Fermi level μ. The resonant width is given by the Kondo temperature T_K. The unitary limit of the conductance, $G \sim 2e^2/h$, can be easily understood in terms of the tunneling through these resonant states. In our model, the overlap between the resonant states of $S = 1$ and $S = 0$ in the dot enhances the Kondo effect. The mean-field calculations have led to a power-law dependence of T_K on Δ in accordance with the scaling calculations.

We have disregarded the Zeeman splitting of the spin-triplet state, E_Z, since this is a small energy scale in the experimental situation, $E_Z \ll T_K$ [15]. When the Zeeman splitting is relevant, another type of the Kondo effect can take place with an even number of electrons, as proposed by Pustilnik *et al.* [34]. They have considered "lateral" quantum dots with an even N, when the ground state is a spin singlet and the first excited state is a triplet. The Zeeman effect can reduce the energy of one component of the triplet state, $|11\rangle$, and finally makes it the ground state. At the critical magnetic field of $E_Z = -\Delta$, the energy of the state $|11\rangle$ is matched with that of the singlet state, $|00\rangle$. The Kondo effect arises from the degeneracy between the two states. A similar idea has been proposed by Giuliano and Tagliacozzo [35]. This type of the Kondo effect has been observed in carbon nanotubes with an even N under high magnetic field ($B \approx 1\text{T}$, $g = 2.0$; $E_Z \gg T_K$). We have also studied this type of Kondo effect using the scaling method and mean-field theory [22]. The mean-field theory is useful to examine the crossover between the regions where the Zeeman effect is irrelevant and relevant.

Acknowledgements

This work was done in collaboration with Yu. V. Nazarov, Delft University of Technology, The Netherlands. The author is indebted to L. P. Kouwenhoven, S. De Franceschi, J. M. Elzerman, K. Maijala, S. Sasaki, W. G. van der Wiel, Y. Tokura, L. I. Glazman, M. Pustilnik, and G. E. W. Bauer for valuable discussions. The author acknowledges financial support from the "Netherlandse Organisatie voor Wetenschappelijk Onderzoek" (NWO) and Japan Society for the Promotion of Science for his stay at Delft University of Technology.

6. Appendix

A Mean-field calculations for $S = 1/2$

The mean-field Hamiltonian, Eq. (32), includes "energy levels" for pseudofermions, $\tilde{E}_\sigma = E_\sigma + \lambda$, which are coupled to the leads with "tunneling amplitude," $\tilde{V} = -\sqrt{2}J\langle\Xi\rangle$. The Green function for the pseudofermions is

$$G_\sigma(\varepsilon) = \frac{1}{\varepsilon - \tilde{E}_\sigma + \mathrm{i}\tilde{\Delta}}, \tag{A.1}$$

where $\tilde{\Delta} = \pi\nu|\tilde{V}|^2$. This represents the resonant tunneling with the resonant width $\tilde{\Delta}$.

The expectation value of the Hamiltonian, Eq. (32), is written as

$$E_{\mathrm{MF}} = \sum_\sigma \left[-\frac{\tilde{\Delta}}{\pi} + \frac{\tilde{E}_\sigma}{\pi}\tan^{-1}\frac{\tilde{\Delta}}{\tilde{E}_\sigma} + \frac{\tilde{\Delta}}{2\pi}\ln\frac{\tilde{E}_\sigma^2 + \tilde{\Delta}^2}{D_0^2} \right] - \lambda + \frac{\tilde{\Delta}}{\pi\nu J}, \tag{A.2}$$

where D_0 is the bandwidth in the leads [9]. We set $\mu = 0$ in this appendix. The constraint of Eq. (30) is equivalent to the condition

$$\frac{\partial E_{\mathrm{MF}}}{\partial\lambda} = \frac{1}{\pi}\sum_\sigma \tan^{-1}\frac{\tilde{\Delta}}{\tilde{E}_\sigma} - 1 = 0. \tag{A.3}$$

This yields $\varepsilon_0 + \lambda = 0$. The minimization of E_{MF} with respect to $\tilde{\Delta}$ (or $|\langle\Xi\rangle|^2$) determines $\tilde{\Delta}$

$$\frac{\partial E_{\mathrm{MF}}}{\partial\tilde{\Delta}} = \frac{1}{2\pi}\sum_\sigma \ln\frac{\tilde{E}_\sigma^2 + \tilde{\Delta}^2}{D_0^2} + \frac{1}{\pi\nu J} = 0. \tag{A.4}$$

For $E_{\mathrm{Z}} = 0$, we find

$$\tilde{\Delta} = D_0 \exp[-1/2\nu J] \equiv \tilde{\Delta}_0. \tag{A.5}$$

This is equal to the Kondo temperature, T_{K}. For $E_{\mathrm{Z}} \neq 0$, Eq. (A.4) yields

$$\tilde{\Delta}^2 + E_{\mathrm{Z}}^2 = \tilde{\Delta}_0^2. \tag{A.6}$$

Using the T matrix, $\hat{T}$, the conductance through the dot, G, is given by

$$\begin{aligned} G &= \frac{e^2}{h}(2\pi\nu)^2 \sum_\sigma |\langle R, k'\sigma|\hat{T}|L, k\sigma\rangle|^2 \Big|_{\varepsilon_k=\varepsilon_{k'}=\mu} \\ &= \frac{e^2}{h}(2\pi\nu)^2 \frac{|V_L|^2|V_R|^2}{(|V_L|^2 + |V_R|^2)^2} \sum_\sigma |\langle\psi_{k'\sigma}|\hat{T}|\psi_{k\sigma}\rangle|^2 \Big|_{\varepsilon_k=\varepsilon_{k'}=\mu} \\ &= \frac{e^2}{h}\frac{4\Gamma_L\Gamma_R}{(\Gamma_L + \Gamma_R)^2} \sum_\sigma \frac{\tilde{\Delta}^2}{(\varepsilon - \tilde{E}_\sigma)^2 + \tilde{\Delta}^2}\Big|_{\varepsilon=\mu}, \end{aligned} \tag{A.7}$$

where $\Gamma_\alpha = \pi\nu|V_\alpha|^2$. This yields Eq. (35) in the text. On the second line in Eq. (A.7), $|\psi_{k\sigma}\rangle = c^\dagger_{k\sigma}|0\rangle = (V_L|L,k\sigma\rangle + V_R|R,k\sigma\rangle)/V$, and the T matrix is evaluated in terms of the Green function, Eq. (A.1), $|\tilde{V}|^2 G_\sigma(\varepsilon = \varepsilon_k)$.

B Mean-field calculations in the present model

For the spin states of the coupling between the spin triplet $S = 1$ in the dot and a conduction electron, we introduce spinors of $S = 1/2$ and $3/2$. Using the Clebsch-Gordan coefficients, they are given by

$$\boldsymbol{\Omega}^{(i)}_{1/2} = \sum_k \begin{pmatrix} (\sqrt{2} f^\dagger_{11} c^{(i)}_{k\uparrow} + f^\dagger_{10} c^{(i)}_{k\downarrow})/\sqrt{3} \\ (\sqrt{2} f^\dagger_{1-1} c^{(i)}_{k\downarrow} + f^\dagger_{10} c^{(i)}_{k\uparrow})/\sqrt{3} \end{pmatrix}, \tag{B.1}$$

$$\boldsymbol{\Omega}^{(i)}_{3/2} = \sum_k \begin{pmatrix} f^\dagger_{11} c^{(i)}_{k\downarrow} \\ (-f^\dagger_{11} c^{(i)}_{k\uparrow} + \sqrt{2} f^\dagger_{10} c^{(i)}_{k\downarrow})/\sqrt{3} \\ (f^\dagger_{1-1} c^{(i)}_{k\downarrow} - \sqrt{2} f^\dagger_{10} c^{(i)}_{k\uparrow})/\sqrt{3} \\ -f^\dagger_{1-1} c^{(i)}_{k\uparrow} \end{pmatrix}. \tag{B.2}$$

The exchange couplings between the triplet state and conduction electrons, Eq. (17), can be rewritten as

$$H^{S=1} = \sum_{i=1,2} J^{(i)} \left[-2\boldsymbol{\Omega}^{(i)\dagger}_{1/2} \boldsymbol{\Omega}^{(i)}_{1/2} + \boldsymbol{\Omega}^{(i)\dagger}_{3/2} \boldsymbol{\Omega}^{(i)}_{3/2} \right]. \tag{B.3}$$

In the same way we define the spinors of $S = 1/2$ to represent the spin couplings between the singlet state $S = 0$ and a conduction electron

$$\boldsymbol{\Psi}^{(i)} = \sum_k \begin{pmatrix} f^\dagger_{00} c^{(\bar{i})}_{k\downarrow} \\ -f^\dagger_{00} c^{(\bar{i})}_{k\uparrow} \end{pmatrix}, \tag{B.4}$$

where $\bar{i} = 2$ and 1 for $i = 1$ and 2, respectively. The conversion between the triplet and singlet states, Eq. (18), is rewritten as

$$H^{S=1\leftrightarrow 0} = -\sqrt{3}\tilde{J} \sum_{i=1,2} \left[\boldsymbol{\Psi}^{(i)\dagger} \boldsymbol{\Omega}^{(i)}_{1/2} + \text{H.c.} \right]. \tag{B.5}$$

In $H^{S=1} + H^{S=1\leftrightarrow 0}$, a mode of the largest coupling with $S = 1/2$ is given by

$$\boldsymbol{\Xi} = \cos\varphi \boldsymbol{\Omega}^{(1)}_{1/2} + \sin\varphi \boldsymbol{\Psi}^{(1)} \tag{B.6}$$

for $J^{(1)} \geq J^{(2)}$, which is Eq. (36) in the text. The corresponding eigenvalue is given by Eq. (38) and φ is determined as in Eq. (39).

The mean-field Hamiltonian, Eq. (37), represents the resonant tunneling through the energy levels for the pseudofermions, $\tilde{E}_S = E_S + \lambda$. The expectation value of Eq. (37), E_{MF}, is evaluated in the same way as in Appendix A. $\partial E_{\mathrm{MF}}/\partial\lambda = 0$ yields

$$\tan^{-1}\frac{\tilde{\Delta}_{11}}{\tilde{E}_{S=1}} + \tan^{-1}\frac{\tilde{\Delta}_{10}}{\tilde{E}_{S=1}} + \tan^{-1}\frac{\tilde{\Delta}_{00}}{\tilde{E}_{S=0}} = \pi, \tag{B.7}$$

where the resonant widths are $\tilde{\Delta}_{11}/\tilde{\Delta}_0 = 2\cos^2\varphi/3$, $\tilde{\Delta}_{10}/\tilde{\Delta}_0 = \cos^2\varphi/3$, and $\tilde{\Delta}_{00}/\tilde{\Delta}_0 = \sin^2\varphi$ with $\tilde{\Delta}_0 = \pi\nu|J_{\mathrm{MF}}\langle\Xi\rangle|^2$. We set $\mu = 0$ here. Minimizing E_{MF} with respect to $\tilde{\Delta}_0$, we obtain

$$\begin{aligned}\frac{2}{3}\cos^2\varphi\ln\frac{\tilde{E}_{S=1}^2+\tilde{\Delta}_{11}^2}{D_0^2} + \frac{1}{3}\cos^2\varphi\ln\frac{\tilde{E}_{S=1}^2+\tilde{\Delta}_{10}^2}{D_0^2}&\\ +\sin^2\varphi\ln\frac{\tilde{E}_{S=0}^2+\tilde{\Delta}_{00}^2}{D_0^2} + \frac{2}{\nu J} = 0.&\end{aligned} \tag{B.8}$$

Equations (B.7) and (B.8) determine λ and $\tilde{\Delta}_0$ (or $|\langle\Xi\rangle|^2$).

The conductance through the dot is given by

$$\begin{aligned}G &= \frac{e^2}{h}(2\pi\nu)^2 \sum_{i,j,\sigma,\sigma'} |\langle R, k'\sigma', j|\hat{T}|L, k\sigma, i\rangle|^2\Big|_{\varepsilon_k=\varepsilon_{k'}=\mu}\\ &= \frac{e^2}{h}(2\pi\nu)^2 \sum_{i,j,\sigma,\sigma'} \frac{\Gamma_R^j}{\Gamma_L^j+\Gamma_R^j}\frac{\Gamma_L^i}{\Gamma_L^i+\Gamma_R^i}|\langle\psi_{k'\sigma'}^{(j)}|\hat{T}|\psi_{k\sigma}^{(i)}\rangle|^2\Big|_{\varepsilon_k=\varepsilon_{k'}=\mu},\end{aligned} \tag{B.9}$$

where $\Gamma_\alpha^i = \pi\nu|V_{\alpha,i}|^2$ and $|\psi_{k\sigma}^{(i)}\rangle = (V_{L,i}|L, k\sigma, i\rangle + V_{R,i}|R, k\sigma, i\rangle)/V_i$. The T matrix can be evaluated, using the Green function for the pseudofermions, $G_{SM}(\varepsilon) = [\varepsilon - \tilde{E}_S + \mathrm{i}\tilde{\Delta}_{SM}]^{-1}$, as in Appendix A. This yields Eq. (43) in the text.

References

[1] *Mesoscopic Electron Transport*, NATO ASI Series E **345**, eds. L. Y. Sohn, L. P. Kouwenhoven and G. Schön (Kluwer, 1997).

[2] D. V. Averin and Yu. V. Nazarov, Phys. Rev. Lett. **65**, 2446 (1990).

[3] L. I. Glazman and M. É. Raikh, Pis'ma Zh. Eksp. Teor. Fiz. **47**, 378 (1988) [JETP Lett. **47**, 452 (1988)].

[4] T. K. Ng and P. A. Lee, Phys. Rev. Lett. **61**, 1768 (1988).

[5] A. Kawabata, J. Phys. Soc. Jpn. **60**, 3222 (1991).

[6] S. Hershfield, J. H. Davies, and J. W. Wilkins, Phys. Rev. Lett. **67**, 3720 (1991); Phys. Rev. B **46**, 7046 (1992).

[7] Y. Meir, N. S. Wingreen, and P. A. Lee, Phys. Rev. Lett. **70**, 2601 (1993).

[8] J. Kondo, Prog. Theor. Phys. **32**, 37 (1964).

[9] A. C. Hewson, *The Kondo Problem to Heavy Fermions* (Cambridge, Cambridge, England, 1993).

[10] K. Yosida, *Theory of Magnetism* (Springer, New York, 1996).

[11] D. Goldhaber-Gordon, H. Shtrikman, D. Mahalu, D. Abusch-Magder, U. Meirav, and M. A. Kastner, Nature (London) **391**, 156 (1998); D. Goldhaber-Gordon, J. Göres, M. A. Kastner, H. Shtrikman, D. Mahalu, and U. Meirav, Phys. Rev. Lett. **81**, 5225 (1998).

[12] S. M. Cronenwett, T. H. Oosterkamp, and L. P. Kouwenhoven, Science **281**, 540 (1998).

[13] F. Simmel, R. H. Blick, J. P. Kotthaus, W. Wegscheider, and M. Bichler, Phys. Rev. Lett. **83**, 804 (1999).

[14] W. G. van der Wiel, S. De Franceschi, T. Fujisawa, J. M. Elzerman, S. Tarucha, and L. P. Kouwenhoven, Science **289**, 2105 (2000).

[15] S. Sasaki, S. De Franceschi, J. M. Elzerman, W. G. van der Wiel, M. Eto, S. Tarucha, and L. P. Kouwenhoven, Nature (London) **405**, 764 (2000).

[16] J. Schmid, J. Weis, K. Eberl, and K. v. Klitzing, Phys. Rev. Lett. **84**, 5824 (2000).

[17] L. P. Kouwenhoven (private communications).

[18] J. Nygård, D. H. Cobden, and P. E. Lindelof, Nature (London) **408**, 342 (2000).

[19] S. Tarucha, D. G. Austing, T. Honda, R. J. van der Hage, and L. P. Kouwenhoven, Phys. Rev. Lett. **77**, 3613 (1996).

[20] L. P. Kouwenhoven, T. H. Oosterkamp, M. W. S. Danoesastro, M. Eto, D. G. Austing, T. Honda, and S. Tarucha, Science **278**, 1788 (1997).

[21] M. Eto and Yu. V. Nazarov, Phys. Rev. Lett. **85**, 1306 (2000).

[22] M. Eto and Yu. V. Nazarov, Phys. Rev. B **64**, 85322 (2001).

[23] M. Eto, Jpn. J. Appl. Phys. **40**, 1929 (2001).

[24] T. Inoshita, Science **281**, 526 (1998).

[25] P. W. Anderson, J. Phys. C **3**, 2436 (1970).

[26] P. Nozières and A. Blandin, J. Phys. (Paris) **41**, 193 (1980).

[27] D. L. Cox and A. Zawadowski, Adv. Phys. **47**, 599 (1998).

[28] F. D. M. Haldane, J. Phys. C **11**, 5015 (1978).

[29] I. Okada and K. Yosida, Prog. Theor. Phys. **49**, 1483 (1973).

[30] M. Pustilnik and L. I. Glazman, Phys. Rev. Lett. **85**, 2993 (2000).

[31] M. Eto and Yu. V. Nazarov, J. Phys. Chem. Solids, in press.

[32] A. Yoshimori and A. Sakurai, Suppl. Prog. Theor. Phys. **46**, 162 (1970).

[33] C. Lacroix and M. Cyrot, Phys. Rev. B **20**, 1969 (1979).

[34] M. Pustilnik, Y. Avishai, and K. Kikoin, Phys. Rev. Lett. **84**, 1756 (2000).

[35] D. Giuliano and A. Tagliacozzo, Phys. Rev. Lett. **84**, 4677 (2000).

Chapter 7

From single dots to interacting arrays

Vidar Gudmundsson, [a] Andrei Manolescu, [a,b] Roman Krahne, [c] and Detlef Heitmann [c]

[a] *Science Institute, University of Iceland, Dunhaga 3, IS-107 Reykjavik, Iceland, E-mail: vidar@raunvis.hi.is*

[b] *Institutul Naţional de Fizica Materialelor, C.P. MG-7 Bucureşti-Măgurele, România*

[c] *Institut für Angewandte Physik und Zentrum für Mikrostrukturforschung, Universität Hamburg, Jungiusstraße 11, D–20355 Hamburg, Germany*

Abstract

We explore the structural changes in the charge density and the electron configuration of quantum dots caused by the presence of other dots in an array, and the interaction of neighboring dots. We discuss what recent measurements and calculation of the far-infrared absorption reveal about almost isolated quantum dots and investigate some aspects of the complex transition from isolated dots to dots with strongly overlapping electron density. We also address the the effects on the magnetization of such dot array.

1. Introduction

Arrays of quantum dots of different shapes and sizes have been explored by far-infrared (FIR) absorption measurements and Raman scattering for a decade by many research groups. The main reason for using arrays has been the need to increase the signal strength of the tiny quantum dots in the weak radiation field applied, whose wavelength is up to 4 orders of magnitude larger than the dots. For lithographically prepared and etched quantum dots no evidence has been found for interaction between the dots on the length scale made available by laser holography for periodic structures. Recently, experiments on field-effect-confined quantum dots in $Al_xGa_{1-x}/GaAs$ heterostructures have yielded signs that have been interpreted as being caused by the periodicity of the confinement potential of the array [1]. Evidence of direct interaction between dots in this same system have also been found [2]. Here we shall review these two cases together with the model calculations used for their interpretation. Such inter-dot interaction had so far only been observed for adjacent large, 20-micron-size, 2D-electron disks in microwave experiments [3].

For the parameters available in field-effect-induced arrays of quantum dots with, typically, a lattice length of few hundred nanometers the interaction effects are expected to be small on the scale of the confinement energy $\hbar\Omega$. As this energy anyways lies in the range of few meV, where the low experimental sensitivity makes measurements challenging, it can be expected that mainly interaction effects leading to changes in the shape of the dots can be detected. With this in mind we explore numerically the subtle effects of the interaction on the ground state of elliptic dots in arrays with a bit shorter lattice length, than is now common in experiments. In addition, we consider the effects on the FIR absorption and the orbital magnetization of the dots. The magnetoplasmon excitations in arrays of circular and noncircular quantum dots have been studied by Zyl et al. in the Thomas-Fermi-Dirac-von Weizsäker semiclassical approximation [4]. They study the deviations from the ideal collective excitations of isolated parabolically confined quantum dots caused by the local perturbation of the confining potential as well as the interdot Coulomb interaction and find the latter indeed to be unimportant unless the interdot separation is of the order of the size of the dots. An analytical model of parabolically confined electrons has been presented with a simplified inter-dot interaction. The model predicts shifts of collective modes and appearance of other modes that are not dipole active [5].

Traditionally, experimental results on the FIR absorption of quantum dots in AlGaAs/GaAs heterostructure have successfully been interpreted in terms of a model of an isolated quantum dot with confinement potential that is parabolic or steeper. The pure parabolic confinement is caused by uniformly distributed ionized donors in the AlGaAs layer that have supplied their electrons to the active dot layer. Furthermore, the extension of the Kohn theorem explains why only center-of-mass modes can be excited in such dots with FIR radiation with wavelength much larger than the dot size [6–8]. Dots satisfying these criteria thus only show a single absorption peak at the frequency of the naked parabolic confinement. In magnetic field the peak

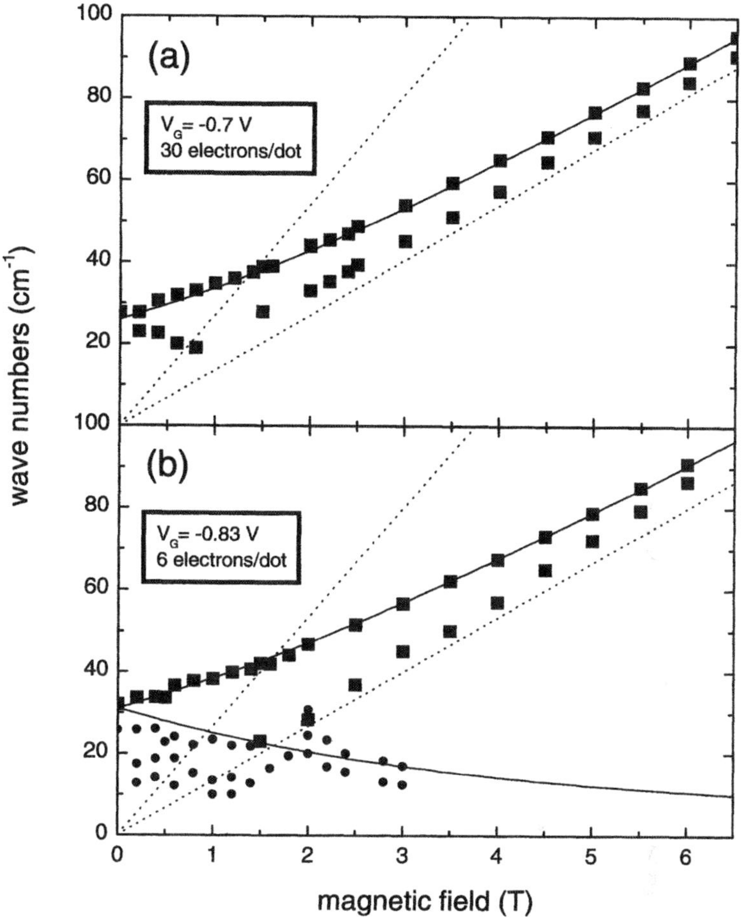

Fig. 1: Experimental dispersion for quantum dots with (a) 30 electrons and (b) 6 electrons. Full lines are fits with the Kohn modes of Eq. (1), the dotted lines are ω_c and $2\omega_c$ extracted from this fit. A new mode, the below-Kohn mode, is observed *below* the upper Kohn mode but clearly *above* ω_c.

splits into two peaks, one approaching the cyclotron frequency from above with increasing magnetic field strength B, and the other decreasing in frequency. The two collective modes are excited with FIR radiation with opposite circular polarization. In accordance with present possibilities in sample preparation, or dot design, the most common deviations form the simple circular parabolic confinement studied have been; elliptic dots, dots with weak square symmetry, and dots with quartic or stronger confinement. Elliptic shape of dots shows up as a simple splitting of the absorption peak at $B = 0$ [9,10], and the square shape produces a characteristic splitting in the upper Kohn mode at finite magnetic field [8,11]. The stronger confinement can produce a trivial blue shift and weaker absorption peaks with magnetic

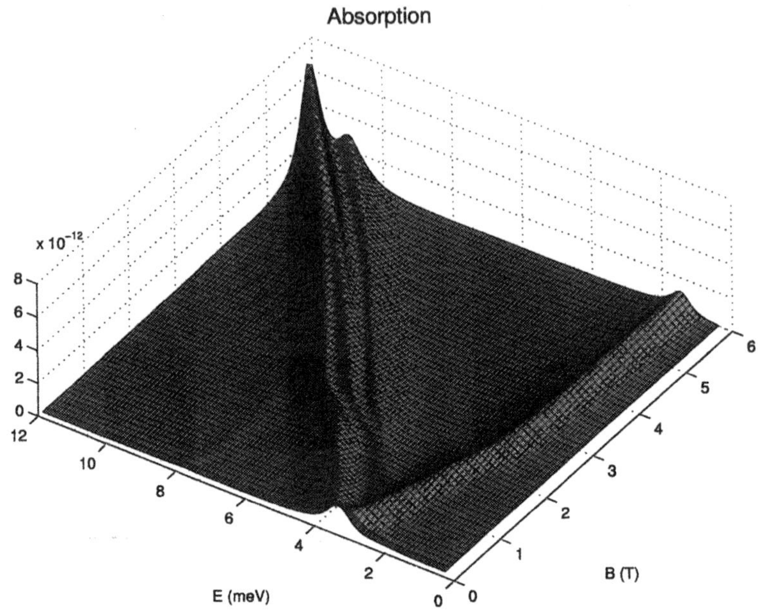

Fig. 2: Calculated dipole absorption for a quantum dot with 5 electrons in a flattened potential described in the text. In addition to the strong Kohn modes new modes below the high-frequency Kohn mode are found also in the calculation. The half-linewidth is 0.3 meV and $T = 1$ K.

dispersion almost parallel to and above the upper Kohn mode [12,13]. In addition, in confinement potentials that do not fullfil the criteria for the Kohn theorem so called Bernstein modes are excited causing characteristic splitting in the upper Kohn mode around higher harmonics of the cyclotron frequency [14,15]. Interestingly enough, researchers have been able to produce ring shaped quantum dots and measure their FIR absorption, but these do not form regular arrays [16–19].

2. Effects of an array

The simplest effects of an array of quantum dots on the confinement potential of an individual dot would be the eventual flattening of the potential imposed by the periodicity of the array. In field-induced dot arrays, where each dot contains only few electrons, it must be possible to have the confinement potential shallow enough that at least electrons in the excited states are affected by this weakening of the confinement. This has been demonstrated by Krahne et al. [1]. The experimental dispersion curves are seen in Fig. 1 for 6 or 30 electrons in each quantum dot. A purely parabolically confined quantum dot has the FIR dispersion of the Kohn modes

$$\omega_{\pm} = \sqrt{\Omega_0^2 + (\omega_c/2)^2} \pm \omega_c/2, \qquad (1)$$

where Ω_0 is the confinement frequency, and $\omega_c = eB/(m^*c)$ is the cyclotron frequency. We have fitted this dispersion with the lower absorption branch and the sharper upper one in the experimental dispersion curves displayed in Fig. 1. In addition to these two branches the experiments show a third branch just below ω_+ however, above the cyclotron frequency ω_c.

It is well known that the energy spectrum of electrons in a periodic lattice can be calculated only for a a magnetic flux commensurable with the unit cell [20]. It is technically difficult to vary the magnetic field continuously to describe the experimental results for an array of dots with interacting electrons [21,22]. We thus resort to a model of an individual quantum dot in the Hartree approximation, but with a potential of the form

$$V(x) = ax^2 + bx^4 + W(x), \tag{2}$$

where $x = r/a_0^*$ is the radial coordinate scaled by the effective Bohr radius $a_0^* = 9.77$ nm in GaAs and

$$W(x) = c\left[1 - f\left(3.9x - 12\right)\right], \tag{3}$$

with $f(x) = 1/(\exp(x) + 1)$. The calculated FIR absorption is shown in Fig. 2 for the parameter choice $a = 0.48$ meV, $b = -1.8^{-3}$ meV, and $c = 6$ meV. These parameters have been selected to give results qualitatively close to the experiment, without performing an actual fit. The model yields a mode just below ω_+ as is seen in the experiment. At low magnetic field the upper Kohn branch, ω_+, has a complex splitting around $\omega = 2\omega_c$ that is dependent on the number of electrons present and for a higher number of electrons develops into a splitting caused by a Bernstein mode [14,15]. At high magnetic field the induced density of the ω_+ mode indeed reflects a center-of-mass mode, but the lower mode, the new one, is the lowest internal mode with one node in the center of the dot [1]. For a confinement stronger than the parabolic one (for example, with $b > 0$ and $c = 0$) this internal mode is usually found above the upper Kohn mode, but here due to the special confinement it has lower energy.

This is clearly an effect of the shape of the confinement potential of an individual quantum dot in an array, but what about a direct interaction between dots?

3. Interaction between dots

3.1 Experimental indications

Indications for interaction between quantum dots have been found in the same type of system when the dots have been prepared to have an elliptic symmetry rather than the circular one [2]. In elliptical quantum dots the rotational symmetry is broken and the degeneracy of the ω_+ and ω_- modes is lifted at $B = 0$. The dispersion of the FIR absorption peaks in a single elliptic quantum dot with parabolic confinement

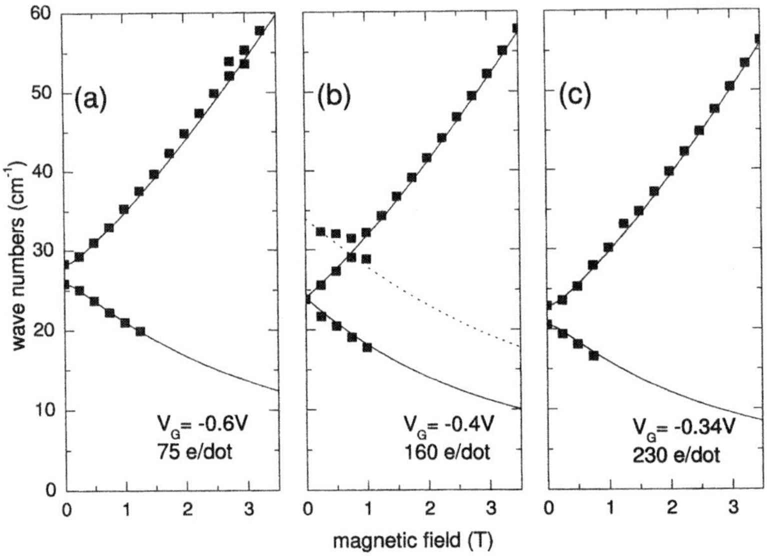

Fig. 3: Magnetic field dispersions for three different gate voltages. The experimental resonance positions extracted from the spectra are depicted by the full squares. The solid lines show a calculation according to Eq. (4) with $\omega_{x(y)}$ as fit parameters. (a) $V_G = -0.6$ V: strong confinement leading to isolated dots. (b) $V_G = -0.4$ V: weaker confinement. Here an anticrossing of the ω_+ mode around $B = 1$ T with another weak resonance, which decreases in energy with increasing B, is observed. This anticrossing is behavior is a characteristic property of square symmetric quantum dots [11]. (c) $V_G = -0.34$ V: weak confinement leading to overlapping dots.

is described by

$$\omega_\pm^2 = \frac{1}{2}\left(\omega_x^2 + \omega_y^2 + \omega_c^2 \pm \sqrt{\omega_c^4 + 2\omega_c^2(\omega_x^2 + \omega_y^2) + (\omega_x^2 - \omega_y^2)^2}\,\right), \tag{4}$$

where ω_x and ω_y are, respectively, the resonance frequencies for the oscillation in the x and y direction at $B = 0$, the two symmetry axis of the dot. Let us consider the long axis of the ellipse to be aligned with the y axis of the coordinate system. The two modes $\omega_+(B = 0) = \omega_x$ and $\omega_-(B = 0) = \omega_y$ are observed with orthogonal linear polarization [2]. Figure 3 shows the magnetic dispersion of the absorption peaks for 3 different values of the gate voltages which is used to control the number of electrons in each quantum dot. For few electrons, Fig. 3(a), the dispersion with two peaks at $B = 0$ reflects the elliptic shape of the quantum dots. Curiously enough, for a higher number of electrons, Fig. 3(b), the peaks at $B = 0$ are degenerate and the curves conform with the dispersion measured [11] and calculated [8] for square shape dots with a characteristic anticrossing in the ω_+ mode at a nonvanishing magnetic field. At even higher electron number, Fig. 3(c), the characteristic traits of the square shape are lost, no anticrossing at finite B and no degeneracy of the modes at $B = 0$ is discernible, but now the dispersion can be well fitted by the formula for elliptic

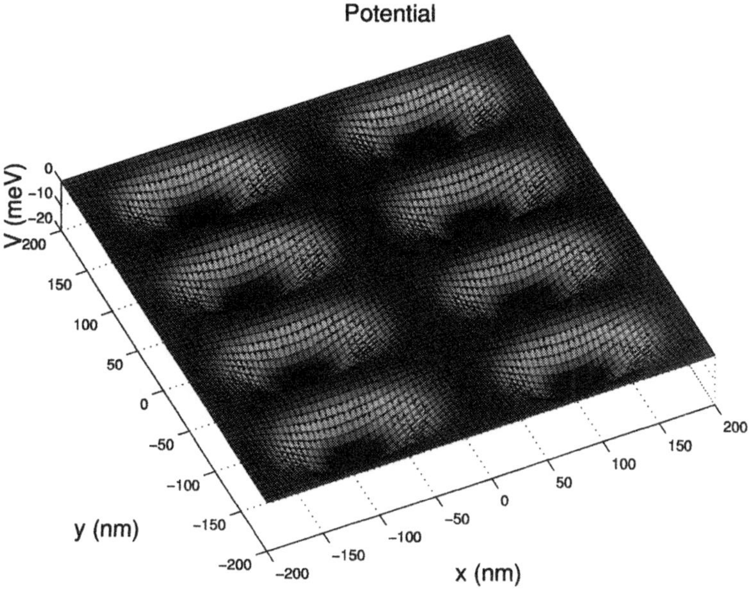

Fig. 4: The periodic confinement potential $V_{\rm QAD}$ for a dot array with aspect ratio 1:2. See equation (5) in text.

dots (4) again. Linearly polarized measurements show that for all gate voltages the energetically higher excitation is polarized in x direction and the energetically lower in y direction. This shows that no rotation of the ellipse out of some electrostatic reason takes place. Thus the actual geometrical shape of the dots must be deformed by some interaction with their neighbors in dependence of the gate voltage.

3.2 Model results for interacting dots: Ground state properties

We model an array of quantum dots as interacting electrons in a periodic potential. We choose to describe the interaction of electrons within a dot at the same level as the interaction of electrons in different dots. In order to distinguish between the effects of different parts of the interaction, the direct one, the exchange, and correlation part, we use density functional theory (DFT) approach in the local spin density approximation (LSDA) to be described below. We describe a simple array of circular or elliptic dots (or antidots) shaped by the potential

$$V_{\rm QAD}(\mathbf{r}) = V_0 \left[\sin\left(\frac{g_1 x}{2}\right) \sin\left(\frac{g_2 y}{2}\right)\right]^2, \qquad (5)$$

where g_i is the length of the fundamental inverse lattice vectors, $\mathbf{g}_1 = 2\pi\hat{\mathbf{x}}/L_x$, and $\mathbf{g}_2 = 2\pi\hat{\mathbf{y}}/L_y$. The Bravais lattice defined by $V_{\rm QAD}$ has period lengths L_x, L_y, and the inverse lattice is spanned by $\mathbf{G} = G_1\mathbf{g}_1 + G_2\mathbf{g}_2$, with $G_1, G_2 \in \mathbf{Z}$. The commensurability condition between the magnetic length $\ell = (\hbar c/(eB))^{1/2}$ and the periods L_i requires magnetic-field values of the form $B = pq\phi_0/L_xL_y$, with $pq \in \mathbf{N}$

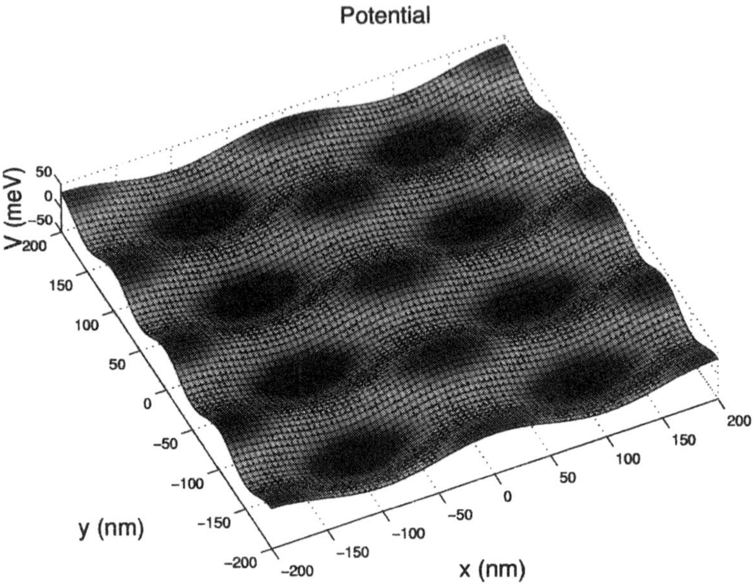

Fig. 5: The simple cosine confinement potential V_{per} with aspect ratio 1:2. See equation (6) in text.

flux quanta, $\phi_0 = hc/e$, in a unit cell [23,21]. Arbitrary rational values can, in principle, be obtained by resizing the unit cell in the Bravais lattice, but numerically this is quite difficult. The term 'circular quantum dot' can of course only describe a dot with few electrons in a square lattice where the electron density is concentrated in the middle of the cell and vanishes well before the cell edge. As the number of electrons increases the electron density must reflect the symmetry of the lattice. The dot potential (5) is seen in Fig. 4. We investigate different confinement potentials later on in order to understand better the interaction between dots. One of interest will be the simple periodic cosine potential

$$V_{\text{per}}(\mathbf{r}) = V_0 \cos(g_1 x) + V_0 \cos(g_2 y), \tag{6}$$

shown in Fig. 5 for an array of elliptic dots.

The exchange-correlation energy per particle, $\epsilon_{xc}(\nu, \xi)$, is parameterized in terms of the total filling factor $\nu = \nu_\uparrow + \nu_\downarrow = 2\pi\ell^2 n$ and the spin polarization

$$\xi = \frac{\nu_\uparrow - \nu_\downarrow}{\nu_\uparrow + \nu_\downarrow} \tag{7}$$

rather than the spin densities $n_\uparrow$ and $n_\downarrow$ [24]. The exchange-correlation potentials are then

$$V_{xc\updownarrow} = \frac{\partial}{\partial \nu}(\nu\epsilon_{xc}) \pm (1 \mp \xi)\frac{\partial}{\partial \xi}\epsilon_{xc}, \tag{8}$$

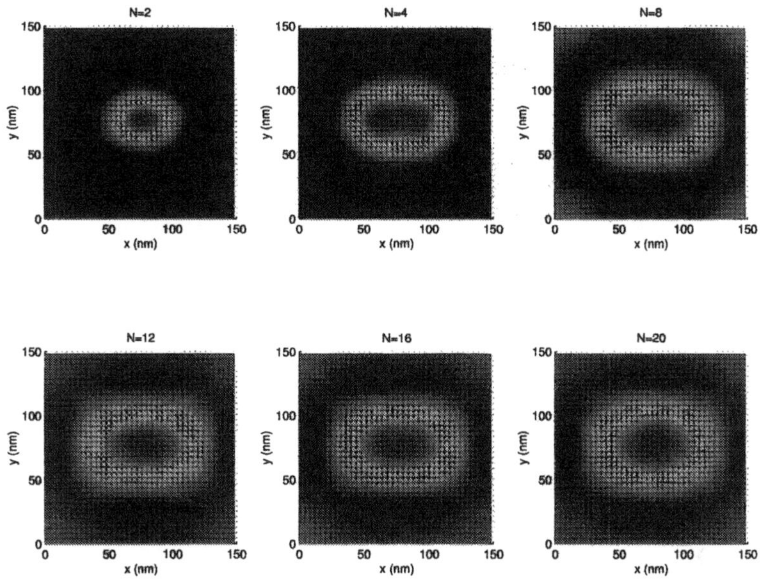

Fig. 6: The electron density distribution for the ground state of interacting elliptical dots in a square lattice. The confinement is according to Eq. (10). $B = 1.654$ T, $T = 1$ K, $V_0 = -16$ meV.

and the exchange-correlation energy is interpolated as

$$\epsilon_{xc}(\nu, \xi) = \epsilon_{xc}^{\infty}(\nu) e^{-f(\nu)} + \epsilon_{xc}^{\mathrm{TC}}(1 - e^{(1-f(\nu))}) \quad \text{with} \quad f(\nu) = 1.5\nu + 7\nu^4 \tag{9}$$

between the infinite magnetic field limit $\epsilon_{xc}^{\infty} = -0.782\sqrt{2\pi n}e^2/\kappa$ and the zero field limit $\epsilon_{xc}^{\mathrm{TC}}$ given by Tanatar and Ceperly [25] generalized to intermediate polarizations [26].

In the numerical calculations we shall assume the magnetic flux density through the relevant unit cell to be constant and set to $B = 1.654$ T leading to a magnetic length $\ell = 19.95$ nm much smaller than the lattice lengths L_x and L_y, which shall be in the range 100 to 200 nm. It should be emphasized here that since the Coulomb interaction is treated equally for all electrons in the system, independent of whether they are in the same dot or not, we turn it totally off when we discuss a noninteracting system.

The resulting Kohn-Sham equations are solved within the symmetric Ferrari basis [27,23,21] and in order to have a large enough basis allowing several electrons in each dot we have to use lattice lengths L_i shorter than the actual ones in experiments. In the present calculations we use upto 16384 basis states. To get back to the experimental results displayed in Fig. 3 we perform a calculation for the ground state properties of an array of dots described by the confinement potential

$$V_{\mathrm{sq}}(\mathbf{r}) = V_0 \left[\sin\left(\frac{g_1 x}{2}\right) \left\{ \sin\left(\frac{g_2 y}{2}\right) \right\}^2 \right]^2 . \tag{10}$$

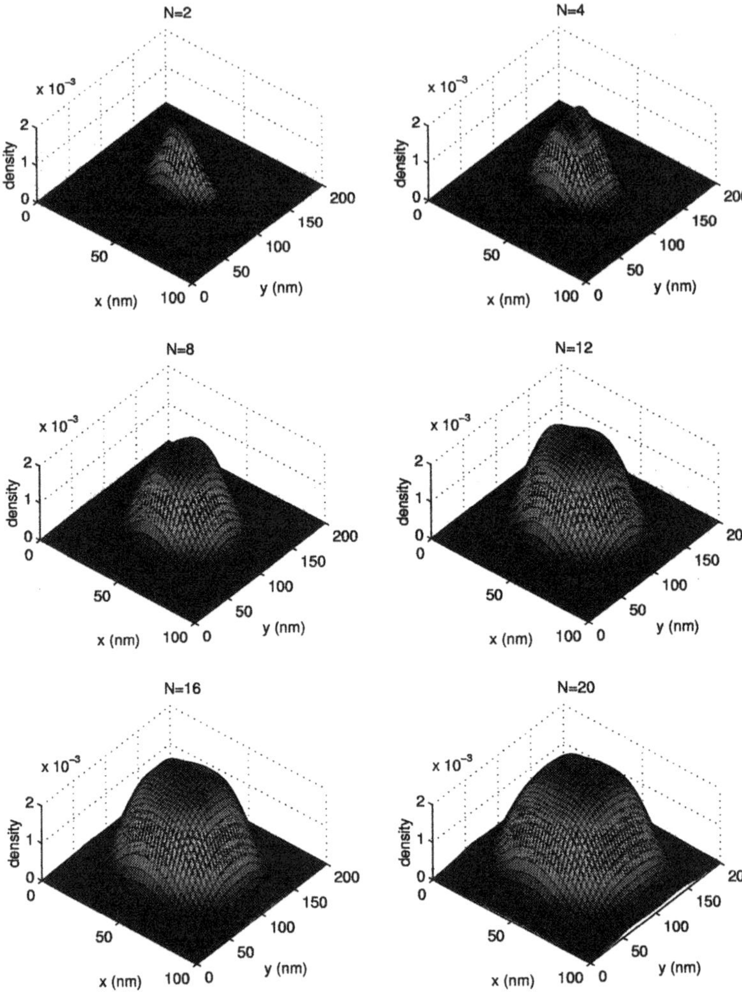

Fig. 7: The electron density distribution for the ground state of *noninteracting* elliptical dots. The confinement is according to Eq. (5). The x and y axis are scaled differently here. $B = 1.654$ T, $T = 1$ K, $V_0 = -16$ meV.

This choice defines a square unit cell, but within each cell the dot confinement is elliptic. The model results are shown as density contours in Fig. 6 for a growing number of electrons. For few electrons the shape of the dots is very close to circular, but with increasing electron number the dots become more elliptic. For still a higher number of electrons the Coulomb repulsion between the narrower edges (their ends) of neighboring dots causes their shape to become more square like. For $N = 20$, when their density clearly overlaps the central region acquires a circular or elliptic form again. Now, we can not maintain that this demonstrates what happens in an absorption experiment, but by a comparison to the noninteracting case we can

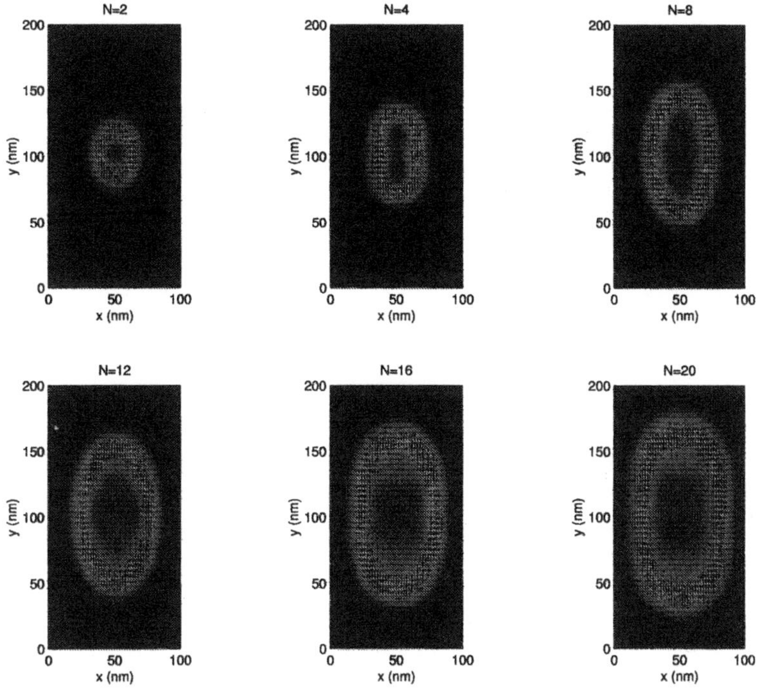

Fig. 8: The electron density distribution for the ground state of *noninteracting* elliptical dots. The confinement is according to Eq. (5). Same case as in Fig. 7. $B = 1.654$ T, $T = 1$ K, $V_0 = -16$ meV.

clearly see that the Coulomb interaction has a strong influence on the the shape of the dots as the number of electrons is increased.

To learn more about these effects we want to compare the results to what happens in the different confinement potentials (5) and (6) introduced above. We start with elliptic dots described by $V_{\rm QAD}$ in Eq. (5) in a rectangular lattice. Although the lattice does not have a square shaped unit cell as the one defined by $V_{\rm sq}$, the distance between the quantum dots measured in the x or the y direction can be expected to be comparable. The density for the noninteracting case can be seen in Fig. 7 with the contours displayed in Fig. 8 with the real aspect ratio between the x and the y axis. Up to $N = 20$ there is only a weak overlapping of the electron density of neighboring cells, but certainly the shape of the dots changes with growing N, even in the absence of the electron-electron interaction. For low N their ellipticity increases as N grows, but for higher electron number the shape slowly approaches the symmetry of the lattice.

The electron density for the interacting case can be seen in Fig. 9 together with the contours in Fig. 10. In contrast to the noninteracting case the Coulomb repulsion is strong enough to push the electron density to already overlap considerably at $N = 12$ or even a bit lower. Interestingly, the repulsion forces the electrons to form

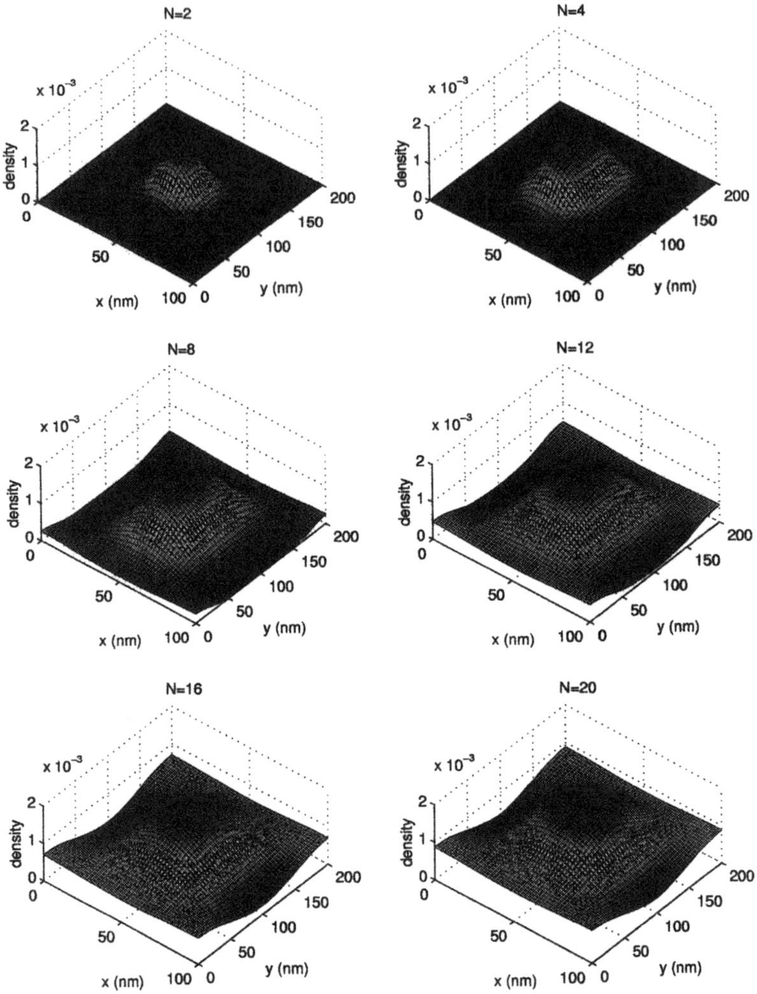

Fig. 9: The electron density distribution for the ground state of *interacting* elliptical dots. The confinement is according to Eq. (5). The x and y axis are scaled differently here. $B = 1.654$ T, $T = 1$ K, $V_0 = -16$ meV.

wires in the direction of the longer axis (y axis) of the ellipses. This behavior is continued to much higher number of electrons as can be verified in Fig. 11, and it can be understood as governed by two facts. First, the repulsion between the dots along the longer edge is stronger than between the shorter edges of the dots. Second, the lower slope of the confinement potential in the y direction than in the x direction determines an asymmetric screening in the electron system. In this connection, it is also clear that the electronic density in the elliptic dots in the square lattice defined by V_{sq} had more space to 'broaden' the dots by occupying the space along the long edge between them than in this system.

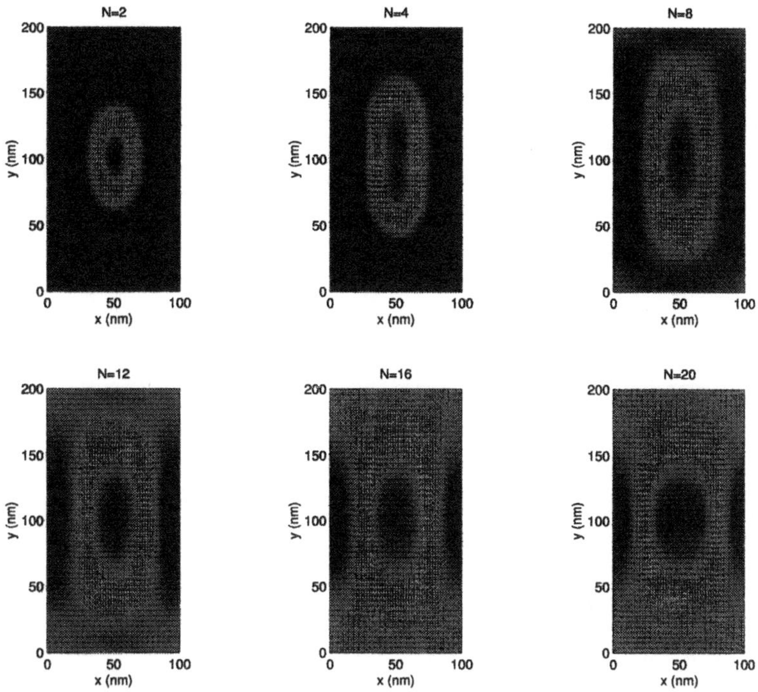

Fig. 10: The electron density distribution for the ground state of *noninteracting* elliptical dots. The confinement is according to Eq. (5). Same case as in Fig. 9. $B = 1.654$ T, $T = 1$ K, $V_0 = -16$ meV.

This comparison opens the question of the role of the steepness of the confining potential itself. To tackle that question we have redone the calculations for the simple cosine potential $V_{\rm per}$ defined by Eq. (6) with a variable strength V_0 but a constant number of electrons $N = 20$. The results for the noninteracting electrons is displayed in Fig. 12. When V_0 is small the overlapping of the electron density into neighboring cells is of the same order of magnitude in both directions, but for strong modulation V_0 the overlapping only takes place between the longer edges of the cell, i.e. modulated wires are formed in the y direction. The curious fact is that absolutely the *contrary* happens for the interacting system shown in Fig. 13, where the wires are formed in the x direction, which is the longer axis of the unit cell. Here several energy bands are occupied in the case of 20 electrons in a unit cell. The single-particle states with high energy are not well localized in the minimum of the potential in the middle of the cell and thus the wave functions of neighboring cells overlap where the distance is the shortest, here in the direction of the short axis L_y. With the interaction turned-on the repulsion in this direction is stronger and the system forms wires in the direction with the shorter interface with the neighboring cell.

The softer confinement potential $V_{\rm per}$ causes more drastic difference between the interacting and the noninteracting electron system, than the more realistic dot con-

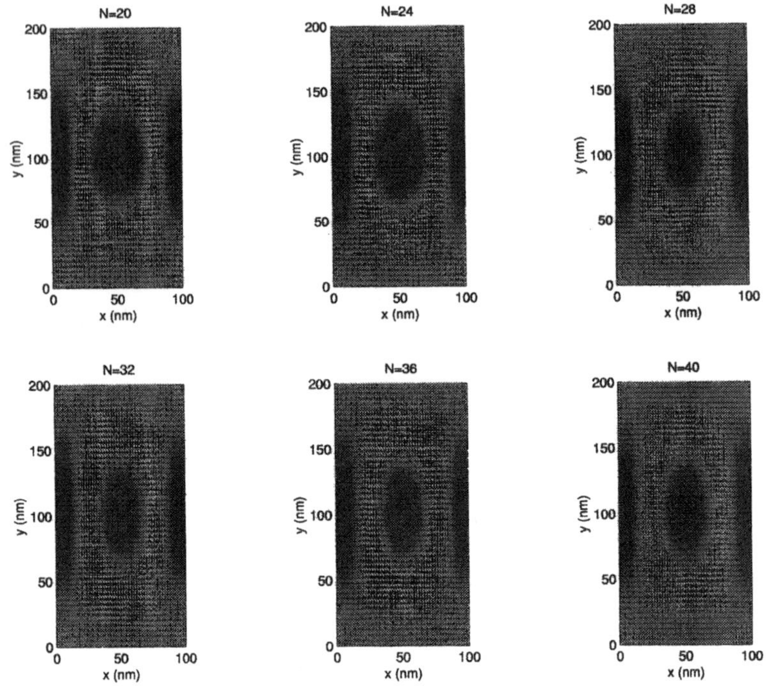

Fig. 11: The electron density distribution for the ground state of *noninteracting* elliptical dots. The confinement is according to Eq. (5). Same system as in Fig. 10, but with a higher number of electrons. $B = 1.654$ T, $T = 1$ K, $V_0 = -16$ meV.

finement $V_{\rm QAD}$, at least more realistic at the lattice lengths and number of electrons considered here.

We have performed the calculations for quantum dots in arrays with $L_x = L_y$ to confirm that no direction for 'wire formation' is preferred in that case, and when $L_y = 1.5L_x$ we already have the wire formation in the preferred directions well developed. To test which parts of the dot interaction are important in influencing the shape we have repeated some of the calculations excluding the exchange and correlation interaction. For the lattice lengths, the electron number, and the confinement selected here the exchange and correlation plays a minor role. There is a fine structure in the density that depends on it, but the overall properties are caused by the direct interaction as could be expected.

3.3 FIR absorption in the model of interacting dots

Due to the large size of the mathematical set of basis functions used in the ground state calculation, in order to describe dots with several electrons and an array with not too short lattice lengths, we are not able to perform a calculation of the FIR absorption for the system with the parameters used above. Instead, we can describe the electrons in the Hartree approximation (HA) without spin and in a smaller lattice

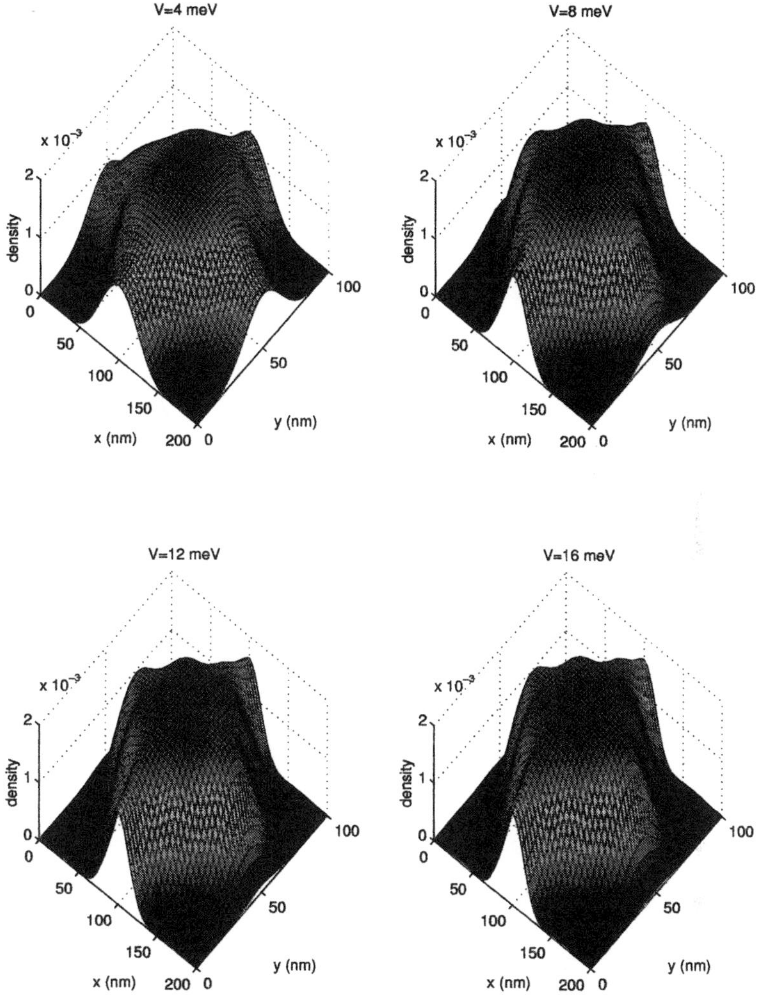

Fig. 12: The electron density distribution for the ground state of a *noninteracting* 2DEG in the simple periodic potential $V_{\rm per}$ described by Eq. (5). The x and y axis are scaled differently here. $B = 1.654$ T, $T = 1$ K, $N = 20$.

with $L_x = 100$ nm and $L_y = 150$ nm in a lower magnetic field strength $B = 1.10$ T.

The FIR absorption is calculated in a self-consistent linear response [22] exciting the system with an external electric field of the type

$$\mathbf{E}_{ext}(\mathbf{r}, t) = -\mathrm{i}\mathcal{E}_0 \frac{\mathbf{k}+\mathbf{G}}{|\mathbf{k}+\mathbf{G}|} \exp\{\mathrm{i}(\mathbf{k}+\mathbf{G})\cdot\mathbf{r} - \mathrm{i}\omega t\}. \tag{11}$$

Here we do not restrict the dispersion relation for the external field, $\omega(\mathbf{k}+\mathbf{G})$, to that of a free propagating electromagnetic wave, but we allow for the more general situation in which the external field is produced as in a near-field spectroscopy or in a Raman scattering set-up. The power absorption is found from the Joule heating of the self-consistent electric field [28,12], $-\nabla\phi_{sc}$, with $\phi_{sc} = \phi_{ext} + \phi_{ind}$,

$$P(\mathbf{k}+\mathbf{G},\omega) = -\frac{\omega}{4\pi}\left[|\mathbf{k}+\mathbf{G}|\phi_{sc}(\mathbf{k}+\mathbf{G},\omega)\phi^*_{ext}(\mathbf{k}+\mathbf{G},\omega)\right]. \tag{12}$$

The induced potential ϕ_{ind} is caused by the density variation $\delta n_s(\mathbf{r})$ due to ϕ_{sc}, which can then in turn be related to the external potential by the dielectric tensor

$$\sum_{\mathbf{G}'} \epsilon_{\mathbf{G},\mathbf{G}'}(\mathbf{k},\omega)\phi_{sc}(\mathbf{k}+\mathbf{G}',\omega) = \phi_{ext}(\mathbf{k}+\mathbf{G},\omega). \tag{13}$$

The dielectric tensor, $\epsilon_{\mathbf{G},\mathbf{G}'}(\mathbf{k},\omega) = \delta_{\mathbf{G},\mathbf{G}'} - \frac{2\pi e^2}{\kappa|\mathbf{k}+\mathbf{G}|}\chi_{\mathbf{G},\mathbf{G}'}(\mathbf{k},\omega)$, is determined by the susceptibility of the electron system,

$$\begin{aligned}\chi_{\mathbf{G},\mathbf{G}'}(\mathbf{k},\omega) = \tfrac{1}{(2\pi L)^2}\textstyle\int d\theta \sum_{\alpha,\alpha'} f_{\alpha,\alpha'}^{\theta,\theta-\tilde{\mathbf{k}}L}(\hbar\omega) J_{\alpha,\alpha'}^{\theta,\theta-\tilde{\mathbf{k}}L-\tilde{\mathbf{G}}L}(\mathbf{k}+\mathbf{G}) \\ \left(J_{\alpha,\alpha'}^{\theta,\theta-\tilde{\mathbf{k}}L-\tilde{\mathbf{G}}L}(\mathbf{k}+\mathbf{G}')\right)^*,\end{aligned} \tag{14}$$

where $\mathbf{k}$ is in the first Brillouin zone, $\tilde{\mathbf{k}} = (k_xL_x, k_yL_y)$ and $\tilde{\mathbf{G}} = (G_1L_x, G_2L_y)$, κ is the dielectric constant of the surrounding medium, $\theta = (\theta_1,\theta_2) \in \{[-\pi,\pi]\times[-\pi,\pi]\}$, and

$$f_{\alpha,\alpha'}^{\theta,\theta'}(\hbar\omega) = \left\{\frac{f^0(\varepsilon_{\alpha,\theta}) - f^0(\varepsilon_{\alpha',\theta'})}{\hbar\omega + (\varepsilon_{\alpha,\theta} - \varepsilon_{\alpha',\theta'}) + i\hbar\eta}\right\}, \tag{15}$$

in which f^0 is the equilibrium Fermi distribution, $\eta \to 0^+$, and

$$J_{\alpha,\alpha'}^{\theta,\theta'}(\mathbf{k}) = (\alpha'(\theta')|e^{-i\mathbf{k}\cdot\mathbf{r}}|\alpha(\theta)). \tag{16}$$

Special care must be taken with respect to the symmetry of the wave functions corresponding to the Hartree states $|\alpha(\theta))$ when translating them across the boundaries of the quasi-Brillouin zones.

In Fig. 14 we see the absorption for an array of electrons with two electrons in each dot (upper right panel) and an array of seven electrons (lower right panel). In the former case the dots are isolated, but in the latter case their densities start to overlap in the x and y directions, almost of the same amount. The structure of the corresponding energy bands is displayed in the left panels of Fig. 14, showing that, indeed, when 7 electrons occupy each dot the chemical potential μ is situated in the continuum states. We fix the polarization of the external field (11) by giving $\mathbf{k}$ a small but finite value, $\mathbf{k}_iL_i = 0.2$ with $i = x, y$, and $\mathbf{G} = 0$ in accordance with FIR radiation.

The FIR absorption of the isolated dots consists of two peaks, one sharp peak and the second broadened and lower. Since the confinement is not parabolic there are higher order peaks at still higher energy that we exclude from our discussion and figure here. At the higher electron number we still can locate the two main peaks, now both split. At an energy below the collective dot modes, which are not

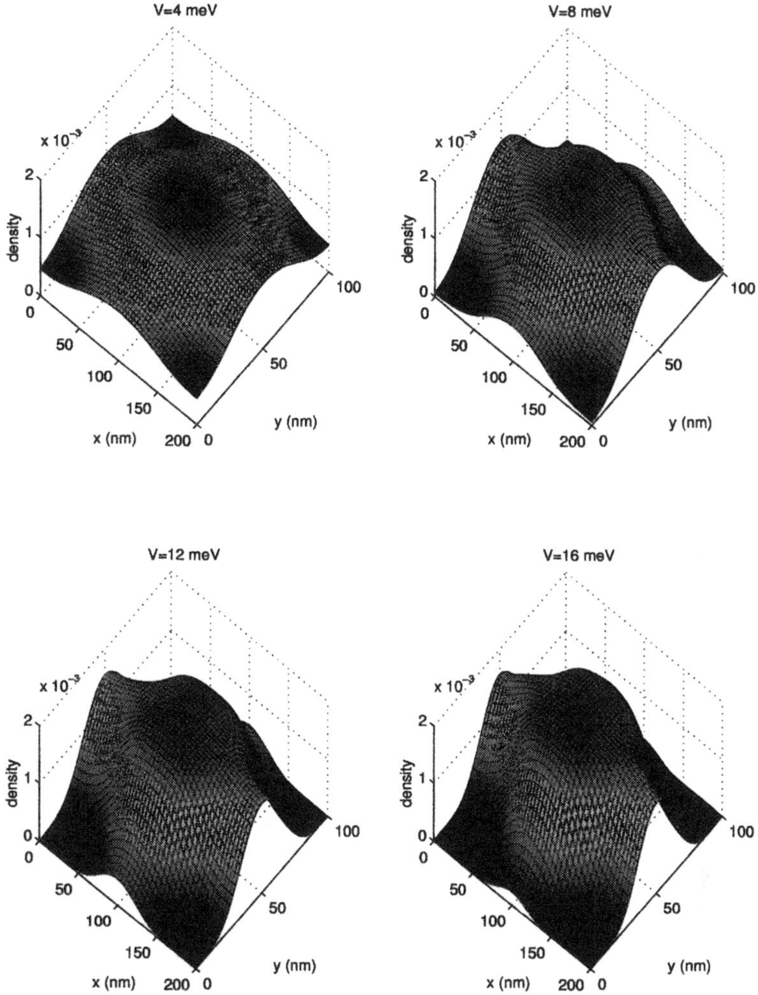

Fig. 13: The electron density distribution for the ground state of an *interacting* 2DEG in the simple periodic potential V_{per} described by Eq. (5). The x and y axis are scaled differently here. $B = 1.654$ T, $T = 1$ K, $N = 20$.

dependent on the direction of $\mathbf{k}$, we find intraband modes caused by transitions in the Landau band where the chemical potential is located. These intraband modes depend on the direction of $\mathbf{k}$ as the structure of the low lying continuum bands reflects the geometry of the dot array. They are generally not seen in experimental spectra since for larger lattice lengths they are at a very low energy range that is not easily accessible.

As we can only consider a very limited system here, we have to be careful about generalizations, but if we analyze the gap between the sets of peaks as a function of the electron number N in the range between 2 and 12 electrons we can see a tendency

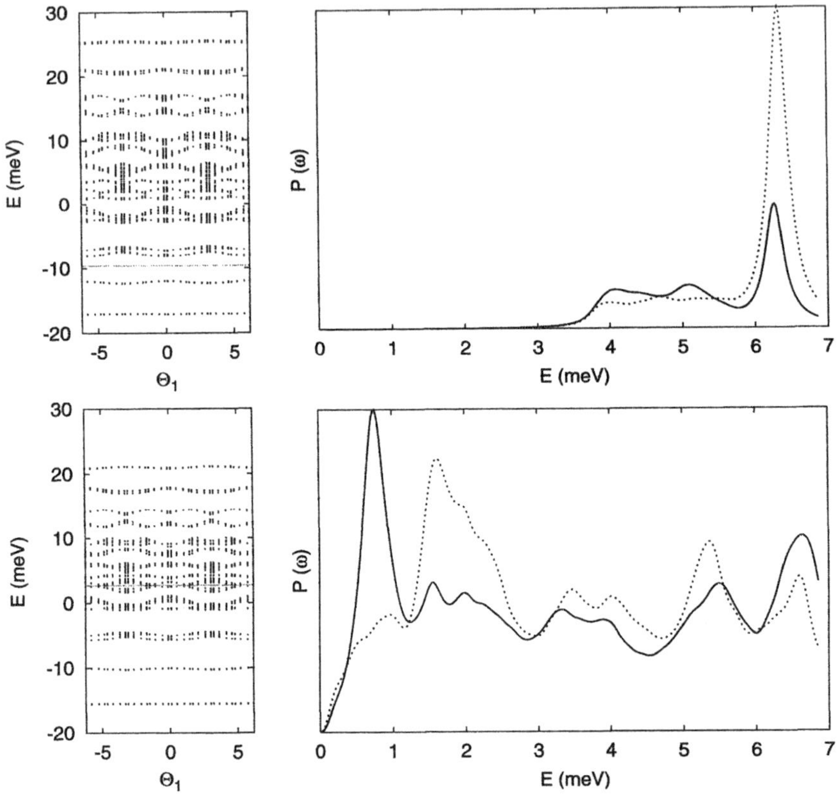

Fig. 14: The FIR absorption for two (upper panel) and seven electrons (lower panel) in the hartree approximation for spinless electrons. $k_xL_x = 0.2$ (solid curve), and $k_yL_y = 0.2$ (dotted curve). The right panels show the band structure projected on the $\Theta_1 = k_xL_x$ axis in the Brillouin zone. The chemical potential μ is indicated by the horizontal solid line.

to a similar evolution as has been reported in experiment [2] and is repeated in Fig. 3. This should only be considered as a very preliminary result and one has to keep in mind that we perform our calculation at a finite magnetic field since the calculation is built on a basis set which has to increase when the magnetic field decreases.

3.4 Effects on magnetization in the ground state

Recently, our attention has been drawn to measurements of the magnetization of a homogeneous two-dimensional electron gas (2DEG) in heterostructures [29,30]. There are efforts underway to extend the experiments on magnetization also to modulated 2DEG's and arrays of dots and antidots. The magnetization is an equilibrium property of the ground state of the electron system so we can calculate it for the system in which we have studied the shape changes of the quantum dots as function of the number of electrons N. The total magnetization can be calcu-

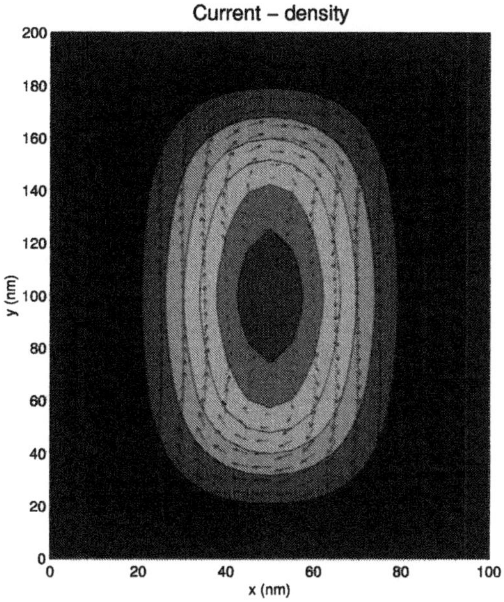

Fig. 15: The interacting electron and current density for 6 electrons in an elliptic quantum dot in the array described by Eq. (5). $B = 1.654$ T, $T = 1$ K, $V_0 = -16$ meV.

lated according to the definition for the orbital M_o and the spin component of the magnetization M_s, [31,32]

$$M_o + M_s = \frac{1}{2c\mathcal{A}} \int_{\mathcal{A}} d^2r \, (\mathbf{r} \times \langle \mathbf{J}(\mathbf{r}) \rangle) \cdot \hat{\mathbf{n}} + \frac{g\mu_B}{\mathcal{A}} \int_{\mathcal{A}} d^2r \langle \sigma_z(\mathbf{r}) \rangle \, , \tag{17}$$

where $\mathcal{A}$ is the total area of the system. The equilibrium local current is evaluated as the quantum thermal average of the current operator,

$$\hat{\mathbf{J}} = -\frac{e}{2}\Big(\hat{\mathbf{v}} |\mathbf{r}\rangle\langle\mathbf{r}| + |\mathbf{r}\rangle\langle\mathbf{r}| \hat{\mathbf{v}} \Big), \tag{18}$$

with the velocity operator $\hat{\mathbf{v}} = [\hat{\mathbf{p}} + (e/c)\mathbf{A}(\mathbf{r})]/m^*$, $\mathbf{A}$ being the vector potential.

A typical current density is shown in Fig. 15 superimposed on the contours of the electron density for one elliptical quantum dot in an array of dots. Even though the density has only one maximum two vortices are seen in the current density. Here again we have used the LSDA described above. The orbital magnetization of arrays of elliptical dots and antidots of different aspect ratio is presented in Fig. 16 and for comparison the last panel shows the magnetization for the electronic system confined by $V_{\rm per}$ (6). The magnetization for the antidot array is almost simply the mirror image of the magnetization for the dot array for the range of N considered here, independent of whether the system forms isolated dots or not. For low N the M_o develops peaks when N equals twice the number of flux quanta pq through the unit cell, i.e. when only the lowest Landau band is completely occupied and all other

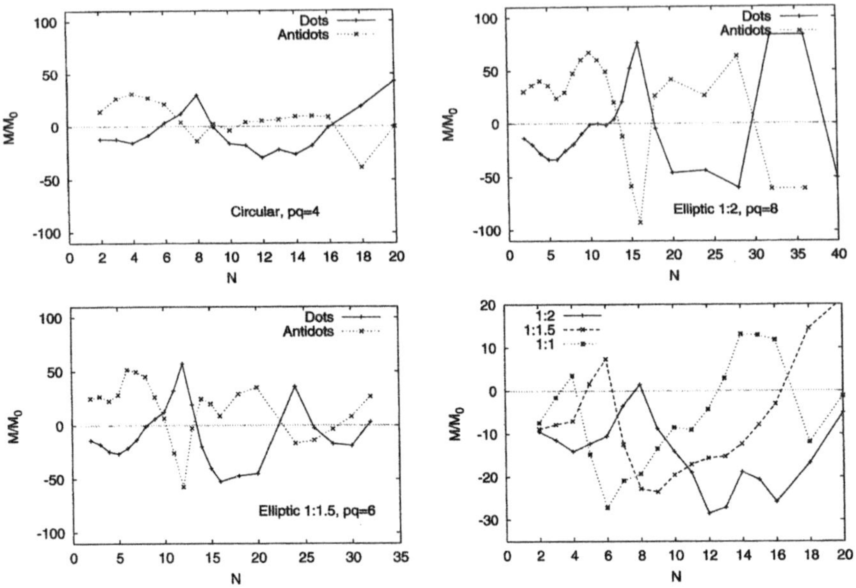

Fig. 16: The orbital magnetization M_o as function of the number of electrons N in a quantum dot or antidot array described by Eq. (5) with aspect ratios 1:1 (upper left panel), 1:1.5 (upper right), 1:2 (lower left), a 2DEG confined by Eq. (6) for all three aspect ratios (lower right panel). $M_0 = \mu_B^*/(L_x L_y)$, $B = 1.654$ T, $T = 1$ K, $|V_0| = 16$ meV.

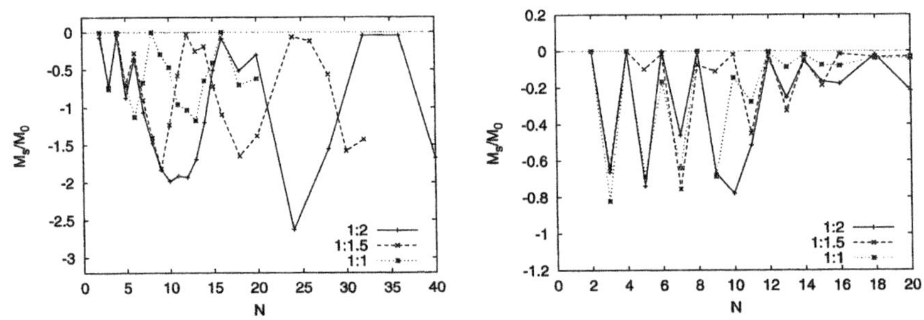

Fig. 17: The spin contribution to the magnetization M_s as a function of the number of electrons N for a dot array described by Eq. (5) (left), and a 2DEG described by Eq. (6). $M_0 = \mu_B^*/(L_x L_y)$, $B = 1.654$ T, $T = 1$ K, $|V_0| = 16$ meV.

bands are empty. The spin contribution to the magnetization, M_s, seen in the left panel of Fig. 17, reflects strong spin polarization as $N = pq$, when the first Landau band is half filled and the exchange energy is thus enhanced. This enhancement of the exchange can also be recognized at higher odd integer multiples of pq.

The situation is a bit different for the electron system confined or modulated by $V_{\rm per}$ (6). Here the splitting of the Landau bands into Hofstadter bands [20,23,21] is stronger than the exchange enhancement of the spin splitting reflected by the fact that the spin contribution to the magnetization M_s in the right panel of Fig. 17 vanishes for even number of electrons in most cases and no strong spin polarization is observed. This happens even when the iteration process of the LSDA has been started with an artificial large g factor that is later reduced to the natural value of 0.44 appropriate for GaAs. The last panel of Fig. 16 shows that the orbital magnetization M_o is also quite different for this system: First, its magnitude does not increase as drastically with the size of the unit cell as for the dots and antidots. Second, the Hofstadter splitting in the lowest Landau band when it is half filled produces a clearer signature than the complete filling of the band. The difference in the magnetization for these two systems has to be connected to their different geometry. At low N the confinement V_{QAD} produces simple dots or antidots, the dots are isolated at first but start to overlap only after the first Landau band has been fully occupied. On the other hand, the electron system in V_{per} forms connected regions for lower N. At this moment we have not discovered any clear signs of the actual geometry of the dots and antidots in the magnetization, and thus we can not distinguish the magnetization of circular or elliptic quantum dots. In order to accomplish this in isolated dots with few electrons we would need to be able to vary the magnetic field continuously for a constant number of electrons [33].

4. Summary

We have reported here on efforts to discern in experiments or predict by model calculations the effects arrays can have on the FIR absorption of quantum dots. There are indications that effects of the periodicity itself have been detected in measured FIR spectra, and even interaction between neighboring quantum dots. Model calculations confirm that the effects of the periodicity are well understood, but the effects of a direct interaction between the electron systems of different dots is very weak and subtle. The direct interaction though seems to be detectable if it can influence the shape of the dots, since the FIR absorption is very dependent on the symmetry of the electron system confined in them.

Acknowledgments

We gratefully acknowledge support from the German Science Foundation DFG through SFB 508 "Quanten-Materialien", the Graduiertenkolleg "Nanostrukturierte Festkörper", the Research Fund of the University of Iceland, and the Icelandic Natural Science Council. We thankfully acknowledge the great help of Birgir F. Erlendsson in parallelizing the execution of the core regions of our programs.

References

[1] R. Krahne, V. Gudmundsson, C. Heyn, and D. Heitmann, Phys. Rev. B **63**, 195303 (2001).

[2] R. Krahne, V. Gudmundsson, C. Heyn, and D. Heitmann, Physica E p. in print (2001).

[3] C. Dahl and J. P. Kotthaus, Phys. Rev. B **46**, 15590 (1992).

[4] B. P. van Zyl, E. Zaremba, and D. A. W. Hutchinson, Phys. Rev. B **61**, 2107 (2000).

[5] M. Taut, Phys. Rev. B **62**, 8126 (2000).

[6] P. A. Maksym and T. Chakraborty, Phys. Rev. Lett **65**, 108 (1990).

[7] D. A. Broido, K. Kempa, and P. Bakshi, Phys. Rev. B **42**(17), 11400 (1990).

[8] D. Pfannkuche and R. Gerhardts, Phys. Rev. B **44**(23), 13132 (1991).

[9] S. K. Yip, Phys. Rev. B **43**, 1707 (1991).

[10] Q. P. Li, K. Karraï, S. K. Yip, S. D. Sarma, and H. D. Drew, Phys. Rev. B **43**(6), 5151 (1991).

[11] T. Demel, D. Heitman, P. Grambow, and K. Ploog, Phys. Rev. Lett. **64**, 788 (1990).

[12] V. Gudmundsson and R. Gerhardts, Phys. Rev. B **43**(14), 12098 (1991).

[13] Z. L. Ye and E. Zaremba, Phys. Rev. B **50**(23), 17217 (1994).

[14] V. Gudmundsson, A. Brataas, P. Grambow, T. Kurth, and D. Heitmann, Phys. Rev. B **51**, 17744 (1995).

[15] I. B. Bernstein, Phys. Rev. **109**(1), 10 (1958).

[16] A. Lorke, R. J. Luyken, A. O. Govorov, J. P. Kotthaus, J. M. Garcia, and P. M. Petroff, Phys. Rev. Lett. **84**, 2223 (2000).

[17] E. Zaremba, Phys. Rev. B **53**(16), R10512 (1996).

[18] J. M. Llorens, C. Trallero-Giner, A. Garca-Cristbal, and A. Cantarero, Phys. Rev. B **64**, 035309 (2001).

[19] A. Emperador, M. Barranco, E. Lipparini, M. Pi, and L. Serra, Phys. Rev. B **62**(23), 4573 (2000).

[20] R. D. Hofstadter, Phys. Rev. B **14**, 2239 (1976).

[21] V. Gudmundsson and R. R. Gerhardts, Phys. Rev. B **52**, 16744 (1995).

[22] V. Gudmundsson and R. R. Gerhardts, Phys. Rev. B **54**, 5223R (1996).

[23] H. Silberbauer, J. Phys. C **4**, 7355 (1992).

[24] M. I. Lubin, O. Heinonen, and M. D. Johnson, Phys. Rev. B **56**, 10373 (1997).

[25] B. Tanatar and D. M. Ceperley, Phys. Rev. B **39**, 5005 (1989).

[26] M. Koskinen, M. Manninen, and S. M. Reimann, Phys. Rev. Lett. **79**, 1389 (1997).

[27] R. Ferrari, Phys. Rev. B **42**, 4598 (1990).

[28] C. Dahl, Phys. Rev. B **41**(9), 5763 (1990).

[29] I. Meinel, T. Hengstmann, D. Grundler, and D. Heitmann, Phys. Rev. Lett. **82**(4), 819 (1999).

[30] I. Meinel, D. Grundler, D. Heitmann, A. Manolescu, V. Gudmundsson, W. Wegscheider, and M. Bichler, Phys. Rev. B **64**, 121306(R) (2001).

[31] J. Desbois, S. Ouvry, and C. Texier, Nucl. Phys. B **528**, 727 (1998).

[32] V. Gudmundsson, S. I. Erlingsson, and A. Manolescu, Phys. Rev. B **61**, 4835 (2000).

[33] I. Magnúsdóttir and V. Gudmundsson, Phys. Rev. B **61**, 10229 (2000).

Chapter 8

Quantum dots in a strong magnetic field: Quasi-classical consideration

A. Matulis

Institute of Semiconductor Physics, Goštauto 11, 2600 Vilnius, Lithuania,
E-mail: amatulis@takas.lt

Abstract

The electron motion in strong magnetic fields (when only the lowest Landau level is populated) is considered. In this case the electron kinetic energy is frozen out and the electrons are guided by a slowly varied potential. Using the adiabatic procedure and expansion in magnetic length series an approximate description is developed. In zeroth order this approximation leads to the classical equations of motion describing the Larmor circle drift in the potential gradient. In the second order the special quantum mechanical description where the electron potential energy plays the role of the total Hamiltonian is constructed. Simple examples of a single and two electrons in the parabolic dot demonstrates that the proposed approximate description gives the main features of the electron system spectrum and the collective phenomena.

1. Introduction

Quantum dots, or *artificial atoms*, have been the subject of intense theoretical and experimental research over the last few years [1]. One useful component of the spectroscopy experiments is a magnetic field applied perpendicular to the quantum dot plane direction which enables one to trace easily the dependence of the quantum dot properties on various parameters. Moreover the strong magnetic field reveals the quantization effects introducing into the electron system the favorable interplay between confining potential and Landau levels.

Recently, the main interest in quantum dots is related to the electron-electron interaction and the collective phenomena, such as the change of the ground state multiplicity, the electron density reconstruction, and the Wigner crystallization. The electron density reconstruction in the finite electron systems was considered in [2]. Now it is known as Shamon-Wen edge — some of the electron density ring around the finite system. Under certain circumstances the ring was reported to become unstable [3] and breaks into separate lumps. Although the possibility to obtain the symmetry braking solutions was argued [4] considering them as an artifact of the approximate methods used, the exact calculations of the electron correlation function [5] undoubtedly indicates that the Wigner crystallization occurs at rather large electron-electron interaction. The electron density plots in [6] show that the strong magnetic field facilitates the electron density edge reconstruction leading to the Wigner crystallization.

Meanwhile, the minimization of the system potential presented in [7] shows that the Wigner crystallization in quantum dots can be successfully considered by classical mechanics. The fact that the strong magnetic field facilitates the Wigner crystallization enables us to suppose that the electron system behavior in very strong magnetic can be described by classical or quasi-classical methods. The purpose of the present article is to show how such methods could be developed. The article is organized as follows. After the formulation of the model in the next Section, in Sections 3 and 4 the main ingradients — fast and slow variables are introduced. Then in Section 5 the adiabatic procedure is discussed and the slow motion Schrödinger equation is considered. In Section 6 the classical equations for the limiting case of a strong magnetic field are derived, and in the next two sections the illustrations of the simplified quantum mechanical description are given. In Appendix A the details of the adiabatic procedure are presented, and in Appendix B the transformation back to the initial coordinates is discussed.

2. Model

We consider the Schrödinger equation

$$i\hbar\frac{\partial}{\partial t}\Psi = \mathcal{H}_T\Psi \tag{1}$$

with the Hamiltonian

$$\mathcal{H}_T = \frac{1}{2m}\left\{\boldsymbol{p} + \frac{e}{c}\boldsymbol{A}(\boldsymbol{r})\right\}^2 + V(\boldsymbol{r}) \tag{2}$$

describing the motion of two-dimensional (2D) electrons in a strong perpendicular homogeneous magnetic field and a slowly varying potential $V(\boldsymbol{r})$. For the sake of simplicity the main equations will be derived for a single electron as the generalization for the system of many electrons is trivial. It will be presented toward the end of the derivation of our formalism.

Choosing the symmetric gauge $\boldsymbol{A} = [\boldsymbol{B} \times \boldsymbol{r}]/2$ we write the main part of the Hamiltonian as

$$\mathcal{H}_0 = \frac{1}{2m}\left\{\left(p_x - \frac{eB}{2c}y\right)^2 + \left(p_y + \frac{eB}{2c}x\right)^2\right\}. \tag{3}$$

We shall consider it as a largest term and treat the remaining potential term $V(\boldsymbol{R})$ as a small perturbation.

3. Landau levels

As in the standard perturbation technique we have to start with the zeroth order problem and solve the following stationary Schrödinger equation

$$\{\mathcal{H}_0 - \varepsilon\}\psi = 0. \tag{4}$$

The solutions are known as Landau levels. The most simple way to obtain these is to introduce the new variables

$$\xi = \frac{\ell_B}{\hbar}p_x - \frac{1}{2\ell_B}y, \quad \eta = \frac{\ell_B}{\hbar}p_y + \frac{1}{2\ell_B}x \tag{5}$$

where $\ell_B = \sqrt{c\hbar/eB}$ is the magnetic length. Using the new variables Hamiltonian (3) can be rewritten as

$$\mathcal{H}_0 = \frac{\hbar\omega_c}{2}(\xi^2 + \eta^2) \tag{6}$$

where $\omega_c = eB/mc$ is the cyclotron frequency, and the new variables obey the following commutation rule

$$[\xi, \eta] = -\mathrm{i}. \tag{7}$$

The zeroth order Hamiltonian is reminiscent of the Hamiltonian of a harmonic oscillator, and it is evident that it has the equidistant discrete spectrum, the Landau levels.

We shall consider the case of a very strong magnetic field when the electrons are in the lowest Landau level. Our task is to reveal how the slowly varying additional potential $V(\boldsymbol{r})$ (as compared with the magnetic length ℓ_B) changes their behavior.

4. Slow variables

We shall treat the variables introduced in the previous Section as *fast* variables because they are included into the main part of the Hamiltonian. But as we are going to solve the 2D problem they are not sufficient to treat the Schrödinger equation (1). We have to introduce two more variables

$$X = \frac{1}{2}x - \frac{\ell_B^2}{\hbar}p_y, \quad Y = \frac{1}{2}y + \frac{\ell_B^2}{\hbar}p_x. \tag{8}$$

We chose them in such way in order to have the most simple commutation relations

$$[\xi, X] = [\xi, Y] = [\eta, X] = [\eta, Y] = 0, \tag{9}$$

and

$$[Y, X] = -\mathrm{i}\ell_B^2. \tag{10}$$

We shall consider those variables as *slow* ones.

Now substituting the initial variables

$$x = X + \ell_B\eta, \quad y = Y - \ell_B\xi \tag{11}$$

into Hamiltonian (2) we arrive at the following expression

$$\mathcal{H}_T = \frac{\hbar\omega_c}{2}(\xi^2 + \eta^2) + V(X + \ell_B\eta, Y - \ell_B\xi). \tag{12}$$

So, we divided the Hamiltonian into two parts. The first major part describing the motion of the electron in the homogeneous magnetic field depends on the fast variables only, while the other part — the slowly varying potential — depends on both fast and slow variables. Thus, we see that the slow and fast variables can not be separated exactly, but the presence of the small parameter (namely, the ratio of the magnetic length ℓ_B and the characteristic potential variation length $l_0 \sim |V/\nabla V|$) enables us to separate them approximately by means of some adiabatic procedure.

5. Adiabatic procedure

In order to develop the adiabatic procedure and apply it to the Schrödinger equation (1), we take the following steps

- we expand the potential into ℓ_B-powers

$$V = V(X, Y) + \ell_B\eta V_X(X, Y) - \ell_B\xi V_Y(X, Y) + \cdots; \tag{13}$$

- divide the Hamiltonian into two parts

$$\mathcal{H} = \mathcal{H}_f + \mathcal{H}_s, \tag{14}$$

$$\mathcal{H}_f = \frac{\hbar\omega_c}{2}(\xi^2 + \eta^2) + \ell_B\eta V_X(X, Y) + \ell_B\xi V_Y(X, Y) + \cdots, \tag{15}$$

$$\mathcal{H}_s = V(X, Y); \tag{16}$$

- present the wave function as the product of its fast and slow parts

$$\Psi = \psi(\eta|X,Y)\Phi(X), \tag{17}$$

- and use the following equation for the fast wave function part

$$\{\mathcal{H}_f - E(X,Y)\}\psi(\eta|X,Y) = 0. \tag{18}$$

In fact, it is the standard adiabatic procedure which has to lead to the Schrödinger equation for the slow electron motion

$$\mathrm{i}\hbar\frac{\partial}{\partial t}\Phi(X) = \mathcal{H}\Phi(X), \tag{19}$$

with the effective slow-motion Hamiltonian

$$\mathcal{H} = V(X,Y) + E(X,Y). \tag{20}$$

However, there are some peculiarities caused by the fact that according to Eqs. (7,10) neither fast nor slow variables commute with each other. That is why both wave function parts in Eq. (17) depend only on a single variable (either η or X), while the other one has to be treated as an operator ($\xi = -\mathrm{i}\partial/\partial\eta$, $Y = -\mathrm{i}l_B^2\partial/\partial X$). Consequently, X and Y variables entering the fast wave function part $\psi(\eta|X,Y)$ and the corresponding eigenvalue $E(X,Y)$ can not be treated as parameters (what is done in a standard adiabatic procedure), but should be considered as operators acting on the slow wave function part. This makes the adiabatic procedure a little bit tricky and cumbersome. Nevertheless due to the presence of the small parameter ℓ_B/l_0 it can be performed. The details of this derivation are presented in the Appendix A. Restricting ourselves up to the order ℓ_B^2 we shall use the following slow-motion Hamiltonian

$$\mathcal{H} = V^{(S)}(\boldsymbol{R}) + \frac{\ell_B^2}{4}\nabla^2 V^{(S)}(\boldsymbol{R}). \tag{21}$$

Here the superscript indicates that the expression should be symmetrized with respect to the permutation of the slow variables X and Y which as we know already do not commute with each other.

The above adiabatic procedure can be easily generalized for the case of the many-electron system. As for the slow motion different electron coordinates $\boldsymbol{R}_i$ commute each with other this generalization reduces to inserting the proper summations into obtained slow motion Hamiltonian and replacing it by the following expression

$$\mathcal{H} = V^{(S)}(\boldsymbol{R}_1, \boldsymbol{R}_2, \cdots) + \frac{\ell_B^2}{4}\sum_{i=1}^{N} \nabla_i^2 V^{(S)}(\boldsymbol{R}_1, \boldsymbol{R}_2, \cdots). \tag{22}$$

Now we consider some simple examples in order to illustrate the application of the proposed simplified description of the electron motion in the case of strong magnetic fields. Let us start with the zeroth order ($\ell_B = 0$) approximation.

6. Classical equations of motion

In zeroth order approximation, we shall take into account only the first term in the Hamiltonian (21) and neglect the commutator (10) between X and Y coordinates. We know that neglecting the commutators leads us to the classical mechanics. However, one should remember that it is not correct just to neglect the commutators. It is necessary to replace them by the corresponding Poisson brackets according to the following rule (note that we inserted ℓ_B^2 instead of $\hbar$)

$$\frac{\mathrm{i}}{\ell_B^2}[A,B] \quad \rightarrow \quad \{A,B\} = \frac{\partial A}{\partial y}\frac{\partial B}{\partial x} - \frac{\partial A}{\partial x}\frac{\partial B}{\partial y}. \tag{23}$$

The most simple way to obtain the classical equations of motion is to use the Heisenberg equations of motion for the operators. Therefore, we write

$$\frac{d}{dt}X = \frac{\mathrm{i}}{\hbar}[H,X] = \frac{\ell_B^2}{\hbar}\frac{\mathrm{i}}{\ell_B^2}[H,X] \rightarrow \frac{\ell_B^2}{\hbar}\{H,X\} = \frac{c}{eB}\frac{\partial V}{\partial Y}, \tag{24}$$

$$\frac{d}{dt}Y = -\frac{c}{eB}\frac{\partial V}{\partial X}. \tag{25}$$

Note in the Heisenberg equations of motion the Plank constant $\hbar$ is used (in spite of the fact that commutator of the variables is proportional to the magnetic length squared), because it has to be in agreement with the slow motion Schrödinger equation (19). Those two equations of motion can be rewritten as a single vector equation

$$\dot{\boldsymbol{R}} = -\frac{c}{eB}[\boldsymbol{e}_z \times \nabla]V(\boldsymbol{R}) \tag{26}$$

where the symbol $\boldsymbol{e}_z$ stands for the unit vector perpendicular to the electron motion plane $z = 0$. It is a well known equation in plasma physics, and it describes the Larmor circle (the rotating electron in a strong magnetic field) drift caused by the gradient of applied additional potential.

Thus, we see that system of 2D electrons in the very strong magnetic field (in the conditions of the fractional quantum Hall effect, when only the part of the lowest Landau level is populated) demonstrates the classical behavior. This classical behavior is rather tricky. They do not behave as electrons. They behave as a system of classical gyroscopes.

Let us now take the ℓ_B^2 order terms into account. In this case the quantum mechanical correction should take place and we have to obtain something like quasi-classical description. In order to understand the main features of such quasi-classical motion let us take the most simple example of the parabolic dot with one and two electrons.

7. Single electron in a parabolic dot

In order to check the correctness of the above described method let us start with the problem of a single electron in a parabolic dot. In this case we have the following potential

$$V(\boldsymbol{r}) = \frac{m\omega_0^2}{2} r^2 \tag{27}$$

with the frequency ω_0 characterizing the strength of the confining potential, and according to Eq. (21) the following slow motion Hamiltonian

$$\mathcal{H} = \frac{m\omega_0^2}{2}(X^2 + Y^2) + \frac{1}{2} m\omega_0^2 \ell_B^2 = \frac{m\omega_0^2}{2}\left\{-\ell_B^4 \frac{\partial^2}{\partial X^2} + X^2 + \frac{1}{2}\ell_B^2\right\}. \tag{28}$$

This the well known Hamiltonian of the harmonic oscillator. Its eigenvalues and the corresponding eigenfunctions are

$$E_n = \frac{\hbar\omega_0}{\gamma}(n+1), \tag{29}$$

$$\Phi_n(X) = \frac{1}{\sqrt{\ell_B 2^n n! \sqrt{\pi}}} \mathrm{e}^{-X^2/2\ell_B^2} H_n(X/\ell_B) \tag{30}$$

where the parameter $\gamma = \omega_c/\omega_0$ characterizes the relative strength of the magnetic field, and H_n stands for the Hermite polynomial.

In order to evaluate the approximation obtained by solving the slow motion Schrödinger equation let us compare it with the exact Fock-Darwin result which is

$$E_{nm} = \hbar\omega_0 \left\{(2n + |m| + 1)\sqrt{1 + \gamma^2/4} + (m-1)\gamma/2\right\} \tag{31}$$

where orbital quantum number m is an integer, and radial quantum number n is an integer and nonnegative. This exact result together with the approximate one (29) are shown in Fig. 1 by solid and dashed curves, correspondingly. We see that in the asymptotic region $\gamma \to \infty$ (shown by the rectangular box) the approximate result is rather close to the rotational levels belonging to the lowest Landau level. Moreover, we may expect the quantitative agreement already at $\gamma \gtrsim 2$.

It is interesting to inspect how the wave function and the corresponding electron density looks like. However, we have to remember that eigenfunction (30) is not the electron wave function itself but it is its slow motion part only. In order to obtain the electron wave function according to Eq. (17) we have to multiply it by the fast motion part. Next we have to go back to the initial variables (11). It can be done using some integral transformation which is described in Appendix B. Using the transformation kernel (B.6) and restricting our consideration to the lowest fast wave function approximation (A.17) we write the total electron wave function in initial x, y variables as

$$\Psi_n(x,y) = \int_{-\infty}^{\infty} d\eta \int_{-\infty}^{\infty} dX \, \langle x, y|\eta, X\rangle \psi_0(\eta)\Phi_n(X)$$

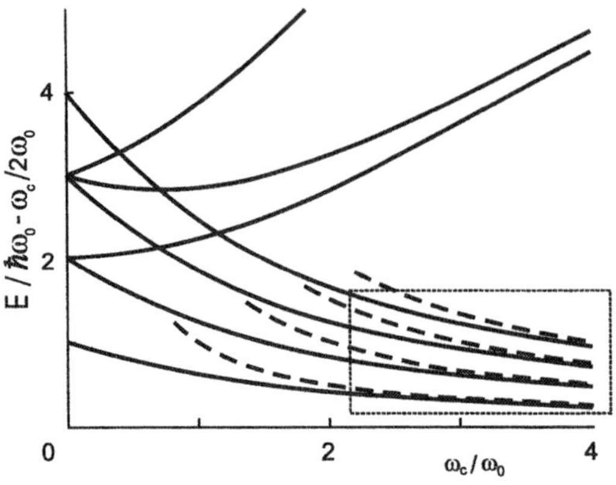

Fig. 1: Electron spectrum in a parabolic dot: solid curves — the exact result according (31), dashed curves — the slow motion approximation (29).

$$= \frac{1}{2\pi\ell_B\sqrt{2^n n!}} \int_{-\infty}^{\infty} d\eta \int_{-\infty}^{\infty} dX\, \mathrm{e}^{-\eta^2/2+\mathrm{i}y(X-\ell_B\eta)/2\ell_B^2-X^2/2\ell_B^2}$$
$$\cdot\,\delta(X+\ell_B\eta-x)H_n(X/\ell_B)$$
$$= \frac{\mathrm{e}^{(\mathrm{i}xy-x^2)/2l_b^2}}{2\pi\ell_B\sqrt{2^n n!}} \int_{-\infty}^{\infty} d\eta\, \mathrm{e}^{-\eta^2+(x-\mathrm{i}y)\eta/\ell_B} H_n(x/\ell_B-\eta). \tag{32}$$

The integral can be evaluated analytically. Using the standard integrals with Hermite polynomials [8] we obtain the following expression for the total electron wave function

$$\Psi_n(x,y) = \frac{1}{\ell_B\sqrt{2^{n+1}n!\pi}}\mathrm{e}^{\mathrm{i}n\varphi}(r/\ell_B)^n\mathrm{e}^{r^2/4\ell_B^2}, \tag{33}$$

and the corresponding electron density in the n-eigenstate

$$\rho_n(\boldsymbol{r}) \sim (r/\ell_B)^{2n}\mathrm{e}^{-r^2/2\ell_B^2}. \tag{34}$$

We see that in the case of large n (in the quasi-classical case) electrons are mainly located on the ring. Equating to zero the derivative of the above density expression we obtain the radius of this ring,

$$r_0 = \ell_B\sqrt{2n}. \tag{35}$$

Now, inserting the n from Eq. (29) we get

$$r_0 = \ell_B\sqrt{\frac{2\gamma E}{\hbar\omega_0}} = \sqrt{\frac{2E}{m\omega_0^2}}, \tag{36}$$

which exactly corresponds to the classical potential energy $E = V(r_0) = m\omega_0^2 r_0^2/2$ of the rotating electron drifting in the confining potential along the circle with the radius r_0. The single difference of quasi-classical electron behavior from the classical one is that now according to Eq. (34) it moves not along the thin trajectory, but it is spread over the ring with the thickness of order ℓ_B.

8. Two electrons in a dot

The other example considered here is the case of two electrons in a parabolic dot where the exact numerically solution (see, for instance, [9]) can be compared with our approximate results. In this case the behavior of electrons is described by the following potential

$$V(\boldsymbol{r}_1, \boldsymbol{r}_2) = \frac{m\omega_0^2}{2}\{r_1^2 + r_2^2\} + \frac{e^2}{|\boldsymbol{r}_1 - \boldsymbol{r}_2|}. \tag{37}$$

Let us introduce the center of mass and relative motion coordinates. We shall do it in a non standard way in order not to spoil the commutations rules for the fast and slow variables which we already used. We use the following definition

$$\boldsymbol{r}_c = \frac{1}{\sqrt{2}}(\boldsymbol{r}_1 + \boldsymbol{r}_2), \quad \boldsymbol{r}_r = \frac{1}{\sqrt{2}}(\boldsymbol{r}_1 - \boldsymbol{r}_2), \tag{38}$$

which leads to the separation of variables as the potential can be presented as a sum of two terms

$$V(\boldsymbol{r}_1, \boldsymbol{r}_2) = V_c(\boldsymbol{r}_c) + V_r(\boldsymbol{r}_r). \tag{39}$$

The potential for the center of mass motion

$$V_c(\boldsymbol{r}_c) = \frac{m\omega_0^2}{2} r_c^2 \tag{40}$$

exactly coincides with the single electron potential (27) which was already considered in the previous section. Consequently, the eigenvalue and eigenfunction of the center of mass motion coincide with those given by Eqs. (29,30). Note that now X and Y have to be replaced by the slow center of mass motion coordinates X_c and Y_c. Following the same procedure as in the previous section, we shall arrive at center of mass motion density given by Eq. (34) with the coordinate $\boldsymbol{r}$ replaced by the center of mass coordinate $\boldsymbol{r}_c$.

The relative motion potential is given as follows

$$V_r(\boldsymbol{r}_r) = \frac{m\omega_0^2}{2} r_r^2 + \frac{e^2}{r_r\sqrt{2}}. \tag{41}$$

According to Eq. (21), this leads to the following slow relative motion Hamiltonian

$$\mathcal{H}_r = \frac{m\omega_0^2}{2}\left(R_r^2 + \frac{\ell_B^2}{2}\right) + \frac{e^2}{R_r\sqrt{2}}\left(1 + \frac{\ell_B^2}{4R_r^2}\right). \tag{42}$$

The symbol $R_{\rm r}^2$ of course has to be replaced by the operator

$$R_{\rm r}^2 \to -\ell_B^4 \frac{\partial^2}{\partial X_{\rm r}^2} + X_{\rm r}^2. \tag{43}$$

The slow-motion Schrödinger equation with Hamiltonian (42) can now be solved by means of Fourier transformation technique presented in Appendix A. However, in the case of two electrons one can find the eigenvalues of the above slow motion Hamiltonian rather easily by paying attention to the fact that the eigenfunctions of $R_{\rm r}^2$ operator diagonalize Hamiltonian (42) as well. We already know the eigenvalues and eigenfunctions of operator $R_{\rm r}^2$ [Eqs. (29,30)]. Thus, in order to obtain the eigenvalues of Hamiltonian (42) we have to make the following replacement just in the above Hamiltonian

$$R_{\rm r}^2 \to \ell_B^2(2n+1). \tag{44}$$

Consequently, the relative slow motion eigenvalue reads

$$E_n^{(r)} = \hbar\omega_0 \left\{ \frac{n+1}{\gamma} + \frac{\lambda\sqrt{\gamma}}{\sqrt{2(2n+1)}} \left[1 + \frac{1}{4(2n+1)} \right] \right\} \tag{45}$$

where the dimensionless parameter of electron-electron interaction $\lambda = l_0/a_B$ is the ratio of the characteristic confining potential length $l_0 = \sqrt{\hbar/m\omega_0}$ and the Bohr radius $a_B = \hbar^2/me^2$.

Addition of the eigenvalue (29) for the center of mass motion, the relative motion eigenvalue (45), and $\hbar\omega_c$ for the lowest Landau level energy provides the final result for the two electron eigenvalues in the parabolic dot in the slow motion approximation

$$E_{N,n} = \hbar\omega_0 \left\{ \gamma + \frac{N+n+2}{\gamma} + \frac{\lambda\sqrt{\gamma}}{\sqrt{2(2n+1)}} \left[1 + \frac{1}{4(2n+1)} \right] \right\}. \tag{46}$$

The dimensionless eigenvalue (in units of $\hbar\omega_0$) dependencies on the relative magnetic field strengths (on parameter $\gamma = \omega_c/\omega_0$) for the case of $N = 0$ and several n values are shown in Fig. 2a. In Fig. 2b these eigenvalues are compared with the exact solution taken from Ref. [9]. We see that, when $n \to \infty$ and $\gamma \to \infty$ (namely, in the asymptotic region) the approximate consideration is in good agreement with the exact one. Moreover, the quasi-classical treatment describes correctly the main features of the electron behavior in strong magnetic fields, namely, the increment of the angular momentum (the quantum number n plays its role) with the increment of the magnetic field strength. Indeed, minimizing relative motion eigenvalue (45) with respect to the magnetic field γ in the case of large quantum numbers n we obtain the ground state orbital momentum

$$n_0 = (\lambda/4)^{2/3}\gamma \tag{47}$$

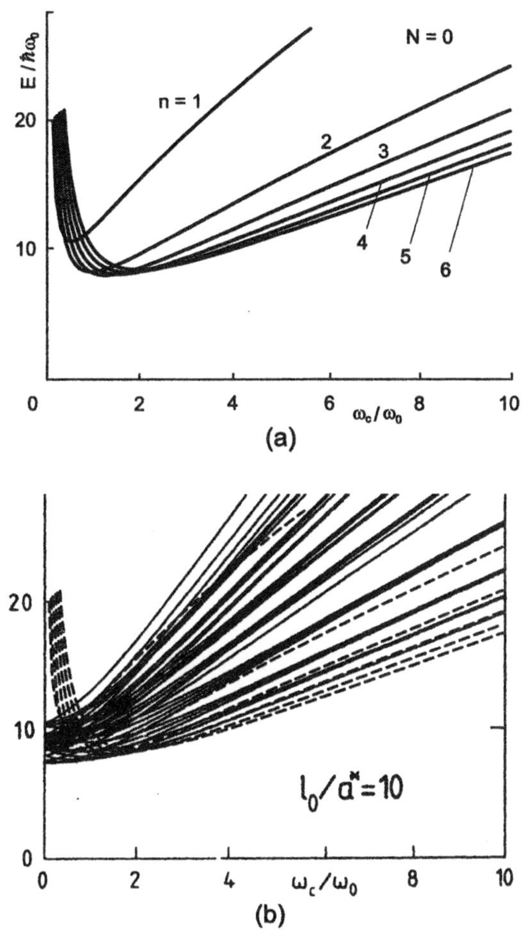

Fig. 2: Spectrum of a two electron system in a parabolic dot: (a) — slow motion approximation results; (b) — the comparison of approximate results (dashed curves) with the exact results (solid curves) taken from [9].

that is proportional to the magnetic field and agrees with the quantum mechanical result obtained in [10].

Let us also look at the electron density given by the quasi-classical approximation. First, we notice that according to what was said above the eigenfunction of Hamiltonian (42) coincides with the eigenfunction of the operator R_{r}^2. Thus, it coincides with wave function (30) with the coordinate X replaced by the relative motion coordinate X_{r}. So, performing the same transformation as it was done in Section 7. we obtain the relative motion density given by expression (34).

Taking into account the fact that $N = 0$ corresponds to the ground two electon state we can write down the two electron distribution function in the following form

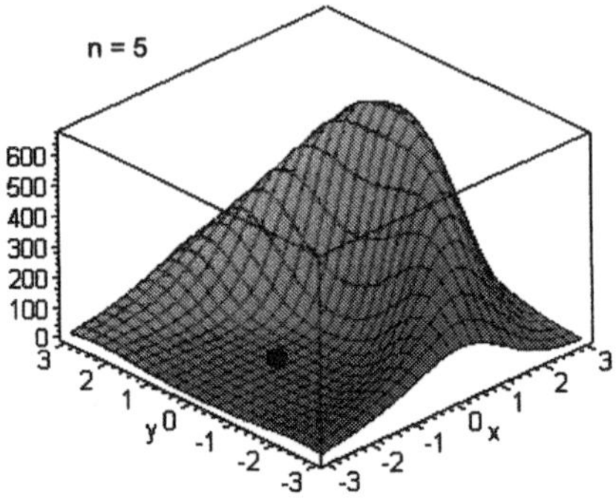

(a)

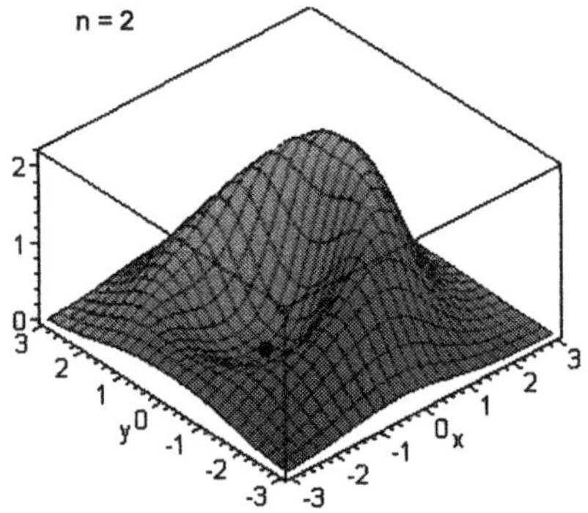

(b)

Fig. 3: Pair-correlation function for various orbital momenta.

$$\rho_n(\boldsymbol{r}_1, \boldsymbol{r}_2) \sim \mathrm{e}^{-r_{\mathrm{c}}^2/2\ell_B^2} \cdot (r_{\mathrm{r}}/\ell_B)^{2n} \mathrm{e}^{-r_{\mathrm{r}}^2/2\ell_B^2} \sim (\boldsymbol{r}_1 - \boldsymbol{r}_2)^{2n} \mathrm{e}^{-(r_1^2+r_2^2)/2\ell_B^2}. \tag{48}$$

In Fig. 3, the above function is plotted as a function of the first electron coordinate with the other electron coordinate fixed at the point corresponding to its classical equilibrium position. This point is indicated by a solid dot. In fact, the plot represents the so called pair correlation function. The plot in Fig. 3 (a) corresponds to $n = 5$ and the plot in Fig. 3 (b) to $n = 2$. We see that for larger n the pair correlation function demonstrates the peak opposite to the fixed electron position which corresponds to the Wigner crystallization of this simple two electron system in the strong magnetic field.

When the magnetic field strength decreases (corresponding to the ground state with the smaller angular momentum value, say, $n = 2$ as shown in Fig. 3 (b)) the Wigner crystal starts to melt — the pair correlation function transforms itself from the peak into ring. Note that it demonstrates the fact that the angular melting precedes the radial one and agrees with quantum mechanical result in [11] obtained for the case without the magnetic field.

9. Conclusions

In the asymptotic region of the strong magnetic fields, when all electrons occupy only the lowest Landau level, their kinetic energy is frozen out, and their behavior is guided by weakly varying (characterized by some characteristic length l_0) additional potential. Applying special fast and slow motion variables, the adiabatic procedure and the expansion in ℓ_B/l_0 powers some simplified approximate description can be developed.

In zeroth order approximation one gets the classical equations which describe the electrons as a system of gyroscopes. Those equations actually are the equations for the Larmor circle drift in the gradient of applied potential.

Applying the expansion up to $(\ell_B/l_0)^2$ order one obtain the self consistent equation set which coincides with the Schrödinger equation where the role of canonical variables play two cartesian slow motion 2D electron coordinates X and Y with the commutator proportional to the magnetic length squared (instead of being proportional to the Plank constant as it is in the standard Schrödinger equation). Two simple examples of a single and two electron in a parabolic dot demonstrate the accuracy and main features of proposed approximate description. The approximate eigenvalues coincide with the exact ones in the asymptotic region when $\gamma = \omega_c/\omega_0 \to \infty$, and even in the intermediate region ($\gamma \gtrsim 2$) one can expect rather good semi-quantitative description.

The electron wave functions can be obtained as a product of the fast wave function part corresponding to the lowest Landau level and slow wave function part obtained by the above specific slow motion Schrödinger equation. After the transformation to the original variables the wave function obtained in this way describes correctly the quantum mechanical ℓ_B/l_0 order correction to the classical electron motion.

The two electron in a dot example shows that the proposed approximate consideration describes such collective phenomena as the Wigner crystallization, the change of the angular momentum in the ground state when the magnetic field strength increases, the phenomena of angular and radial melting of the Wigner crystal. We hope that this approximate method can be useful for the consideration of more sophisticated many electron systems when the straightforward solution of quantum mechanical equation meets computational difficulties.

Acknowledgements

I would like to acknowledge Prof. François Peeters and Dr. Bart Partoens from the Antwerp University. Most my ideas on quantum dots appeared during my numerous visits there, due to the close collaboration and the discussions with them. I would like to thank Dr. Egidijus Anisimovas for drawing my attention to various representations of electron wave functions in magnetic field.

10. Appendix

A Slow motion Hamiltonian

As it was mentioned in Section 5. for performing the adiabatic procedure steps we have to pay attention to the fact that the variables X and Y do not commute with each other. That is why, instead of using the straightforward expansion (13) we apply the following Fourier transformation

$$V(x,y)=\frac{1}{2}\int_{-\infty}^{\infty}\frac{dk}{2\pi}\int_{-\infty}^{\infty}\frac{dq}{2\pi}\left\{\mathrm{e}^{ikx}\mathrm{e}^{iqy}+\mathrm{e}^{iqy}\mathrm{e}^{ikx}\right\}V(k,q), \tag{A.1}$$

$$V(k,q)=\int_{-\infty}^{\infty}dx\int_{-\infty}^{\infty}dy\mathrm{e}^{-ikx}\mathrm{e}^{-iqy}V(x,y). \tag{A.2}$$

These two expressions can be considered as a definition of the operator function $V(x,y)$. Thus in the first expression the symbols x and y will be considered as the operators, while in the second one x and y are just the dummy integration variables. The main advantage of such potential representation is that the operators x and y are moved from the general potential function $V(\boldsymbol{r})$ to more simple exponent functions. Although the old x and y variables commute we used the symmetric exponent product which will be necessary in further derivation.

Now we substitute variables (11) into exponentials and expand them into ξ,η-powers

$$\begin{aligned}
&\mathrm{e}^{ik(X+\ell_B\eta)}\mathrm{e}^{iq(Y-\ell_B\xi)}=\mathrm{e}^{ikX}\mathrm{e}^{iqY}\mathrm{e}^{i\ell_Bk\eta}\mathrm{e}^{-i\ell_Bq\xi}\\
&\quad=\mathrm{e}^{ikX}\mathrm{e}^{iqY}\left\{1+i\ell_Bk\eta-\frac{1}{2}\ell_B^2k^2\eta^2\right\}\left\{1-i\ell_Bq\xi-\frac{1}{2}\ell_B^2q^2\xi^2\right\}\\
&\quad=\mathrm{e}^{ikX}\mathrm{e}^{iqY}\left\{1+i\ell_B(k\eta-q\xi)-\frac{1}{2}\ell_B^2k^2\eta^2-\frac{1}{2}\ell_B^2q^2\xi^2+\ell_B^2kq\eta\xi\right\},
\end{aligned} \tag{A.3}$$

$$\begin{aligned}
&\mathrm{e}^{iq(Y-\ell_B\xi)}\mathrm{e}^{ik(X+\ell_B\eta)}=\mathrm{e}^{iqY}\mathrm{e}^{ikX}\mathrm{e}^{-i\ell_Bq\xi}\mathrm{e}^{i\ell_Bk\eta}\\
&\quad=\mathrm{e}^{iqY}\mathrm{e}^{ikX}\left\{1+i\ell_B(k\eta-q\xi)-\frac{1}{2}\ell_B^2k^2\eta^2-\frac{1}{2}\ell_B^2q^2\xi^2+\ell_B^2kq\xi\eta\right\}.
\end{aligned} \tag{A.4}$$

Taking into account the equality

$$a\eta\xi+b\xi\eta=\frac{1}{2}a(\eta\xi+\xi\eta+\mathrm{i})+\frac{1}{2}b(\xi\eta+\eta\xi-\mathrm{i}) \tag{A.5}$$

$$=\frac{1}{2}(a+b)(\xi\eta+\eta\xi)+\frac{\mathrm{i}}{2}(a-b) \tag{A.6}$$

we write the following expansion of the symmetric product of exponentials

$$\begin{aligned}
&\mathrm{e}^{ik(X+\ell_B\eta)}\mathrm{e}^{iq(Y-\ell_B\xi)}+\mathrm{e}^{iq(Y-\ell_B\xi)}\mathrm{e}^{ik(X-\ell_B\eta)}\\
&\quad=\left\{\mathrm{e}^{ikX}\mathrm{e}^{iqY}+\mathrm{e}^{iqY}\mathrm{e}^{ikX}\right\}\\
&\qquad\times\left\{1+i\ell_B(k\eta-q\xi)-\frac{1}{2}\ell_B^2(k\eta-q\xi)^2\right\}
\end{aligned}$$

$$+\frac{i}{2}\ell_B^2 kq\left\{e^{ikX}e^{iqY} - e^{iqY}e^{ikX}\right\}$$
$$= 2\left[e^{ikX}e^{iqY}\right]^{(S)} L(\xi,\eta,k,q) + i\ell_B^2 kq\left[e^{ikX}e^{iqY}\right]^{(A)}. \tag{A.7}$$

Note how the symmetric and antisymmetric exponent products and function $L(\xi,\eta,k,q)$ are defined.

Inserting the above expansion into Fourier transformation (A.1), changing the parameters k and q by the operators $i\partial/\partial x$ and $i\partial/\partial y$ acting on exponentials, and performing the integration by parts we arrive at the following potential expansion

$$V(X+\ell_B\eta, Y-\ell_B\xi) = \int_{-\infty}^{\infty} dx \int_{-\infty}^{\infty} dy \int_{-\infty}^{\infty} \frac{dk}{2\pi} \int_{-\infty}^{\infty} \frac{dq}{2\pi} V(x,y)$$
$$\times \left\{L(\xi,\eta,k,q)\left[e^{ik(X-x)}e^{iq(Y-y)}\right]^{(S)} + \frac{i}{2}\ell_B^2 kq\left[e^{ik(X-x)}e^{iq(Y-y)}\right]^{(A)}\right\}$$
$$= \int_{-\infty}^{\infty} dx \int_{-\infty}^{\infty} dy \int_{-\infty}^{\infty} \frac{dk}{2\pi} \int_{-\infty}^{\infty} \frac{dq}{2\pi} V(x,y)$$
$$\times \left\{L(\xi,\eta,i\partial/\partial x, i\partial/\partial y)\left[e^{ik(X-x)}e^{iq(Y-y)}\right]^{(S)}\right.$$
$$\left. - \frac{i\ell_B^2}{2}\frac{\partial^2}{\partial x\partial y}\left[e^{ik(X-x)}e^{iq(Y-y)}\right]^{(A)}\right\}$$
$$= \int dx \int dy \int \frac{dk}{2\pi} \int \frac{dq}{2\pi}\left\{\left[e^{ik(X-x)}e^{iq(Y-y)}\right]^{(S)} L(\xi,\eta,-i\partial/\partial x,-i\partial/\partial y)\right.$$
$$\left. - \frac{i\ell_B^2}{2}\left[e^{ik(X-x)}e^{iq(Y-y)}\right]^{(A)}\frac{\partial^2}{\partial x\partial y}\right\}V(x,y)$$
$$= \int dx \int dy \int \frac{dk}{2\pi} \int \frac{dq}{2\pi}\left\{\left[e^{ik(X-x)}e^{iq(Y-y)}\right]^{(S)}\left\{1 + \ell_B\left(\eta\frac{\partial}{\partial x} - \xi\frac{\partial}{\partial y}\right)\right.\right.$$
$$\left.\left. - \frac{\ell_B^2}{2}\left(\eta\frac{\partial}{\partial x} - \xi\frac{\partial}{\partial y}\right)^2\right\} - \frac{i\ell_B^2}{2}\left[e^{ik(X-x)}e^{iq(Y-y)}\right]^{(A)}\frac{\partial^2}{\partial x\partial y}\right\}V(x,y). \tag{A.8}$$

Actually it is the definition of the operator function expansion which we have to use instead of expression (13). It can be rewritten in more simple way if we use the following operator function definition

$$F^{(XY)}(X,Y) = \int dx \int dy \int \frac{dk}{2\pi} \int \frac{dq}{2\pi} e^{ik(X-x)}e^{iq(Y-y)}F(x,y). \tag{A.9}$$

It defines the function with ordered operators — all operators X stand on the left side of the operators Y in all terms of its Taylor expansion, or Fourier transform.

Defining the symmetric and antisymmetric operator function as

$$F^{(S)}(X,Y) = \frac{1}{2}\left\{F^{(XY)}(X,Y) + F^{(YX)}(X,Y)\right\}, \tag{A.10}$$

$$F^{(A)}(X,Y)=\frac{1}{2}\left\{F^{(XY)}(X,Y)-F^{(YX)}(X,Y)\right\} \tag{A.11}$$

we rewrite the potential expansion in the following formal simple form

$$V=\left\{1+\ell_B\left(\eta\frac{\partial}{\partial X}-\xi\frac{\partial}{\partial Y}\right)-\frac{\ell_B^2}{2}\left(\eta\frac{\partial}{\partial X}-\xi\frac{\partial}{\partial Y}\right)^2\right\}V^{(S)}+\frac{\mathrm{i}\ell_B^2}{2}V_{XY}^{(A)}. \tag{A.12}$$

Now we are ready to perform the next step of our adiabatic procedure, namely, to insert the obtained potential expansion into the fast Hamiltonian (15) and solve the fast eigenvalue problem (18). It can be easily performed using the standard perturbation technique.

Indeed, using the modified fast Hamiltonian

$$\mathcal{H}_f=\mathcal{H}_0+\mathcal{H}_1+\mathcal{H}_2, \tag{A.13}$$

$$\mathcal{H}_0=\frac{\hbar\omega_c}{2}(\xi^2+\eta^2), \tag{A.14}$$

$$\mathcal{H}_1=\ell_B\left(\eta\frac{\partial}{\partial X}-\xi\frac{\partial}{\partial Y}\right)W^{(S)}, \tag{A.15}$$

$$\mathcal{H}_2=-\frac{\ell_B^2}{2}\left(\eta\frac{\partial}{\partial X}-\xi\frac{\partial}{\partial Y}\right)^2W^{(S)} \tag{A.16}$$

we obtain the zeroth order eigenvalue and function

$$E_0=\frac{\hbar\omega_c}{2},\quad \psi_0(\eta)=\pi^{-1/4}\mathrm{e}^{-\eta^2/2}. \tag{A.17}$$

Next, due to the zeroth order function symmetry we get $E_1=0$, and solving the first order equation

$$\{\mathcal{H}_0-E_0\}=-H_1\psi_0=-\ell_B(V_X^{(S)}-\mathrm{i}V_Y^{(S)})\eta\psi_0 \tag{A.18}$$

we define the first order correction to the wave function

$$\psi_1(\eta|X,Y)=-\frac{\ell_B(V_X^{(S)}-\mathrm{i}V_B^{(S)})}{\hbar\omega_c}\eta\psi_0. \tag{A.19}$$

Then from the second order equation one can easily get the following second order eigenvalue correction

$$\begin{aligned}E_2(X,Y)&=\int_{-\infty}^{\infty}d\eta\,\psi_0(\eta)H_2\psi_0(\eta)+\int_{-\infty}^{\infty}d\eta\,\psi_0(\eta)H_1\psi_1(\eta|X,Y)\\&=\frac{\ell_B^2}{4}\left\{V_{XX}^{(S)}+V_{YY}^{(S)}\right\}-\frac{\ell_B^2}{2\hbar\omega_c}\left\{V_X^{(S)2}+V_Y^{(S)2}\right\}.\end{aligned} \tag{A.20}$$

Now having the fast motion problem eigenvalue calculated we can proceed with the adiabatic procedure. For this purpose we present the total wave function as

$$\Psi(\eta,X,t)=\mathrm{e}^{-\mathrm{i}E_0t}\left\{\psi_0(\eta)+\psi_1(\eta|X,Y)\right\}\Phi(X,t), \tag{A.21}$$

and inserting it into Eq. (1), we obtain the following expression

$$\left\{ \mathrm{i}\frac{\partial}{\partial t} + E_0 - H_f - V^{(S)} - \frac{\mathrm{i}\ell_B^2}{2} V_{XY}^{(A)} \right\} \cdot \left\{ \psi_0(\eta) + \psi_1(\eta|X,Y) \right\} \Phi(X,t) = 0. \tag{A.22}$$

Multiplying the above equation by function $\psi_0(\eta)$ from the left side, integrating it over all $-\infty < \eta < \infty$ interval and taking into the action of the fast Hamiltonian H_f on the fast wave function part we arrive at the slow motion equation (19) with the effective Hamiltonian

$$\mathcal{H} = V^{(S)} + \frac{\mathrm{i}\ell_B^2}{2} V_{XY}^{(A)} + \frac{\ell_B^2}{4} \nabla^2 V^{(S)} - \frac{\ell_B^2}{2\hbar\omega_c} \{\nabla V^{(S)}\}^2. \tag{A.23}$$

The performed procedure is consistent at least with the accuracy up to ℓ_B^2. That is why we shall omit the second and the last terms in (A.23) as they are of order ℓ_B^4. In the last term this dependence appears due to the additional factor ω_c in the denominator, while in the second term it is caused by slow variable commutator (10). Omitting these terms we arrive at the final slow motion Hamiltonian (21).

B Coordinate transformation

According to the ideas of the quantum mechanics we can change the wave function variables by means of the following transformation

$$\Psi(x,y) = \int_{-\infty}^{\infty} d\eta \int_{-\infty}^{\infty} dX \, \langle x,y|\eta,X\rangle \Psi(\eta,X) \tag{B.1}$$

where the transformation function $\langle x,y|\eta,X\rangle$ has to be chosen as the eigenfunction of operators $\hat{x}$ and $\hat{y}$ with the corresponding eigenvalues x and y. Namely, this transformation function has to obey the following equations

$$\{\hat{x} - x\}\langle x,y|\eta,X\rangle = 0, \tag{B.2}$$
$$\{\hat{y} - y\}\langle x,y|\eta,X\rangle = 0, \tag{B.3}$$

or

$$\{X + \ell_B\eta - x\}\, \langle x,y|\eta,X\rangle = 0, \tag{B.4}$$
$$\left\{ -\mathrm{i}\ell_B^2 \frac{\partial}{\partial X} + \mathrm{i}\ell_B \frac{\partial}{\partial \eta} - y \right\} \langle x,y|\eta,X\rangle = 0. \tag{B.5}$$

It can be checked straightforwardly that the following transformation function satisfies both equations

$$\langle x,y|\eta,X\rangle = \frac{1}{2\sqrt{\pi}\ell_B} e^{\mathrm{i}y(X-\ell_B\eta)/2\ell_B^2} \delta(X + \ell_B\eta - x). \tag{B.6}$$

The normalization factor is chosen in agreement with the condition

$$\int_{-\infty}^{\infty} dx \int_{-\infty}^{\infty} dy\, \langle x, y|\eta, X\rangle\langle x, y|\eta', X'\rangle = \delta(\eta - \eta')\delta(X - X'). \tag{B.7}$$

References

[1] L. Jacak, P. Hawrylak, and A. Wójs, *Quantum dots* (Springer-Verlag, Berlin, 1998).

[2] C. de C. Chamon and X. G. Wen, Phys. Rev. B **49**, 8227 (1994).

[3] E. Goldmann and S. R. Renn, cond-mat/9909071 (1999).

[4] K. Hirose and N. S.Wingreen, Phys. Rev. B **59**, 4604 (1999).

[5] P. A. Maksym, Phys. Rev. B **53**, 10871 (1996).

[6] S. M. Riemann, *et all*, Phys. Rev. Lett. **83**, 3270 (1999).

[7] V. M. Bedanov and F. M. Peeters, Phys. Rev. B **49**, 2667 (1994).

[8] I. S. Gradshteyn and I. M. Ryzhik, *Table of Integrals, Series and Products* (Academic Press, New York, 1994), chap. 7.3, p. 843.

[9] U. Merkt, J. Huser and M. Wagner, Phys. Rev. B **43**, 7320 (1991).

[10] A. Matulis, F. M. Peeters, Solid St. Comm., **117**, 655 (2001).

[11] A. V.Filinov, M. Bonitz, and Yu. E. Lozovik, Phys. Rev. Lett., **86**, 3851 (2001).

Chapter 9

Micro-Hall-magnetometry

M. Rahm, J. Biberger, and D. Weiss*

Institut für Experimentelle und Angewandte Physik, Universität Regensburg, Germany
**E-mail:dieter.weiss@physik.uni-regensburg.de*

Abstract

Micro-Hall sensors are sensitive tools to examine magnetization patterns on a nanoscale. These Hall sensors can either be used as non-invasive 'tips' which are scanned across a magnetic surface and deliver spatially resolved information on sub-micron magnetic stray field distributions or, alternatively, as miniaturized magnetometers to study the magnetization reversal of individual nanomagnets. Micro-Hall-magnetometry together with complementary imaging techniques such as, Lorentz- and magnetic force microscopy provide important insights in the magnetic switching process of 'mesoscopic' magnets. Below we give a brief survey of these techniques applied to magnetic nanopillars, micro-rhombs and nanodisks.

1. Introduction

The past two decades have shown the development of what is now called 'mesoscopic physics'. Due to advanced fabrication techniques the range of magnitude of experimentally accessible feature sizes has been reduced to dimensions, which allowed the discovery of many phenomena unknown before. The transition from classical to mesoscopic behavior takes place as soon as the size of the structures to be examined is comparable to some characteristic lengths of the system including electron mean free path, phase relaxation length and Fermi wavelength [1]. Miniaturizing the device dimensions beyond phase breaking scales provided, for example, experimental evidence of the Aharonov-Bohm effect in the solid state [2–4]. Other mesoscopic features are conductance fluctuations [5] and electrical properties determined by the exact geometry of the devices (see [6,7] and other examples below). However, not only the research concerned with electrical transport has been promoted by the reduction of feature size, but also, for example, the large area of magnetism. An important characteristic length in this field is represented by the exchange length $l_{ex} = \sqrt{A/K}$ [8]. Here A represents the exchange stiffness constant, which gives a measure of the strength of the exchange interaction trying to keep magnetic moments parallel. The anisotropy constant K describes the impact of magnetic anisotropy attempting to align the spins in a specific direction of the ferromagnetic body. It arises from different origins such as crystal structure or shape of the particle. The latter point becomes of crucial importance if the size of ferromagnetic particles is reduced to the micrometer and sub-micrometer range. As long as the dimensions of the ferromagnet exceed l_{ex} significantly the magnetization configurations typically show multi domain patterns [9–11]. Scaling down the magnets to sizes of typical domain wall widths results in the occurrence of coherent spin structures, which can be described as continuous, 'flowing' magnetization patterns [12]. In this transition regime the magnetic properties of these so called nanomagnets can be tailored by varying their shape, size, material and structure [13–18]. This makes them an interesting subject not only for current fundamental research, but also for many economically relevant applications – mainly in memory and data storage technology [19–22]. The purpose of this article is to give an introductory survey of micro-Hall-magnetometry, which represents a method capable of providing insight into the magnetic behavior of small ferromagnetic particles. The method employs mesoscopic Hall junctions to gather information about nanoscale magnets.

2. Principles of Hall-magnetometry

A variety of methods has been developed to detect the characteristic magnetic properties of magnetic nanoparticles. While some of them are sensitive to the magnetization of the whole sample others just probe its surface magnetization. Lorentz Transmission Electron Microscopy (LTEM, a short description of this method is given below) and optical methods using the Faraday effect, for instance, belong to the former category, whereas Kerr microscopy and Scanning Electron Microscopy with Polarization Analysis (SEMPA) are examples for the latter techniques. Other methods utilize the particles' stray field. The sinks and sources of the stray field are the so called surface and volume charges caused by a divergence of the magnetization which appear for example at domain walls. Typical stray fields of magnetic particles with sub-micrometer dimensions are in the range of some 10 mT at a distance of about 50 nm. Therefore only a few techniques are able to provide the necessary sensitivity. Micro-SQUIDs, extensively used by Wernsdorfer et al. [23], are powerful tools among them, but are restricted to low temperatures. Another useful method to investigate magnetic stray fields on a sub-micron scale is Magnetic Force Microscopy (MFM, see text below) where a tiny magnetic tip is scanned over the sample's surface. The interaction between tip and stray field is recorded. As will be explained below, this method is invasive and can disturb the magnetization of the particle. Micro-Hall-magnetometry, in contrast, imposes only a negligible perturbation on the nanomagnet during the magnetization reversal process. The magnetic field caused by the sensor current is only on the order of 10 μT. An important advantage of this method is that it can be employed over a wide range of temperatures i. e., from cryogenic temperatures up to ambient temperature. The principle of conventional macroscopic Hall sensors is generally known. Charge carriers that carry a current in a cross shaped Hall device are deflected by the Lorentz force caused by the normal component of the stray field. Therefore a voltage can be detected perpendicular to the current path, between the voltage probes on either side. In principle this procedure is applicable also for very small magnetic (stray) fields to be measured. The Hall voltage U_{H} is described by the expression

$$U_{\mathrm{H}} = \frac{IB}{n_{\mathrm{S}}\mathrm{e}}$$

where I is the applied current, B is the magnetic field, n_{S} stands for the carrier density and e is the elementary charge. The Hall coefficient $1/n_{\mathrm{S}}\mathrm{e}$ is sensor specific. As the Hall voltage is proportional to the current, the maximum signal is restricted by the current that the miniaturized crosses can sustain. The way to provide a sufficiently high Hall coefficient is to reduce the carrier density. Therefore one needs materials with a small carrier density and low resistivity. Metallic systems are not ideal in this respect, because the areal electron densities of thin metal films (assuming a thickness of 10 nm) are of order 10^{17} cm^{-2}. An enhancement of the voltage signal by orders of magnitude can be obtained by applying semiconductor Hall devices. Extremely sensitive Hall sensors can be fabricated from GaAs/AlGaAs semiconductor heterostructures providing a two dimensional electron gas (2DEG) only some ten nanometers below the surface. The mobility and the density of 2DEG electrons are typically several 10^5 cm$^2/Vs$ and some 10^{11} cm^{-2}, respectively.

3. Sample fabrication, operational modes and limitations

In a first fabrication step, these devices are patterned by conventional optical lithography involving wet-chemical mesa etching and alloying ohmic contacts. In a second

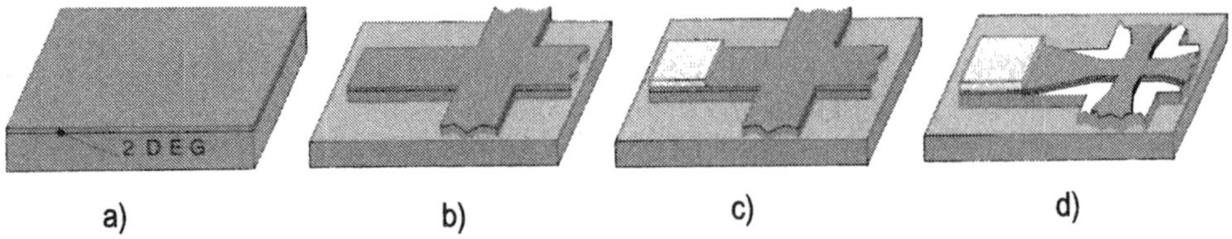

Fig. 1: Scheme of the fabrication steps of a micro-Hall sensor. a) The whole structure is based on a semiconductor heterojunction with a 2DEG very close to the surface. b) Wet chemical etching defines the cross shaped mesa structure. c) Thermally evaporated metal contacts are alloyed into the semiconductor material in order to form ohmic contacts. d) Electron beam lithography and dry etching are used to confine the final sensor geometry.

step the actual cross shaped Hall sensor with lateral dimensions of only some hundred nanometers is defined. This step is done by e-beam lithography followed by chemically assisted ion beam etching (CAIBE) or reactive ion etching (RIE). Because the electron gas in GaAs/AlGaAs is depleted on the edges of the mesa structure this final restriction of the crosses results in sensors with even smaller effective dimensions. Figure 1 sketches the different steps of the preparation of a Hall sensor.

One possible approach to apply the Hall sensors for probing stray fields on a submicron scale is Scanning Hall Probe Microscopy (SHPM) [25–27]. The method is similar to magnetic force microscopy (described below), but instead of a magnetic tip, a micro-Hall sensor is scanned across the surface to probe the local magnetic stray field by measuring the resulting Hall voltage. The guiding of the micro-sensor is

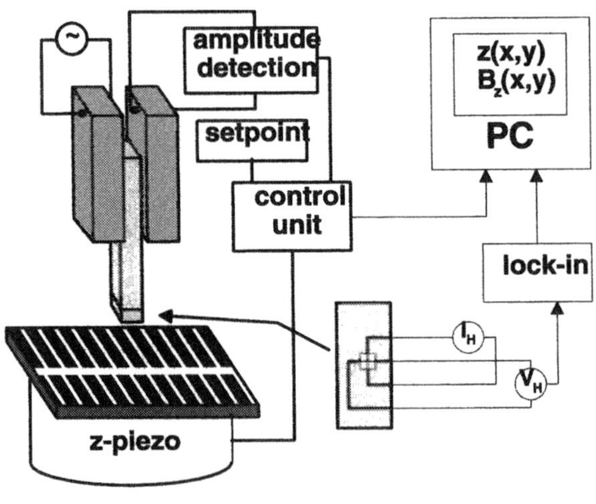

Fig. 2: Schematic view of the shear force detection assembly. The cantilever which is sandwiched between two piezo plates (gray) oscillates at its resonance frequency driven by one of the piezos. The amplitude is detected by the second piezo plate and serves as the control signal for the scanner z-piezo element to maintain constant sensor-sample distance. Taken from [24].

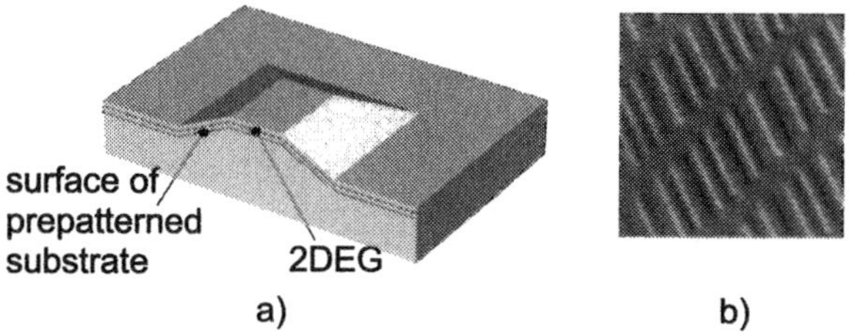

Fig. 3: a) Non planar 2DEGs are used to raise the sensitive area of the Hall sensors. The section illustrates the prepatterned substrate which is overgrown by the epitactic semiconductor heterojunction. b) Stray field map of a magnetic hard disk taken by scanning a Hall probe over the disk's surface. The micrograph shows a scanning area with a size of $(48\ \mu m)^2$. Taken from [25].

accomplished by the scanning unit of a scanning microscope the probe is attached to. The sample sensor distance can be adjusted e. g., by a shear force distance control as described in [24] and shown in Fig. 2. Recording the Hall voltage across the scanned area gives a complete map of the magnetic field distribution of a magnetic surface. In order to adopt this method for magnetic fields that fluctuate on a sub-micron length scale, some extra features have to be implemented. The distance between the sample's surface and the sensor becomes crucial, because of the rapid spatial decay of the stray field. The structure of the sensor should be optimized by minimizing the distance between the surface and the sensitive area of the probe. This can be achieved by raising the central area of the Hall probe by employing non planar 2DEGs [25]. Figure 3 a) illustrates the experimental realization. The heterostructure is grown by molecular beam epitaxy on a prepatterned frustum of pyramid on the GaAs substrate. A sub-micron Hall sensor is patterned on top of the mesa. Hence, the sensitive area of the Hall sensor is the most elevated part of the whole probe. Figure 3 b) shows the stray field map of a magnetic hard disk recorded by SHPM, the single bits can be recognized very well. The current lateral resolution is ≈ 200 nm and limited by the depletion at the lateral boundaries of the GaAs based 2DEG [25].

Miniaturization of the Hall sensors into the few nanometer regime is limited by the number of electrons in the sensor. For typical carrier densities between 10^{11} cm^{-2} and 10^{13} cm^{-2}, 3 to 250 electrons can be found in a 50 nm $\times$ 50 nm area. Such small numbers of electrons involved in the charge transport lead to increased noise and therefore to worthless signals if the amplitude of the noise exceeds the actual Hall voltage. Thus the highest accessible lateral resolution of nanoscale Hall sensors based on 2DEG systems is estimated to be of order ≈ 50 nm.

Instead of moving micro-Hall sensors as scanning probes they can be utilized as miniaturized magnetometers to study magnetization reversal of nanomagnets. The basic idea is to pattern the magnetic particle to be examined directly onto the Hall cross sensor. This technique, which is illustrated by the sketch in Fig. 4, has become a powerful method for the investigation of micron and sub-micron size magnetic particles, as will be demonstrated by several examples in the following sections. If

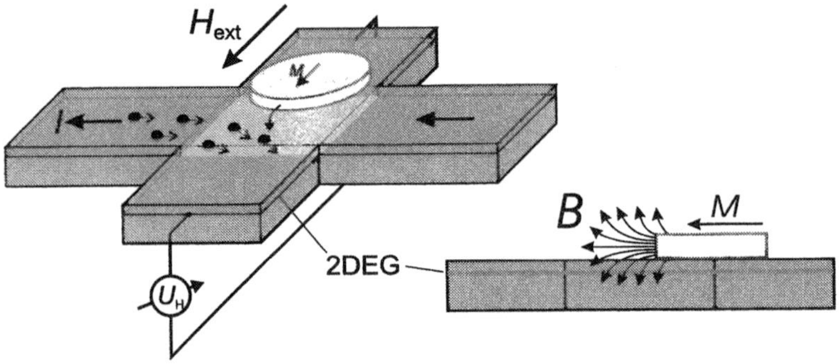

Fig. 4: Schematic sketch of a Hall sensor including a disk-shaped magnetic particle on its top. 2DEG-electrons entering the junction area are deflected by the inhomogeneous stray field emanating from the micro-magnet. In the ballistic transport regime the stray field is averaged over the light gray shaded region in the center of the cross.

the magnet to be investigated has a size comparable to the size of the active area of the Hall cross, the stray field is distributed inhomogeneously across the active area. Thus it is crucial to know how the Hall voltage depends on the inhomogeneous magnetic field penetrating the sensitive area of the sensor. This problem has been solved both for the ballistic and the diffusive transport regime.

The ballistic motion of electrons in a mesoscopic Hall bar under the influence of a local inhomogeneous magnetic field has been studied numerically [28,29]. These theoretical investigations are based on a classical approach [7], which is justified in the low-magnetic-field-regime where no quantized Hall conductance appears. The model used is comparable to an electron billiard. As the Fermi wavelength of the electrons is usually much smaller than the size of the Hall cross, the electrons are treated as classical particles, which are specularly reflected at the 2DEG boundaries. For a given inhomogeneous distribution of the magnetic field, classical trajectories are calculated for a large number of electrons injected into the cross junction region. These calculations are used to determine the transmission and reflection probabilities within the Landauer-Büttiker formalism applied to calculate the Hall voltage.

To generate an inhomogeneous magnetic field across the active area the calculations were performed for a disk-shaped magnetic anti-dot placed on the junction [28,29]. Underneath the anti-dot the field is set to zero while it is assumed to be homogeneous outside. This situation is depicted in the inset of Fig. 5. The calculated curves plotted in Fig. 5 show the dependence of the Hall factor $\alpha = R_{\mathrm{H}}/B$ on the diameter of the disk (with $R_{\mathrm{H}} = U_{\mathrm{H}}/I$, Hall resistance). While α decreases with increasing diameter, $\alpha^* = R_{\mathrm{H}}/\langle B\rangle$ appears to be constant. Here $\langle B\rangle$ represents the magnetic field averaged over the central cross junction area. As further investigations revealed, α^* is also independent of the exact position of the magnetic anti-dot on top of the junction region. The same results were obtained when the calculations were performed for a magnetic dipole instead of an anti-dot. This demonstrates that R_{H} is independent of the detailed distribution of the magnetic field penetrating the central area. This theoretical result is important for the practical use of the Hall

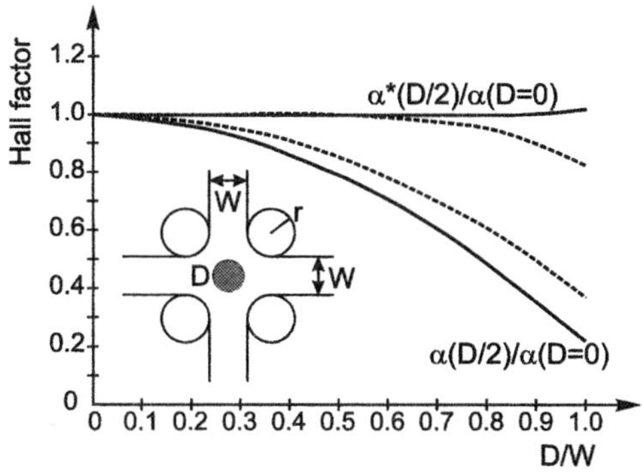

Fig. 5: The inset sketches the Hall cross sensor with width W of the current and voltage probes. The radius of curvature of the corners is represented by r. The junction area is penetrated by a homogeneous magnetic field outside the disk (diameter D), whereas no field emanates from the interior (magnetic anti-dot). The full lines of the graph display the Hall factor for sharp corners (r=0), the dashed lines are calculated for r/W=0.1. Data taken from [28].

cross sensor, because it connects the measured Hall voltage in a quantitative way with the magnetic flux density in the cross area.

However, some restrictions to that rule have to be considered. For example, a very strong local magnetic field can deflect the incident electrons and prevent them from reaching the central, strongest part of the field distribution. In this case the electrons do not explore the whole junction area. This leads to deviations from the picture given above [30]. Another restriction concerns the lithographic quality of the employed Hall sensors. Samples fabricated by optical or electron beam lithography often have rounded corners. Curves calculated for circularly rounded corners are plotted as dashed lines in Fig. 5. For a radius $r/W = 0.1$ the Hall factor α^* is constant for $D/W < 0.7$. This means that the low field Hall resistance is still determined by the average magnetic field.

Furthermore it has been reported that the Hall voltage in some cases does not depend linearly on an applied magnetic field at low B [6,7,31–33]. Among the observed magnetoresistance anomalies are the quenching of the Hall effect, a negative Hall resistance and the appearance of the last Hall plateau. These phenomena were explained by collimation and scattering of ballistic electrons dependent on the exact geometry of the junction. In our experiments these anomalies were not observed. Quantum interference effects can be suppressed by using high currents (up to 30 μA) which heat the electron system to higher temperatures where the quantum fluctuations wash out [34].

In the diffusive regime, the transport properties of a 2DEG in a four-terminal Hall junction have been examined by Ibrahim et al. [35], who studied the Hall voltage response to different nonuniform axially symmetric magnetic field profiles numerically. For example, the inhomogeneous magnetic distributions of a dot and an anti-dot, a Gaussian and a dipole field profile were treated. For low magnetic fields they found that the Hall resistance is insensitive to the detailed field profile. Instead, it rather depends on the total flux through the cross junction region. However, an essential difference with respect to the ballistic regime exists: While in the ballistic regime the measured Hall voltage depends on the magnetic flux averaged over the immediate junction region, in the diffusive regime the relevant area is twice as large. The reason for the different areas lies in the fact that a part of the transport current is spread into the voltage probes, when the Hall sensor is operated in the diffusive regime. In Ref. [36] and [37] it is demonstrated numerically, that the Hall response generated in the regions outside the central cross area is smaller than the contribution originating from inside. As long as the inhomogeneous magnetic field is sufficiently weak, the incoming electrons can reach any part of the junction area, and the Hall voltage depends on the average field through the effective area. For locally strong magnetic fields however, there might be areas which are not reached by electron trajectories. In this case the field averaging mechanism breaks down.

Whether transport is ballistic or diffusive depends on the size of the Hall junction in respect to the mean free path l_e of the electrons. If W is much smaller than l_e, the transport is ballistic and if l_e is significantly smaller than W it is diffusive. Thus increasing temperatures induce a change from ballistic to diffusive transport. With increasing temperature, increased phonon scattering reduces the electron mean free path. How this transition affects the micro-Hall measurements is illustrated in the following by means of a magnetic particle with a rectangular stray field hysteresis loop. An individual pillar shaped Nickel dot is placed on the sensor by using e-beam lithography and electroplating [39]. With this aim, the Hall cross is covered by a 10 nm thin Cr/Au gate electrode deposited by thermal evaporation and subsequent lift off. Figure 6 shows a SEM image of such a Hall cross device with an electroplated nickel pillar on top of the crossing area [38]. The electroplated dots with diameters of about 150 nm show aspect-ratios (height/diameter) up to 3. This means that, due to shape anisotropy, the dots behave like single domain particles for magnetization reversal along the axis of the cylinders. In Fig. 7, the Hall voltage is depicted for

Fig. 6: Nickel pillar (height: 370 nm, diameter: 170 nm) in the center of a micro-Hall sensor (width W: 850 nm). Taken from [38].

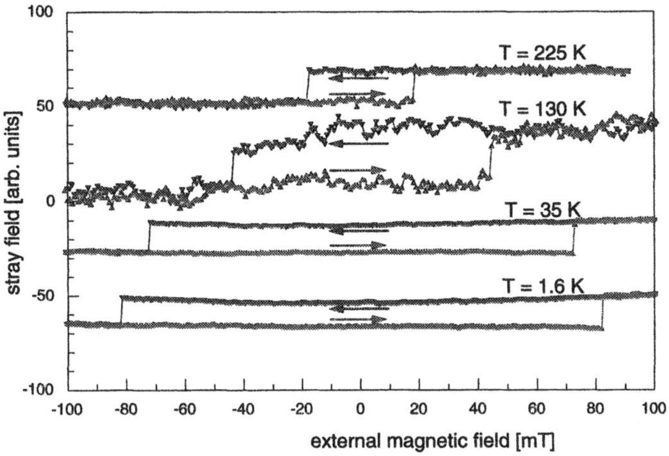

Fig. 7: The four loops which are offset vertically for clarity, show hysteresis loops of the same Ni pillar taken at different temperatures. With increasing temperature the coercive field decreases. The transition from the ballistic to the diffusive transport regime (T = 130 K) is characterized by strikingly strong noise. Taken from [39].

different temperatures during magnetization reversal. The first thing to recognize is a significant decrease of the switching field for rising temperatures. This dependency is almost linear for the Ni dots in these measurements. This is in disagreement with the Néel-Brown model of thermally assisted magnetization reversal over a single potential barrier [40] expected and measured for single domain particles [41–43].

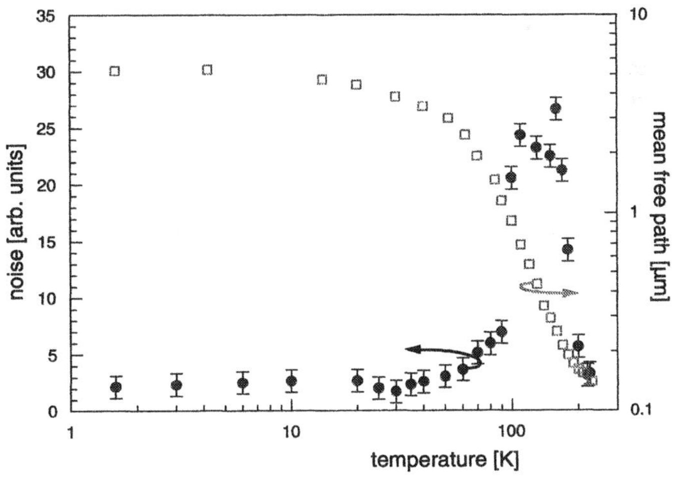

Fig. 8: The graph shows the temperature dependent noise (black dots) of the Hall measurements in Fig. 7 and the mean free path of the 2DEG electrons in the Hall cross. For temperatures between 100 K and 180 K, where the mean free path is in the order of the lateral dimensions of the sensor, the noise reveals a distinct peak. Taken from [38].

The temperature dependence as well as a large variation in switching fields for dots with similar shape and size suggest that oxide layers (capable of pinning the magnetization) on the surface of the pillars strongly influence the magnetization reversal process. Another interesting aspect in Fig. 7 concerns the change of the transport regime. Although for the lower and the highest temperatures the noise is comparatively low, the curve measured at 130 K reveals much more noise. This fact can be associated with the temperature dependence of the mean free path l_e. Figure 8 shows corresponding data together with the noise level for the Hall measurements. With l_e becoming shorter the noise becomes maximal, when l_e is in the order of the lateral dimensions of the Hall sensor, i. e., when the transport regime changes. As described above, this transition is associated with an increase of the sensitive area of the cross. Therefore one might expect a drop in the amplitude of the Hall signal for higher temperatures. This is true in principle, but as a matter of fact the increasing temperature is also connected with a slight change of the effective dimensions (depletion length) and the electron density (Hall coefficient) of the sensor and thus the amplitude of the Hall signal is not reduced.

4. Measurement technique

In the last section some Hall measurements were anticipated. This section is concerned with the way these measurements were actually carried out and how the hysteresis loops are recorded. An ac current of fixed amplitude between 1 μA and 30 μA is applied to the current channel of the Hall cross while the Hall voltage is measured by standard lock-in technique. It is useful to maximize the voltage signal by controlling the electron density with the help of a voltage applied to the metallic gate that covers the whole structure.

The variable homogeneous magnetic field is provided by a commercial ^{4}He cryostat with a superconducting magnet. The system is also equipped with a variable temperature insert (VTI) allowing to adjust the temperature between 1.4 K and almost the room temperature. In the experiment displayed in Fig. 6, the externally applied magnetic field was oriented along the axis of the nickel cylinder perpendicular to the 2DEG. This magnetic field is used to switch the magnetization direction of the nickel pillar. Both the externally applied magnetic field and the perpendicular component of the stray field of the nanomagnet contribute to the measured Hall voltage. To extract the signal caused by the particle, we subtract the Hall voltage resulting from the external field. This is done by subtracting either the Hall signal of an empty reference Hall junction or the linear Hall voltage obtained when the magnetic particle is saturated. The linear part of the curve also serves for calibrating the probe for quantitative stray field measurements. A possible offset in the Hall voltage due to geometrical imperfections of the sensor is also eliminated.

For a full hysteresis loop, the particle is saturated in a strong (some Tesla) applied field which is then swept slowly from about 0.2 T to -0.2 T with a rate of typically 0.5 mT/s. Then the particle is saturated in the negative field direction before the second branch of the loop (from -0.2 T to 0.2 T) is recorded. In order to gain additional information about the mechanism of the magnetization reversal process it is often useful to sweep the external field in different directions relative to the sample [44,45]. By means of a tilted field experiment, e. g., it was shown that an acicular Ni particle behaving as single domain particle, if magnetized along the easy axis, is not single domain [46]. Tilting the sample in the external field is enabled by a pivoted sample stage. Although the stage is equipped with only one axis of rotation the sample can be fastened to the stage with the rotation axis perpendicular or parallel to the 2DEG. The first orientation is especially suited for the examination of oblate nanomagnets, which will be the topic of the following sections.

5. Complementary methods of investigation

In trying to understand the hysteresis loops measured by micro-Hall-magnetometry a major problem connected to any non-imaging method of investigation becomes evident: It is difficult to draw the right conclusions from stray field measurements alone. However, the application of a method, which enables the visualization of the magnetic configuration during magnetization reversal, can help to interpret the hysteresis loops correctly. In this section, two techniques capable of imaging magnetic structures in micron and sub-micron size particles will be introduced, namely Lorentz Transmission Electron Microscopy (LTEM) and Magnetic Force Microscopy (MFM).

In order to use a TEM for magnetic investigations, some modifications in the imaging system are necessary. The major problem in the conventional operation mode consists of the high magnetic field at the sample's position generated by the objective lens. As this field would always saturate the particle unintentionally, the TEM is equipped with a special lens system, called Lorentz lens, which causes a negligible magnetic field at the position of the sample. The objective lens itself is operated at low current to produce the magnetic field needed to perform in-situ magnetization reversal experiments [47,48]. In order to guarantee sufficient transparency for the electron beam, the thin magnetic particles are patterned on a Si_3N_4 membrane typically 15 nm to 30 nm thick.

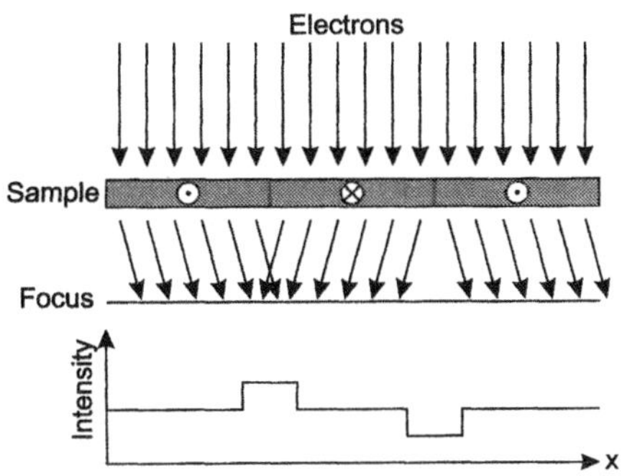

Fig. 9: Imaging of domain walls in the Fresnel mode (see text).

Figure 9 shows schematically a magnetic sample which is split into three domains separated by 180° walls. The incoming electrons are deflected by the Lorentz force, as soon as they pass through regions of non-vanishing magnetic flux. As the domains are magnetized in different directions, the electrons also get deflected in different directions, which leads to the partial superposition of electrons emerging from adjacent domains. Defocusing the Lorentz lens therefore has the effect that, the electron density at the position of a domain wall appears to be increased or decreased (see Fig. 9). Hence, the domain walls are visualized as bright and dark lines (Fresnel mode). The Fresnel imaging mode provides information on the direction of the magnetic flux perpendicular to the path of the electron beam.

The magnetic force microscope belongs to the family of scanning probe microscopes [49–52]. An atomic force microscope can be employed for high resolution magnetic imaging, when the conventional tip is replaced by a tip, that is covered by a hard magnetic film. This means that the tiny magnetic sensor is positioned at the end of the cantilever, which can be scanned across the surface of the magnetic sample using piezo elements. The oscillation of this cantilever can be detected by a laser deflection system. There are two different mechanisms that result in the exertion of a mechanical force on the cantilever. First, the roughness of the sample's surface causes the cantilever to bend, as it is also the case in the conventional AFM measurement mode. Second, the magnetic stray field of the sample, which arises from magnetic surface and volume charges, exerts an additional force resulting from the interaction with the magnetic moment of the tip. In order to distinguish between both contributions, the apparatus is operated in a special mode, called Liftmode (developed by Digital Instruments). It is characterized by scanning every line twice: During the first run information about the topography is gathered, so that the second scan can be performed at a fixed height above the surface. This scan, therefore, serves to extract the magnetic information. It is the out of plane component of the stray field which influences the oscillation of the cantilever during the lifted scan, whereas other effects of the surface-tip interaction play essentially no role. Although MFM provides a high spatial resolution, there are two severe disadvantages inherent in this method. The magnetic moment of the tip can switch the magnetic configuration during the scanning process. Moreover, the external magnetic field, needed for magnetization reversal, does not only influence the sample, but also the magnetic coating of the tip, which can seriously disturb the measurement.

It is important to emphasize that MFM detects magnetic charges, which act as the sources of the sample's stray field. MFM and micro-Hall-magnetometry detect the same magnetic property of the sample. However, MFM is an imaging technique, whereas micro-Hall-magnetometry enables quantitative stray field measurements. Comparing LTEM and MFM, the first can be used to visualize magnetic configurations, which do not generate stray fields. In contrast, MFM is extremely sensitive to domain patterns that do produce magnetic fields. Moreover, the LTEM can not detect any out of plane component of the magnetization, aligned in the direction of the electron beam. MFM, however, is highly sensitive to this out of plane component. Thus, combining the three methods of investigation results in a powerful set of tools for the examination of magnetic nanoparticles.

Fig. 10: Rhombic Ni particle placed on top of a Hall cross. The widths of the current and voltage paths are 800 nm. The micro-rhomb is 2 μm long, 300 nm broad and 70 nm high. The Maltese crosses served as alignment marks. Taken from [46].

6. In-plane measurements on rhombic particles

Apart from interesting questions concerning fundamental physics, ferromagnetic particles were proposed to be employed as storage elements in hard drives and as memory cells in MRAM devices [19–22]. The latter unify the advantage of a non-volatile magnetic memory with the high speed of a storage device operated by electrical currents [53]. However, application in memory cells in MRAM devices requires a high stability of the memory state, a high repeatability of read/write cycles and uniformity of the switching fields for different, but nominally identical nanoparticles. The last point means that magnetization reversal has to take place in a well defined manner. In the following section we demonstrate how effectively the mechanism of magnetization reversal can be influenced by just varying the geometrical shape of a magnetic particle (see also [13,15,54,55]).

For a micron or sub-micron size particle of well defined shape, the energy stored in the stray field strongly depends on the direction in which the particle is magnetized. In this way the stray field energy leads to the shape anisotropy mentioned above, which relates the magnetization pattern with the geometrical shape of the micromagnet. For example, it is much easier to magnetize a prolate particle along its longitudinal rather than the perpendicular direction. In oblate particles, the shape anisotropy forces the magnetization to lie in the plane of the sample. For most Hall measurements on such particles, therefore, the external field used for magnetization reversal is also applied in this plane parallel to the 2DEG. From the standpoint of measuring accuracy, these in-plane measurements offer the advantage of the particle's signal not being superimposed by the external field, as it is the case for out of plane measurements (cf. Ni pillars above).

In detail, we investigated the switching behavior of rhombic Ni elements, which were thermally evaporated to a final height of 70 nm. While the length of 2 μm of the long axis of the rhombs was kept fixed, the length of the short axis was varied in steps of 100 nm between 100 nm and 700 nm. The micro-rhombs were positioned on top of the Hall sensors with one end of the long axis being placed just at the center of the cross (see Fig. 10). In order to obtain the stray field hysteresis loop, the external field H was applied in the plane of the 2DEG parallel to the long axis of the rhomb. Figure 11 shows the hysteresis measurements of three micro-rhombs with widths of 700 nm (top), 400 nm (middle) and 100 nm (bottom).

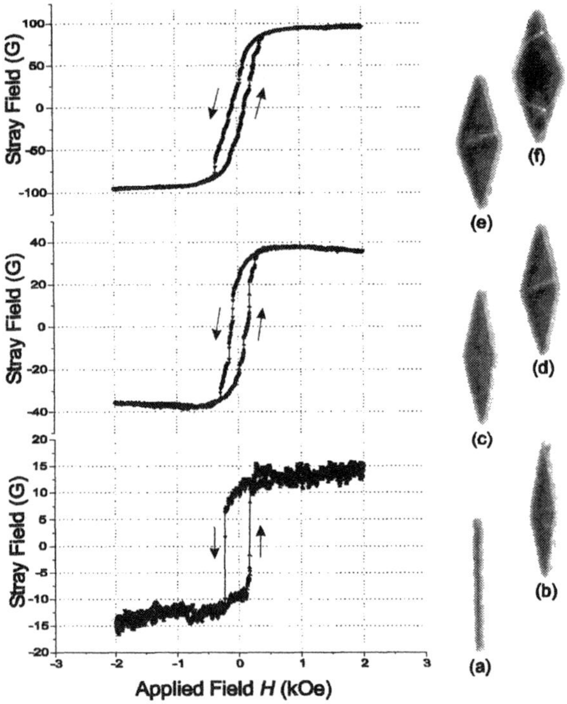

Fig. 11: LEFT: The hysteresis loops were measured with the external field H applied parallel to the long axis of the rhombs. The width of the rhombs was varied from 700 nm (top) over 400 nm (middle) to 100 nm (bottom). Note that the number of observable jumps decreases in this succession. RIGHT: Lorentz images taken at remanence. Widths of the particles: 100 nm in figure (a), 300 nm up to 700 nm in steps of 100 nm for figure (b) to (f) successively. In rhombs with widths larger than or equal to 500 nm domain walls (in white) can clearly be observed. In contrast, no domain walls seem to occur in narrower rhombs. Taken from [46].

These hysteresis loops can be interpreted by additionally using LTEM images (see right side of Fig. 11). The bottom loop shows only one distinct jump during magnetization reversal, which occurs after a slight decrease of the stray field, i. e. the particle is not fully saturated at remanence ($H = 0$). Because of the shape anisotropy, small elongated particles tend to be magnetized parallel to their long directions, which means that there are only two stable states of magnetization in the absence of an external field. During magnetization reversal the magnetization of the particle switches from one stable state with all spins in one direction, parallel to the long axis, to the other one with all spins aligned in the opposite direction, as soon as the coercive field is exceeded. This behavior was also observed by LTEM. The fact, that the switching is preceded by a gradual decrease of the measured stray field can be explained by the alignment of the spins along the edges of the rhomb and, for this particular particle, also by lithographic imperfections.

The hysteresis loop of the broadest rhomb (width = 700 nm) seems to describe a relatively complex magnetization reversal process (see Fig. 11, top). It alternately reveals sections characterized by a smooth increase of the stray field and sharp jumps. Loops exhibiting these features are typical of magnetization reversal accompanied by the existence of magnetic domains. The jumps in the curves are ascribed to sudden changes of the configuration of the domains or to the abrupt depinning of walls from pinning centers of various possible kinds. Sections of smooth increase of magnetization might be related to the continuous movement of domain walls between energy barriers or to the rotation of magnetization within domains. The LTEM-image corresponding to the 700 nm rhomb indeed shows domain walls (only the white ones can be seen clearly), which form a multiple vortex pattern in remanence. Vortex structures avoid the existence of magnetic charges by closing the magnetic flux. So, on the one hand, the stray field energy can be distinctly lowered. However, the spins are not aligned fully parallel, with the angle between them increasing by approaching the center of the vortex structure. That is why, on the other hand, the exchange energy usually is comparatively high in these magnetic patterns. In our examination only the magnetization reversal of the broadest rhomb with 700 nm width took place via this multiple vortex structure.

The number of observable jumps in the measured hysteresis loops increases with increasing width of the rhombs, whereas at the same time the coercive field is decreasing. Although in the LTEM-pictures only the three broadest rhombs reveal clearly visible domain walls, as can be seen on the right side of Fig. 11, the hysteresis loops of the rhombs with widths of 400 nm (see middle of Fig. 11) and 300 nm still show several jumps. This suggests that magnetization reversal is still accompanied by magnetic domains, although the walls can hardly be discovered in the Lorentz images.

As far as the application in data storage devices is concerned, single domain particles characterized by hysteresis loops similar to the one shown at the bottom of Fig. 11 have been proposed as storage elements in hard drives. With its two stable states of magnetization, every particle can store the information of one binary bit. Particles revealing complex magnetization reversal processes like the rhomb with a width of 700 nm can hardly be used in practice. However, it must not be concluded that the magnetization reversal via a magnetic vortex structure inevitably results in uncontrollable magnetic behavior, as will be demonstrated in the next section concerned with magnetic nanodisks.

7. Magnetic nanodisks

Because of their highly symmetric shape, nanodisks represent a very interesting mesoscopic magnetic system. Shape anisotropy only keeps the magnetization in the plane of the sample, but does not put any further restrictions on the direction of magnetization as it is the case, for example, in acicular elements like the rhombs described above. Further, the existence of corners or artificially patterned edge structures [56] can make a particle behave magnetically more complex. Such problems are also excluded by applying the simple geometry of a circle. We investigated nanodisks fabricated by thermal evaporation of Permalloy to a final height of 60 nm. The diameters of the disks range from 450 nm up to 850 nm. In order to obtain a maximum magnetic signal, the disks were placed on top of the Hall crosses with one half lying above the active area, while the other one was located on the mesa path of the voltage probe. As an example, Fig. 12 displays the system consisting of the Hall sensor and the magnetic disk, which is placed on the cross in the described manner. During magnetization reversal the external magnetic field H was applied parallel to the voltage probe.

Two typical kinds of hysteresis loops measured by micro-Hall-magnetometry are shown in Fig. 13. The shape of the loops drastically deviates from the ones characteristic of single domain behavior or of reversal processes accompanied by magnetic domains. One striking feature is that the stray field vanishes in the remanent state. Over an extended range of several hundred Oe around zero the loop is closed. This means that the detected stray field does not depend on the magnetic history in this field range. Another feature is represented by the open parts of the loop which occur at the beginning and the end of the reversal. Sometimes they come into existence by a single jump [see Fig. 13 a)], but a more complex process (see Fig. 13 b)) is also possible. Their elimination, however, always takes place in one distinct jump,

Fig. 12: SEM-picture of a Permalloy disk (height: 60 nm, diameter: 850 nm) placed on a Hall cross of width 1 μm. The white arrow indicates the direction of the externally applied magnetic field.

which is preceded by a decrease of the slope of the measured curve coming from lower fields. This behavior can be ascribed to a vortex magnetization structure.

Again, we use the Fresnel mode of Lorentz transmission electron microscopy, which provides additional information about the in-plane magnetization. Figure 14 shows images of 43 nm high Permalloy disks in remanence with diameters of about 200 nm [57]. Most remarkably, they exhibit bright and dark spots in the center of the disks, but no domain walls are observable. This can be explained by a circular closed-flux pattern. Figure 15 demonstrates that a magnetic vortex structure acts, due to the Lorentz force, like a focusing or diverging lens for the incoming parallel electron beam. In this way the direction of rotation of the magnetic vortex determines whether the electrons are deflected towards the center or away from it leading to the bright and dark spots.

The energy of a magnetic vortex structure is typically composed of a low contribution of stray field energy and a high contribution of exchange energy. Approaching the center of the magnetic vortex, the angle between adjacent spins increases more

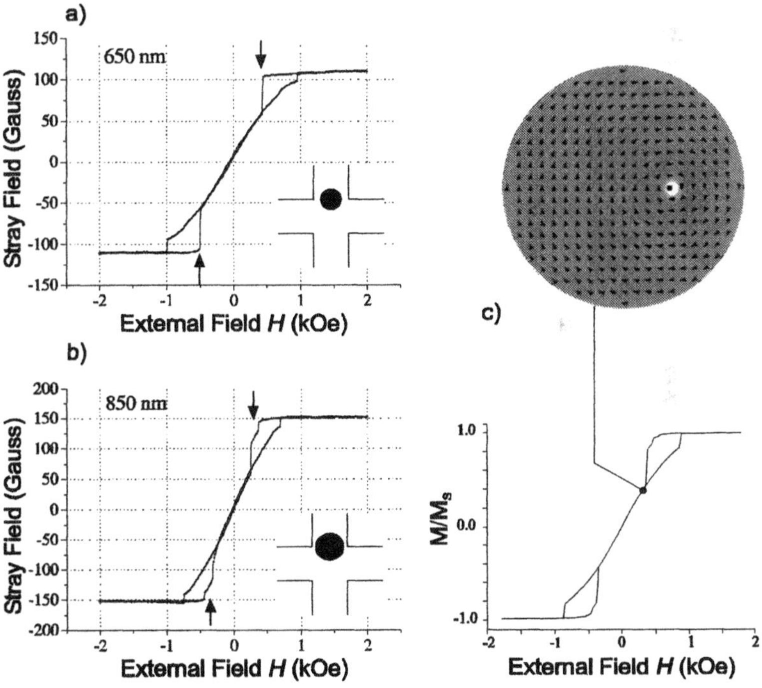

Fig. 13: Hysteresis loops of 60 nm high Permalloy disks with two different diameters. Note that in a) the vortex is formed in one distinct jump, whereas in b) the transition from the nearly saturated to the flux-closed state occurs in a more complex process (see arrows). In c) a calculated hysteresis loop of a Permalloy disk (height 50 nm, diameter 600 nm) is shown, together with a 'snapshot' of the magnetization configuration during the reversal process. The arrows indicate the in-plane components of the magnetization, whereas the gray scale background represents the out of plane component.

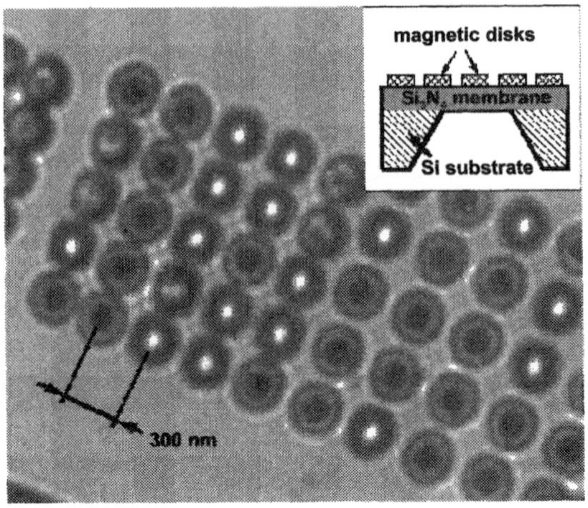

Fig. 14: Fresnel image of 43 nm high Permalloy disks with diameters of 200 nm on a Si_3N_4 membrane. The dark or bright spot in the center of the disks reflects the vortex structure of the magnetization oriented clockwise or counterclockwise. The outer ring structures are due to Fresnel fringes. Taken from [57].

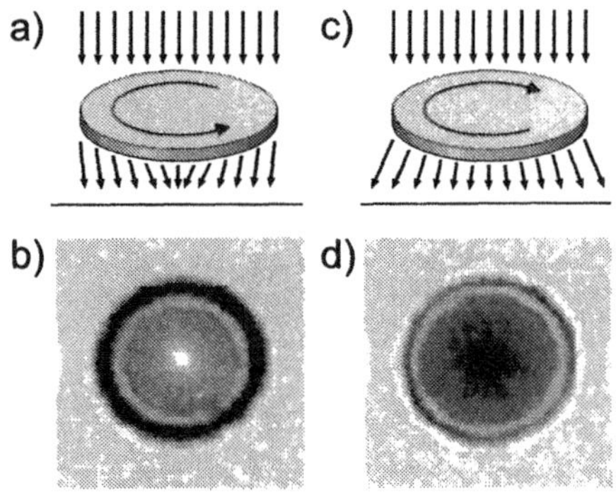

Fig. 15: Origin of the bright and dark spots in the center of the Lorentz micrographs. While in a) the counterclockwise orientation of the magnetization vortex focuses the incoming electron beam (black arrows), the electron beam is defocused below the specimen for a clockwise orientation c). The corresponding intensity distribution in a defocused plane is shown in b) and d), respectively. According to [57].

and more, leading to a sharp rise of exchange energy density at the vortex core. Thus it was supposed that the magnetization at the center of the flux-closed structure turns perpendicular to the surface of the flat sample [12]. This perpendicular component, whose existence was also supported by micro-magnetic simulations, can not be detected by LTEM, because the Lorentz force disappears for the sample's magnetization parallel to the incoming electron beam. Therefore, we used a magnetic force microscope to sense the small magnetic stray field caused by the central out of plane component of the magnetization.

Figure 16 displays a characteristic MFM-picture of the Permalloy disks having a diameter of 900 nm and a height of 50 nm [58]. In contrast to the Lorentz micrographs, bright and dark spots in this image correspond to a component of magnetization pointing in the positive or negative direction normal to the plane of the disk. In Ref. [59] it was supposed that the orientation of the out of plane component is not correlated to the sense of rotation of the magnetic vortex. This could be corroborated by MFM-measurements published in [58], where the direction of the central perpendicular component was switched independently of the orientation of the vortex.

Summing up the results collected by the LTEM and MFM investigations, a clear picture of the magnetization reversal process in circular Permalloy elements emerges and explains the hysteresis loops of Fig. 13. Figure 17 shows the upper half loop of Fig. 13 a) sweeping the external field from the remanent state to positive saturation

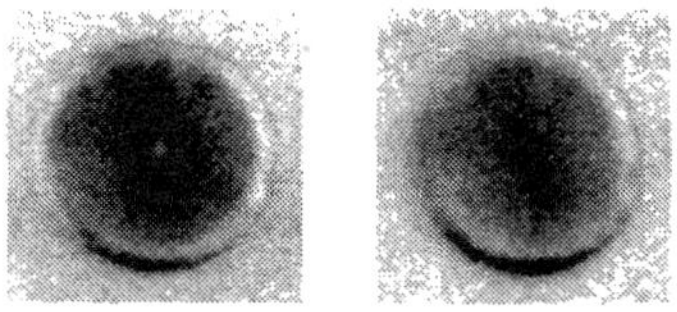

Fig. 16: MFM-images of two Permalloy disks, which are 50 nm high and have a diameter of 900 nm, in remanence. The left disk shows a bright spot in its center, whereas the center of the right one appears dark. This means that the out of plane component of magnetization, which is placed at the center of the disk, is oppositely aligned in these particles. Taken from [58].

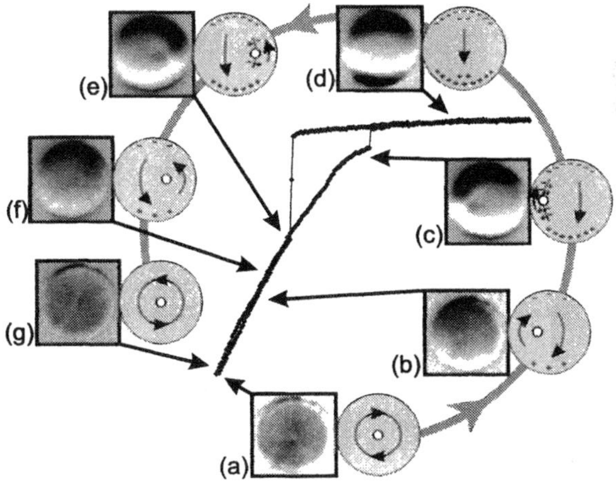

Fig. 17: Upper part of the hysteresis loop shown in Fig. 13 a). The curve is illustrated by MFM-images, whose magnetization configurations (including magnetic charges) are additionally sketched next to the micrographs. The external field, which starts at H=0 (a) and ends at H=0 (g) is directed from top to bottom during the whole process, only the field strength is varied in between. According to [58].

and back again. The sketches attached to the curve illustrate the corresponding magnetization configurations measured by MFM and depicted next to them. (a) At remanence a symmetric vortex structure avoids magnetic charges and – accordingly – any measurable stray field apart from the tiny central component. (b) Driving H slowly to higher field values (up-sweep) moves the central core of the vortex perpendicularly to the direction of H. Thus the section of the disk in which the magnetization is aligned parallel to the external field grows at the expense of the other part being magnetized mainly in the opposite direction. This process results in magnetic charges appearing at the edges of the disk. They are responsible for the gradual rise of the Hall voltage signal. (c) When the vortex core finally approaches the edge of the disk, it transforms into a wall-like structure running parallel to the edge. As magnetic charges of opposite signs are built on both sides of this wall, the emanating stray field only weakly influences the Hall voltage, which is proportional to the magnetic field averaged over the whole active area. Therefore the slope of the hysteresis curve decreases. Finally, the annihilation of the vortex pattern causes the jump in the loop. (d) Now the disk is almost fully saturated. After reversing the sweep direction of H (down-sweep) this nearly saturated state is stable beyond the annihilation field. (e) During the magnetization process in this particle, the transition to the vortex state occurs abruptly. Sometimes, however, vortex formation takes place via metastable intermediate states. A typical example of a hysteresis loop presenting this behavior is shown in Fig. 13 b). (f) Further reduction of H is accompanied by a continuous decrease of the detected stray field, which results from the movement of the vortex core back to the center of the disk (g). It is remarkable that the sense of rotation of the flux-closed pattern can be reversed after saturation in the external in-plane field as is the case here. In contrast to TEM-investigations, where the sense of rotation can be revealed immediately by the brightness of the central spot, the MFM offers a different way of distinguishing between both states: After the application of an external in-plane field, the sense of rotation can be extracted from the direction in which the vortex core is shifted.

8. Conclusions

In view of the experiments mentioned above, some conclusions concerning the value of micro-Hall-magnetometry can be drawn. Micro-Hall-magnetometry offers significant advantages compared to other measurement techniques. For example, it provides sufficient sensitivity to examine single nanoparticles without influencing their magnetic behavior during the experiment. In addition, a high lateral resolution, which is limited by the minimal width of the active area, can be achieved. Operated in the ballistic transport regime, Hall sensors can be used to perform quantitative measurements of the stray field over a wide range of temperatures and external magnetic fields. As Hall-magnetometry does not visualize the magnetic configuration of a particle, there is always the demand for additional methods capable to image the magnetization configuration. While Lorentz transmission electron microscopy and magnetic force microscopy have demonstrated their usefulness in the examples given

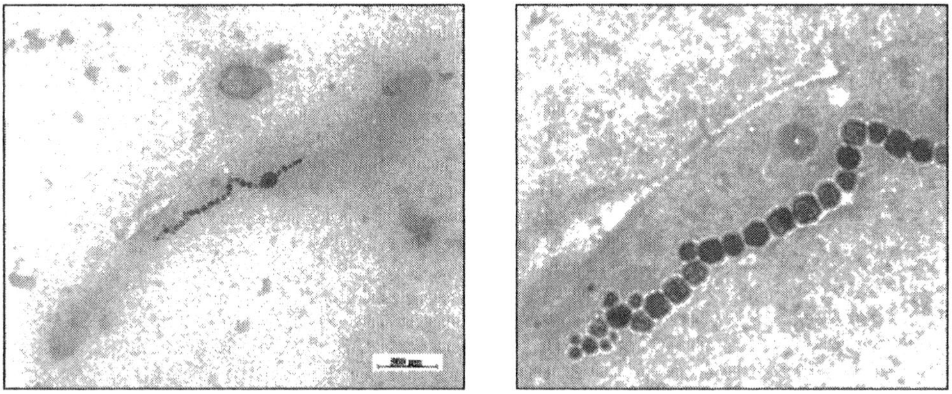

Fig. 18: LEFT: TEM-image (150 kV) of the bacterium Magnetospirillum magnetotacticum, which is about 4 μm long and contains a chain of 28 magnetosomes. RIGHT: The zoom into the left picture provides a more detailed view of the magnetosomes that consist of magnetite and have diameters of 40 nm to 50 nm. Taken from [60].

above, computer simulations have also taken an important place in recent years.

Although micro-Hall-magnetometry has meanwhile reached a level, which allows its application in many areas, the development of this single particle measurement technique is far from being complete. Some tasks still represent a real experimental challenge. An example is the quantitative measurement of the out of plane component of the magnetic vortex core in a nanodisk. Stacks of nanodisks have been proposed to be employed as memory cells in future MRAM devices [61], but the magnetic singularity at the center of the disks was supposed to affect their functionality [62,20]. Hall sensors might also be used to investigate biological systems. One example, a magnetic bacterium, which could be examined individually with respect to its magnetic properties [60], is shown in Fig. 18. In order to find solutions, the development of Hall sensors with improved characteristics is of crucial importance.

Acknowledgements

We thank W. Wegscheider (Universität Regensburg and Walter Schottky Institut München) and V. Umansky (Weizmann Institute of Science, Israel) for providing the GaAs-AlGaAs heterojunction material. The work was supported by the DFG Forschergruppe 370: Ferromagnet-Halbleiter-Nanostrukturen: Transport, magnetische und elektrische Eigenschaften.

References

[1] S. Datta, *Electronic Transport in Mesoscopic Systems* (Cambridge University Press 1997).

[2] Y. Aharonov, D. Bohm, Phys. Rev. **115**, 485 (1959).

[3] S. Washburn, R. A. Webb, Adv. Phys. **35**, 375 (1986).

[4] C. J. B. Ford, T. J. Thornton, R. Newbury, M. Pepper, H. Ahmed, D. C. Peacock, D. A. Ritchie, J. E. F. Frost, G. A. C. Jones, Appl. Phys. Lett. **54**, 21 (1989).

[5] S. Washburn, R. A. Webb, Rep. Prog. Phys. **55**, 1311 (1992).

[6] C. J. B. Ford, S. Washburn, M. Büttiker, C. M. Knoedler, J. M. Hong, Phys. Rev. Lett. **62**, 2724 (1989).

[7] C. W. J. Beenakker, H. van Houten, Phys. Rev. Lett. **63**, 1857 (1989).

[8] J. Shi, S. Tehrani, M. R. Scheinfein, Appl. Phys. Lett. **76**, 2588 (2000).

[9] S. McVitie, J. N. Chapman, IEEE Trans. Magn. **24**, 1778 (1988).

[10] R. D. Gomez, T. V. Luu, A. O. Pak, K. J. Kirk, J. N. Chapman, J. Appl. Phys. **85**, 6163 (1999).

[11] M. Hehn, K. Ounadjela, J.-P. Bucher, F. Rousseaux, D. Decanini, B. Bartenlian, C. Chappert, Science **272**, 1782 (1996).

[12] A. Hubert, R. Schäfer, Magnetic Domains: The Analysis of Magnetic Microstructures (Springer-Verlag, Berlin/Heidelberg, 1998).

[13] R. P. Cowburn, J. Phys. D: Appl. Phys. **33**, 1 (2000).

[14] R. D. McMichael, J. Eicke, M. J. Donahue, D. G. Porter, J. Appl. Phys. **87**, 7058 (2000).

[15] K. J. Kirk, S. McVitie, J. N. Chapman, C. D. W. Wilkinson, J. Appl. Phys. **89**, 7174 (2001).

[16] W. Wernsdorfer, K. Hasselbach, D. Mailly, B. Barbara, A. Benoit, L. Thomas, G. Suran, J. Magn. Magn. Mat. **145**, 33 (1995).

[17] J. Yu, U. Rüdiger, L. Thomas, S. S. P. Parkin, A. D. Kent, J. Appl. Phys. **85**, 5501 (1999).

[18] M. Kläui, J. Rothman, L. Lopez-Diaz, C. A. F. Vaz, J. A. C. Bland, Z. Cui, Appl. Phys. Lett. **78**, 3268 (2001).

[19] S. Tehrani, E. Chen, M. Durlam, M. DeHerrera, J. M. Slaughter, J. Shi, G. Kerszykowski, J. Appl. Phys. **85** 5822 (1999).

[20] J.-G. Zhu, Y. Zheng, G. A. Prinz, J. Appl. Phys. **87**, 6668 (2000).

[21] M. Todorovic, S. Schultz, J. Wong, A. Scherer, Appl. Phys. Lett. **74**, 2516 (1999).

[22] S. Y. Chou, P. R. Krauss, L. Kong, J. Appl. Phys. **79**, 6101 (1996).

[23] W. Wernsdorfer, D. Mailly, A. Benoit, J. Appl. Phys. **87**, 5094 (2000).

[24] T. Schweinböck, D. Weiss, M. Lipinski, K. Eberl, J. Appl. Phys. **87**, 6496 (2000).

[25] T. Schweinböck, Raster-Hall-Mikroskopie, Ph.D. thesis, Universität Regensburg (2001).

[26] A. M. Chang, H. D. Hallen, L. Harriot, H. F. Hess, H. L. Kao, J. Kwo, R. E. Miller, R. Wolfe, J. van der Ziel, T. Y. Chang, Appl. Phys. Lett. **61**, 1974 (1992).

[27] A. Oral, S. J. Bending, Appl. Phys. Lett. **69**, 1324 (1996).

[28] F. M. Peeters, X. Q. Li, Appl. Phys. Lett. **72**, 572 (1998).

[29] X. Q. Li, F. M. Peeters, A. K. Geim, J. Phys.: Condens. Matter **9**, 8065 (1997).

[30] S. V. Dubonos, A. K. Geim, K. S. Novoselov, J. G. S. Lok, J. C. Maan, M. Henini, Physica E **6**, 746 (2000).

[31] M. L. Roukes, A. Scherer, S. J. Allen, H. G. Craighead, R. M. Ruthen, J. P. Beebe, E. D. Harbison, Phys. Rev. Lett. **59**, 3011 (1987).

[32] H. U. Baranger, A. D. Stone, Phys. Rev. Lett. **63**, 414 (1989).

[33] A. M. Chang, T. Y. Chang, H. U. Baranger, Phys. Rev. Lett. **63**, 996 (1989).

[34] A. K. Geim, I. V. Grigorieva, J. G. S. Lok, J. C. Maan, S. V. Dubonos, X. Q. Li, F. M. Peeters, Y. V. Nazarov, Superlattices and Microstructures **23**, 151 (1998).

[35] I. S. Ibrahim, V. A. Schweigert, F. M. Peeters, Phys. Rev. B **57**, 15416 (1998).

[36] S. J. Bending, A. Oral, J. Appl. Phys. **81**, 3721 (1997).

[37] S. Liu, H. Guillou, A. D. Kent, G. W. Stupian, M. S. Leung, J. Appl. Phys. **83**, 6161 (1998).

[38] H.-D. Schuh, Mikro-Hall-Magnetometrie, Ph.D. thesis, Universität Regensburg (2000).

[39] D. Schuh, J. Biberger, A. Bauer, W. Breuer, D. Weiss, IEEE Trans. Magn. **37**, 2091 (2001).

[40] L. Néel, Ann. Geophys. **5**, 99 (1949).

[41] W. Wernsdorfer, K. Hasselbach, A. Benoit, B. Barbara, B. Doudin, J. Meier, J.-P. Ansermet, D. Mailly, Phys. Rev. B **55**, 11552 (1997).

[42] W. Wernsdorfer, E. Bonet Orozco, K. Hasselbach, A. Benoit, B. Barbara, N. Demoncy, A. Loiseau, Phys. Rev. Lett. **78**, 1791 (1997).

[43] J. G. S. Lok, A. K. Geim, U. Wyder, J. C. Maan, S. V. Dubonos, J. Magn. Magn. Mat. **204**, 159 (1999).

[44] G. Meier, D. Grundler, K.-B. Broocks, C. Heyn, D. Heitmann, J. Magn. Magn. Mat. **210**, 138 (2000).

[45] E. Bonet, W. Wernsdorfer, B. Barbara, A. Benoit, D. Mailly, A. Thiaville, Phys. Rev. Lett. **83**, 4188 (1999).

[46] M. Rahm, J. Bentner, J. Biberger, M. Schneider, J. Zweck, D. Schuh, D. Weiss, IEEE Trans. Magn. **37**, 2085 (2001).

[47] K. J. Kirk, J. N. Chapman, C. D. W. Wilkinson, J. Appl. Phys. **85**, 5237 (1999).

[48] J. N. Chapman, M. R. Scheinfein, J. Magn. Magn. Mat. **200**, 729 (1999).

[49] Y. Martin, H. K. Wickramasinghe, Appl. Phys. Lett. **50**, 1455 (1987).

[50] P. Grütter, H. J. Mamin, D. Rugar, in: Springer Series in Surface Sciences, (Springer-Verlag, 1992), vol. 28.

[51] A. Wadas, Magnetic Force Microscopy, in: S. Amelinckx (Ed.), Handbook of Microscopy. Applications in Materials Science, Solid-State Physics and Chemistry. Methods II, VCH, Weinheim u.a., 1997.

[52] S. Porthun, L. Abelmann, C. Lodder, J. Magn. Magn. Mat. **182**, 238 (1998).

[53] G. A. Prinz, J. Magn. Magn. Mat. **200**, 57 (1999).

[54] T. Schrefl, J. Fidler, K. J. Kirk, J. N. Chapman, J. Magn. Magn. Mat. **175**, 193 (1997).

[55] K. J. Kirk, J. N. Chapman, C. D. W. Wilkinson, Appl. Phys. Lett. **71**, 539 (1997).

[56] M. Herrmann, S. McVitie, J. N. Chapman, J. Appl. Phys. **87**, 2994 (2000).

[57] J. Raabe, R. Pulwey, R. Sattler, T. Schweinböck, J. Zweck, D. Weiss, J. Appl. Phys. **88**, 4437 (2000).

[58] R. Pulwey, M. Rahm, J. Biberger, D. Weiss, IEEE Trans. Magn. **37**, 2076 (2001).

[59] T. Shinjo, T. Okuno, R. Hassdorf, K. Shigeto, T. Ono, Science **289**, 930 (2000).

[60] W. Cebulla, Magnetometrie an magnetischen Bakterien, Master's thesis, Universität Regensburg (Oktober 2000).

[61] K. Bussmann, G. A. Prinz, S.-F. Cheng, D. Wang, Appl. Phys. Lett. **75**, 2476 (1999).

[62] K. Bussmann, G. A. Prinz, R. Bass, J.-G. Zhu, Appl. Phys. Lett. **78**, 2029 (2001).

Chapter 10

Stochastic optimization methods for biomolecular structure prediction

T. Herges, H. Merlitz and W. Wenzel*

Institute for Nanotechnology, Forschungszentrum Karlsruhe,
Postfach 3640, D-76021 Karlsruhe, Germany
**E-mail: Wolfgang.Wenzel@int.fzk.de*

Abstract

We discuss the use of stochastic optimization methods for biomolecular structure prediction, in particular in application to protein structure prediction and receptor ligand docking. After a brief discussion of our motivation for these approaches, we present a brief overview of the dominating physical effects that are important for protein structure prediction and outline our strategy to address this problem. We discuss the strength and weaknesses of several possible optimization methods, including the stochastic tunneling method. Finally we give examples of applications of this methodology both for protein structure prediction and receptor ligand docking.

1. Introduction

1.1 Motivation

Biomolecular structure prediction remains one of the main outstanding problems of theoretical biophysical chemistry [1]. One of its primary goals is the prediction of the three-dimensional, tertiary structure of proteins on the basis of their amino acid sequence [protein structure prediction (PSP)]. Proteins are the building blocks of the cellular machinery of all living organisms and plants [2]. Their enzymatic capabilites often surpass the efficiency of competing chemical processes by orders of magnitude. The light harvesting mechanism of chlorophyll and the synthesis of the penicillin family of antibiotics as two impressive examples in this context [3].

Experimental methods to determine the sequence of a particular protein, either directly through protein sequencing, or indirectly through the use of genetic information, have made enormous progress in the last decade, resulting in a pool of several hundred thousand sequenced proteins for various organisms. Unfortunately, sequence information alone is often insufficient to elucidate the biological function or mechanism of a protein. Most proteins spontaneously assume a unique three dimensional "native" conformation, which is a prerequisite for their proper function. Failure to assume this structure often has catastrophic consequences for the metabolism of the cell (e.g., BSE or Jacob-Kreutzfeld disease). To understand the mechanism of a protein, knowledge of this three-dimensional structure is very helpful. Experimental methods for protein structure determination are orders of magnitude more involved and more expensive than sequencing techniques. Although their number is steadily growing, the protein database (PDB) presently contains about 13,000 spatially resolved structures [4]. Therefore there is a large gap between the pool of available sequences and the set of structurally resolved proteins, which is likely to contain a wealth of important biological and biomedical information. There are entire families of important proteins, in particular transmembrane proteins, that are almost impossible to resolve structurally with present day experimental techniques. Theoretical methods for PSP may be helpful to close this gap.

In addition there are many questions regarding the details of the function of proteins, in particular those of dynamical nature, that are difficult to address experimentally at the present time. This is because time-resolved x-ray crystallography and NMR techniques are not readily available with the required resolution. Many of these problems, as well as questions regarding protein-protein association or protein-ligand interactions, would benefit from accurate theoretical methods for PSP. In particular simulation techniques addressing protein-ligand interactions would contribute significantly to the development of new pharmaceutical agents for a variety of diseases [5].

Closely related to the PSP challenge is the protein folding problem [6–8], where one tries to understand the thermodynamic mechanisms of protein folding. The very fact why molecules of such complexity have a unique three-dimensional structure and how they attain it is presently not well understood. Obviously an efficient solution to the protein folding problem would also provide a mechanism for PSP, but presently this seems to be far out of reach. Even though a solution to the protein folding problem may not be directly useful for PSP, investigations into this problem provide crucial information for the design of PSP methods that attempt to predict the native structure without recourse to the folding dynamics.

1.2 Characteristics of the PSP problem

As for their chemical composition, proteins are linear chains composed of varying sequences of 21 naturally occurring amino acids connected with peptide bonds. The peptide bond polymerizes the chain when the RNA encoding the protein is read by the ribosome in the cell. The amino acids are differentiated by their side chains that vary in their chemical composition. From the perspective of PSP their most important feature is the absence of covalent binding between the sidechains of most amino acids. Only methionine and cysteine may form disulfide bridges with one another. Therefore, only few disulfide bridges occur in most proteins and the number of possible connections is limited. From the perspective of PSP, it is thus sensible to consider a given pairing of these amino-acids as given and to solve the PSP problem under the resulting additional constraints. In the absence of covalent bonding, the amino acids are differentiated by the non-covalent characteristics of their sidechains, such as partial charges and hydrophobicity [9].

Only non-covalent interactions determine the structure of the protein. The most important are hydrogen bonding, electrostatic interactions, van-der-Waals interactions and solvent interactions. The energy scale of each of these interactions is of the order of a few tenth to a few kcal/mol, i.e., much smaller than typical covalent energy scales. Therefore one can consider the covalent structure of the protein as a constraint on the possible protein folds, neither bond length nor bond angles can vary significantly in low-energy conformations of the protein. Only rotations of the backbone dihedral angles or dihedral angles of the sidechains are low-energy degrees of freedom that can be explored in the folding process. Consequently these degrees of freedom also define the search space in PSP.

Since the protein backbone has one hydrogen-bond donor and acceptor in each amino-acid, it is not surprising that configurations in which all of these backbone donors and acceptors pair with other backbone donors and acceptors are particularly favorable. It has been observed since the early 60's that arrangements of the backbone dihedral angles that lead to the satisfaction of the backbone-backbone hydrogen bonds are very prevalent in naturally occurring proteins. These arrangements, e.g., left- and right turning helices and beta sheets have been termed the 'secondary structure' of the protein. The three dimensional arrangement of these motifs into the native structure of the protein is termed the tertiary structure. The association of the various protein chains to fully functional complexes is termed quartary structure.

1.3 State of the art

Early attempts for PSP focused on secondary structure prediction, which remains an active area of research to date. The best methods available today, most based on neural network models, are able to assign between 60-80% of the secondary structure of a given protein with reasonable reliability. Unfortunately, from the point of tertiary structure prediction, the most successful methods available today are based on neural networks, so that little information about the underlying physical mechanisms can be derived from these predictions.

Tertiary structure prediction is much less developed. Most methods competing in the CASP structure prediction contest are based on homologies to structurally resolved proteins and heuristic input. As with secondary structure prediction, methods that are based on non-physical input are difficult to evolve systematically. Physics based models, on the other hand are presently neither very reliable, nor efficient.

Considering the interactions described as relevant to protein folding above, one may hope that quantum effects play only a minor role in the description of the folding process and the discrimination of good folds. Therefore an atomistic approach to PSP may be successful using classical molecular forcefields as the discriminating function [10,11]. One approach to PSP is thus to model the folding process with molecular dynamics and use the resulting structure as the prediction. There are two major obstacles that must be overcome in this approach: First, the protein folding process is slow compared to typical MD time scales: typical protein folding times range from 10 ms to 1 s. In order to resolve the trajectory of the system with sufficient accuracy, most presently available MD methods operate on a femtosecond timescale. An atomistic simulation of a 10 ms folding process would thus require $O(10^{13})$ simulation steps. A typical protein has of order $O(10^4)$ atoms, each of which interacts via long-range interactions with a significant number of neighbors, say $O(1000)$. Folding the protein thus requires of $O(10^{20})$ force evaluations. Since the solvent has significant impact on the folding process, the entire system has to be embedded in a sufficiently large bath of water, which may increase the total CPU requirement by another order of magnitude. Using 1 GFLOP processors, folding a single protein can thus be expected to consume about 30000 CPU years and great efforts are being made to make hardware resources at that scale available to this problem [12]. Even so, its appears presently unlikely that a PSP method that simply extrapolates this strategy will be able to significantly augment experimental procedures in the foreseeable future [13].

One may therefore ask whether it is possible to devise alternate simulation strategies that circumvent the folding bottleneck described above [14]. An obvious starting point is the elimination of the explicit treatment of the solvent molecules [15], which often consumes the majority of the numerical effort associated with the simulation of the overall system. Upon closer inspection of this approximation, we find that the introduction of an implicit solvent model has far deeper implications on PSP than the obvious reduction of the computational effort resulting from the reduction of the degrees of freedom of the simulation. We note that the overwhelming majority of the entropic contribution to the folding process are solvent contributions, mediated by the hydrophobic and hydrophilic effects of the different amino acid side chains. Incorporating these terms into an implicit solvent model we obtain in conjuction with the internal energy of the protein a good model for the total *free energy* of the system [9]. As indicated above most proteins attain a unique stable native structure. If the protein is in thermodynamic equilibrium with its environment, this structure must therefore correspond to the global minimum of its free energy surface. As is well known from the simulation of many physical systems with complex dynamics, it is possible to locate the thermodynamically stable state of the system using *stochastic optimization methods* without recourse to its dynamics *orders of magnitude faster* than in an simulation approach [16,17].

To implement this approach to PSP, several important questions must be answered:

- Is there a suitable classical forcefield to describe the internal energy of a protein?
- Are there adequate implicit solvent models?
- Are there suitable and efficient global optimization methods that are able to reliably locate the global minimum of the resulting free energy landscape of the protein?

In the past several years we have implemented this strategy, developing both forcefields and stochastic optimization methods suitable to this task. In the following sections we describe the ingredients of this approach and give an overview of our results. A final section deals with the application of this methodology to a related problem in rational drug design [5,18].

2. Biomolecular forcefield

Over the last decades many classical forcefields [19–22] have been developed to investigate numerous phenomena in physical, organic and inorganic chemistry. The difficulties encountered in PSP justify the development of specific forcefields for the following reasons: Their molecular building blocks, i.e. the amino acids, are well defined and limited in number. The chemical complexity associated with the design of a forcefield specific to peptides and proteins is therefore less than that of generic organic substances. By exploiting the fact that only a limited number of building blocks will occur, their ingredients may be specifically adapted to provide a more accurate description of the system. Secondly, we are interested only in the low-energy conformations of the model. As a result, many degrees of freedom that are associated with covalent interactions, e.g., bond stretching, may be neglected in the description of the system.The accuracy of this approximation can be checked against the available data in the PDB. The remaining degrees of freedom are thus only rotations about the dihedral angles of the backbone and of freely rotatable single bonds of the sidechains. This reduction of the number of degrees of freedom leads to a dramatic increase in the efficiency of the simulation.

Generically the intramolecular forcefield should include Lennard-Jones interactions that provide for steric repulsion, van-der-Waals attraction, electrostatic interactions, and hydrogen-bonding interactions. The Lennard-Jones Potential is parameterized as

$$V(r) = V_0 \left[\left(\frac{r}{r_0} \right)^{12} - 2 \left(\frac{r}{r_0} \right)^6 \right] \quad (1)$$

with interaction strength V_0 and equilibrium distance r_0. The INT force field represents all atoms except apolar CH_n individually. CH_n groups are approximated by a single sphere comprising both the carbon and the hydrogen atoms (united atom approach).

The Lennard-Jones potential rises steeply for $r < 0.8\, r_0$. Comparison with PDB data shows that many proteins have such clashing configurations when Lennard-Jones parameters of standard organic forcefields are employed. We therefore developed a strategy to fit the LJ radii in accordance with the experimental data and fitted the LJ radii of our model to a subset of 134 proteins of the PDB database. The associated LJ interaction strength were taken from the OPLS forcefield [23]. Note that the choice of LJ interaction strength should not be crucial in the PSP problem. Proteins in native or near-native configurations almost always assume relatively dense configurations which attain a packing density of up 75% of hexagonal close packing. As a result, all near-native configurations have similar attractive LJ contributions. Unfolded contributions, for which the number of LJ contacts may be significantly different are much higher in energy than folded configurations. Since we are only interested in the former, errors in describing the latter can be tolerated (in contrast to folding studies, where temperature dependent equilibria between folded and unfolded configurations must be considered). Also note that in simulations with explicit solvent molecules there are LJ interactions between peptide and solvent atoms. This atom-dependent effect must be incorporated into the implicit solvent model.

Coulomb interactions in proteins are complicated, in particular regarding screening effects of the solvent. In the INT forcefield we have implemented an approach which models this effect with group-dependent and interaction dependent effective dielectric constants [24].

In many forcefields, hydrogen bonding is modeled by dipolar interactions of 8/12 potentials similar to Eq. (1). Using only the electrostatic interactions (including the dipolar interactions) in [24,25] we noted significant deviations in the backbone dihedral angles in the modeling of long helices. Since such hydrogen bonds are very important in the stabilization of the native structure we have parameterized the backbone-backbone hydrogen bonds as:

$$V(r, \phi, \psi) = f_r(r) f_\phi(\phi) f_\psi(\psi) \quad (2)$$

where each component of the potential is modeled by a superposition of Gaussian functions that are fitted to reproduce the experimentally observed structures.

For the implicit solvent model, the simplest conceivable choice is to assign a free energy of solvation proportional to the effective contact area each atom of the protein/peptide has with the solvent. We have subdivided the atom types of the

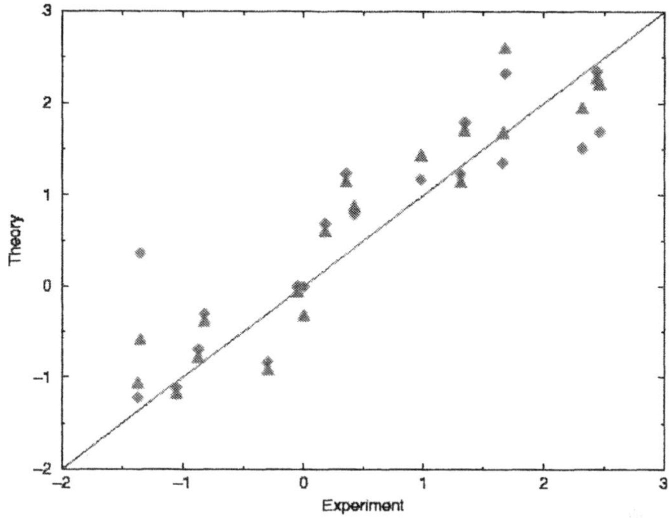

Fig. 1: Correlation between the free energies of solvation between experimental data for Gly-X-Gly and two solvent accessible surface area based models (in units of kcal/mol) that differ in the number of atom groups used in the fit. The INT forcefield uses the fit indicated by the triangles with an RMS error of less than 0.5 kcal/mol.

forcefield into suitable subgroups and fitted the resulting model to the available experimental Gly-X-Gly data [15]

3. Optimization methods

Stochastic optimization methods are now being used in a multitude of applications, ranging from circuit design on silicon wafers to airline flight schedules. In these and many other applications the objective is to minimize a given cost function that depends on a large number of discrete or continuous variables [7,26]. In analogy to physical problems, the cost function describes a potential energy surface (PES) in the parameter space and its global minimum optimizes the desired objective. Stochastic optimization methods are applied when enumerative methods are too costly. This is generically the case in high-dimensional optimization problems, where the total number of possible configurations grows exponentially with the number of variables.

Stochastic optimization methods successively improve one or several configurations of the underlying model to obtain an approximant of the global optimum of the PES. The optimization process thus maps onto a fictitious dynamical process of one or several configurations that move in the configurations space. The process stops when either a certain previously defined amount of computational resources has been spent or when the dynamical process terminates in a stable configuration. In either case there is no guarantee that the stochastic process has found the global optimum of the PES. Indeed, due to the stochastic nature of the process there can be no guarantee of finding the global optimum. In the absence of perfection it is important to differentiate two possible goals in stochastic optimization: in many applications (e.g. circuit design) the quality of the solution is only measured by its energy difference to the global optimum, the "distance" of the configuration obtained to the global optimum is completely irrelevant. In other problems, such as PSP, this distance is crucial. Since we do not seek to "optimize" the folding energy, but to derive useful information from the three-dimensional structure obtained, a low-lying metastable state that has a large RMSD to the true native state may contain virtually no useful information.

The computational challenge in stochastic optimization methods depends strongly on the number of degrees of freedom and the complexity of the PES. The latter depends on the total number of low-lying metastable states, the ability to efficiently explore the configuration space and the average height of transition states that separate low-lying metastable states.

3.1 Simulated annealing

The fundamental challenge in stochastic optimization is to balance the number of moves of the dynamical process in which the energy of the system increases against those in which the number of systems decreases. In high-dimensional problems the number of metastable states often grows exponentially with the system size. The simplest stochastic optimization method, repeated local optimization starting from random initial conditions, will therefore also require an exponentially large number of steps. To significantly reduce the computational effort, stochastic optimization methods must therefore also move uphill. In simulated annealing [27] this challenge is met by simulating the finite temperature dynamics of the system. Starting from a configuration r with energy $E(r)$ one generates a new configuration r' with energy $E(r')$ which replaces the original configuration with probability

$$P = \begin{cases} \exp\left(-\beta[E(r') - E(r)]\right) & \text{if } \mathrm{E(r') > E(r)} \\ 1 & \text{otherwise,} \end{cases} \tag{3}$$

where $\beta = 1/(kT)$ is the fictitious inverse temperature. At any given temperature such an (ergodic) Monte-Carlo process [28] samples the configurations r of the PES according to their thermodynamic probability. Therefore, at high temperatures, moves with or against the gradient are accepted with almost equal probability. At low temperature only downhill moves are accepted. In simulated annealing one thus starts with high temperature simulation and gradually cools the system to zero temperature. If *ergodicity is not lost* during the cooling schedule, the simulation will stop in the global minimum of the PES with probability one.

For locally smooth PES the search is greatly improved by locally minimizing the new configuration after its generation (basin hopping technique) [26]. The particle then travels only among the local minima of the PES, eliminating the costly exploration of intermediate states altogether.

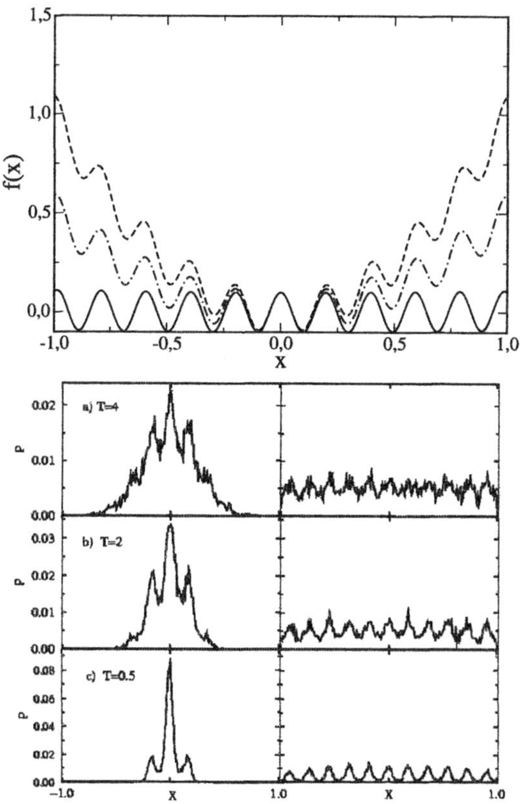

Fig. 2: (a) Top panel: Schematic potential energy surfaces $f(x) = Ax^2 + cos(x/n)$ that differ in their ruggedness, i.e., the ratio of the energy difference of nearby local minima to the height of the intervening transition state. (b) Bottom panel: Distribution of 10,000 SA processes started at random initial positions for the potential with A=1 (left) and A=0.1(right) at the given temperatures respectively.

In many rugged PES simulated annealing suffers from the so-called freezing problem. As illustrated schematically in Fig. 2 (a), the ruggedness of the PES depends on the ratio of the energy difference of adjacent local minima to the height of the intervening transition state. Figure 2 (b) traces the distribution of two sets of SA processes for a smooth (left) and (rugged) PES respectively. In the latter case the particles remain trapped in their respective local minima, because when the temperature is low enough to thermodynamically differentiate between adjacent minima, the probability of crossing the transition state is already exponentially suppressed [1].

[1] For optimization problems, where the distance between the global optimum and its approximant is irrelevant, SA can be considered successful even for the rugged PES illustrated here. One should also note that for smooth potentials, basin hopping eliminates the freezing problem.

3.2 Parallel tempering

For PES in which the progress of simulated annealing (SA) is slow, the freezing problem may be circumvented by allowing a trapped particle to escape from a local minimum by increasing the temperature of its simulation. Following this idea the parallel tempering method replaces the unidirectional cooling of SA by a set of concurrent simulations at different temperatures $\{T_i | i = 1 \dots n\}$, which occasionally exchange configurations with probability

$$p = \exp\left(-(\beta_1 - \beta_2)(E_1 - E_2)\right), \tag{4}$$

where β_i and $E_i (i = 1, 2)$ are the inverse temperatures and energies of the two simulations/configurations respectively. This mechanism permits each particle to alternate between low temperature simulations where only the closest local minimum is explored and high temperature simulations where it diffuses freely across potential barriers. The specific choice in Eq. 4 allows all simulations to remain in thermal equilibrium so that thermal averages can be computed at a variety of temperatures simultaneously (detailed balance). Compared to straightforward SA, parallel tempering (PT) incurs an n-fold increase in cost for a given total simulation length. On rugged or glassy PES, however, where the escape time from a given local minium can be exponentially long, this overhead may be more than compensated for. Recently a number of methods have been proposed that provide similar mechanisms by generalizing the Monte-Carlo method [29–31] to simulate ensembles other than the canonical. However, the efficiency of at least some of these techniques has been questioned for glassy PES [32].

3.3 Stochastic tunneling

The stochastic tunneling (STUN) method [33] incorporates the ability to escape metastable states by letting the particle in the minimization process "tunnel" forbidden regions of the PES. As in SA we retain the idea of a biased random walk, but apply a non-linear transformation to the potential energy surface:

$$E_{\mathrm{STUN}}(x) = 1 - \exp\left[-\gamma(E(x) - E_0)\right] \tag{5}$$

where E_0 is the lowest minimum encountered by the dynamical process so far. Alternately a suitable upper bound for the global minimum can be used for E_0. This effective potential preserves the locations of all minima, but maps the entire energy space from E_0 to the maximum of the potential onto the interval $[0, 1]$.

At a given finite temperature of O(1), the dynamical process can therefore pass through energy barriers of arbitrary height, while the low energy-region is resolved even better than in the original potential. The degree of steepness of the cutoff is controlled by the tunneling parameter γ. Figure 3 (b) illustrates the STUN potential energy surface for a 1D model potential (see below) at a hypothetical point in the simulation where the minimum indicated by the arrow has been found as the present best estimate for the ground state. Obviously, there are many possible transformations that have similar broad characteristics. However, as we argue in the following,

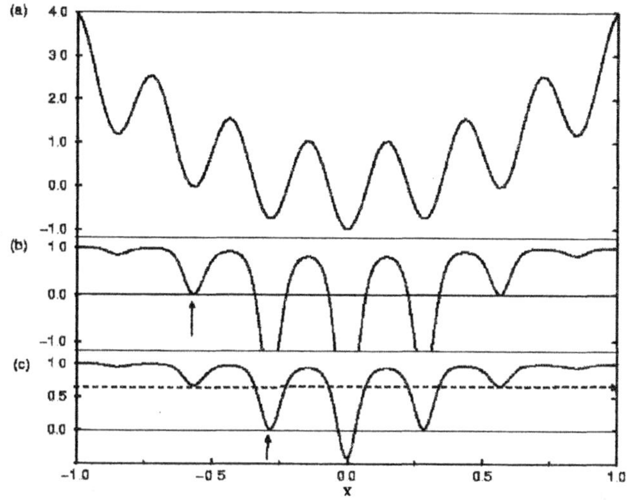

Fig. 3: Schematic one dimensional potential energy surface and its transformations under the STUN procedure, provided that the local minima indicated by the arrows have been found. Part (a) shows the original potential energy surface as in Fig. 2, parts (b) and (c) the transformed PES under the assumption that the minima indicated by the arrow are the best configurations found so far in the simulation, respectively.

a generic physical mechanism is responsible for the advantage of the STUN method over its traditional stochastic cousins.

If we consider a Monte-Carlo (MC) process at some inverse temperature β on the STUN PES, a MC-step from x_1 to x_2 with $\Delta = E(x_2) - E(x_1)$ is accepted with probability

$$\tilde{w}_{1\to2} \approx \exp\left(-\tilde{\beta}\Delta\right) \qquad \text{for} \qquad \gamma\Delta \ll 1 \tag{6}$$

with an effective, energy dependent temperature

$$\tilde{\beta} = \beta\gamma e^{\gamma(E_0 - E(x_1))} \leq \beta\gamma. \tag{7}$$

In this limit the dynamical process on the STUN potential energy surface can be interpreted as an ordinary MC process with an energy dependent temperature which rises with the local energy relative to E_0. For large $E_1 \gg E_0$ the effective temperature becomes infinite and the particle diffuses (or tunnels) freely through potential barriers of arbitrary height. As better and better minima are found, ever larger portions of the high-energy part of the PES are flattened out. Comparing a STUN simulation on the transformed PES with a MC simulation on the original one, the transformation can be viewed as regulatory mechanism for the temperature of the simulation.

One can exploit this realization to use the fixed energy-scale of the effective potential to broadly classify the dynamics of the minimization process into phases corresponding to a local search and to "tunneling" phases, simply by comparing

E_{eff} with some fixed pre-defined threshold E_{thresh}. We can then vary β such that the particle spends approximately the same amount of time in both optimization modes. Figure 3 illustrates an effective PES just after the local minimum indicated by the arrow has been found. At low effective temperature there is almost no probability density at the edges of the present well, i.e., very little escape probability – running the process at this temperature corresponds to a local search. At high temperature the particle escapes the well with relative ease as is required to find the global minimum. In order to switch between search and tunneling phases β is changed by some fixed factor whenever a moving average of E_{eff} crosses a predefined threshold E_c. If $E_{\text{eff}} > E_c$ (tunneling phase) β is reduced by some fixed factor, otherwise it is increased.

While the transformation in Eq. 5 is not the only possible functional, we believe that there are a number of features that constrain its construction: (i) The transformation must be strongly nonlinear in the high-energy regime, as only such a transformation will lead to a nearly constant effective PES for high energies and true "tunneling". (ii) There must be a parameter that modulates the degree of compression (γ), since the ratio the energy differences of adjacent local minima to the transition state energy separating them varies from problem to problem. (iii) Requiring an essentially flat PES at high energy (for typically unbounded PES) requires a transformation that maps the interval $[E_0, \infty]$ onto some finite interval, which can be chosen as $[0, 1]$ without loss of generality. (iv) It is possible to use a fixed inverse temperature β as a second parameter and to quench the configuration whenever a configuration with an energy lower than E_0 is encountered. The optimization of this additional parameter can be avoided, when one adopts the self-adjusting cooling schedule introduced above. While we believe that the transformation we chose in equation 5 is a natural and minimal candidate that has these features, it is possible that more efficient transformations exist that adapt specifically to the particular problem under study.

4. Results

4.1 Protein structure prediction

We have first investigated the folding of small peptide fragments that are believed to assume a unique three dimensional structure even when removed from their environment in the protein. Figure 4 shows the overlay of the crystal structure of a helical 13 amino-acid residue fragment of the 1HRC protein with the structure we have obtained in STUN simulations. Encouragingly, the backbone configurations of these two structures are identical to better than experimental resolution. Figure 5 (a) shows the evolution of the total energy of the structure from an unfolded configuration to the folded configuration as a function of the number of energy evaluations. Figure 5 (b) shows the effective energy and the effective temperature. Several heating an cooling cycles were required to fold the helix fragment and "tunneling phases" that occur when the effective energy is relatively high significantly aided the search process. In these phases the original energy of the system undergoes significant

Fig. 4: Overlay of the crystal structure of a 13 residue helical fragment of 1HRC (Residues 92-105) with the structure obtained in the simulation.

fluctuations that are much larger in magnitude than the difference in energy of two successive metastable states. Circumnavigating these energy barriers in a traditional simulation would significantly slow the optimization process. We conducted several dozen STUN runs for this, as well as for other fragments that were investigated to verify that the structure we had obtained corresponds to the global optimum of the system. For 1HRC we found no competing structures with either PT or SA. We noted that in SA the helix could not be folded even with a tenfold increase of the computational effort. Hence STUN appears to present a viable and efficient optimization strategy to optimize peptide fragments of this length. Helical segments are stabilized by the short range hydrogen bonds. We found that it is possible to artificially destabilize the helical structure if the prefactor of the solvent interactions is increased to unphysical values.

An example for a non-helical 12 amino-acid fragment of the 1UBQ protein is shown in Fig. 6. In this structure hydrogen bonding interactions that attempt to stabilize a helix compete with longer range hydrogen bonding and solvent interactions to form a structure that is part helix part bend. The figure again illustrates the good overlap that was found in our STUN simulations for the simulated configuration and the corresponding crystal structure. A prerequisite for this success is a good balance between hydrogen bonding terms and solvent interactions in the force field.

We have also attempted to fold the 36 residue headpiece of the villin protein that was recently simulated with molecular dynamics [13]. The best configuration obtained with about a CPU week on a single PC is shown in Fig. 7 (b) in comparison with the NMR structure. The fraction of native contacts was similar in both studies, although more than 85 years of CPU time were invested in the MD simulation on a 256 node CRAY-T3E supercomputer. This comparison illustrates the increase in efficiency that can be obtained through the use of stochastic optimization methods, even though both simulations failed to reach the NMR structure. We find however that the structure obtained in our simulation has a lower energy that that of the NMR structure, indicating that this failure is not due to a failure of the optimiza-

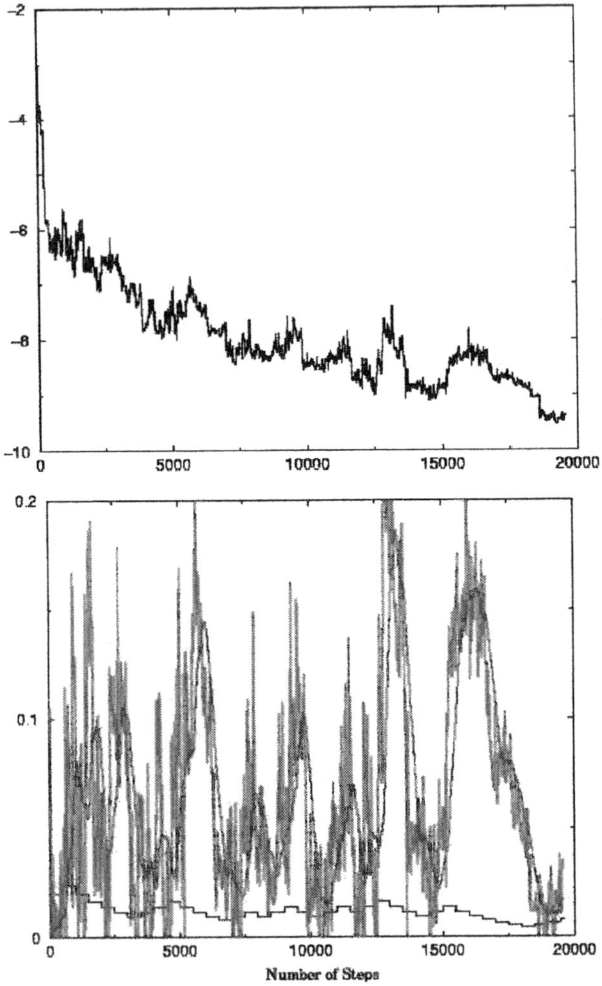

Fig. 5: Application of the stochastic tunneling method to the folding of a 13 amino acid helix fragment of 1HRC (Residues: 92-105). The top of the figure shows the total energy of the system as a function of the number of simulation steps. The lower part shows the effective energy, its moving average (dashed) and the effective inverse temperature of the STUN procedure. Both tunneling and local search phases are relevant to determine the native structure of the peptide, note that tunneling phases with relatively high effective energy correspond to large fluctuations of the original energy in the upper part.

tion strategy, but is attributable to a shortcoming of the forcefield. This suggests a rational decoy strategy to systematically improve the forcefield the we presently implement. We generate a large set of "good" candidates that compete with the NMR structure. As long as one of these decoys has a better energy than the native configuration, the forcefield must be modified to stabilize the native configuration in comparison to all other decoys. When this is achieved we generate new decoys by

Fig. 6: Overlay of the crystal structure of a helical bend in 1UBQ with the simulated structure.

refolding the peptide, generating either new configurations that are yet again better in energy than the NMR structure or ultimately folding the peptide. This strategy is presently implemented in our ongoing work.

4.2 Receptor ligand docking

A related low-dimensional optimization problem of considerable practical interest is the receptor-ligand docking problem, where suitable ligands must be selected for a given, structurally characterized receptor [5,18]. In order to select suitable ligands large chemical databases must be screened *in-silico* and for each ligand the best possible fit between ligand and receptor must be determined. Even in the most simple atomistic model, where both protein and ligand are treated as inflexible molecules, efficient numerical techniques to screen large databases in any reasonable timeframe are still lacking.

The reason for this difficulty lies in the competition between two vastly different energy scales in the problem, where steric repulsion competes with attractive electrostatic forces and hydrogen bonding to determine the global minimum of the PES. The tight fit between receptor and ligand (key-lock-principle) complicates the optimization problem significantly because it is almost impossible to reorient the ligand within the receptor, while there are few specific interactions between ligand and receptor outside the receptor pocket. Here we illustrate the performance of STUN in comparison with PT and SA for two receptor-ligand pairs, dihydrofolate reductase (4dfr) with methotrexate and the retinol binding protein (1rbp) with retinol respectively.

Fig. 7: Comparison of the (a) NMR structure and the (b) simulated structure of 1VII.

For the simulations discussed below we used a scoring function:

$$S = \sum_{\text{Protein}} \sum_{\text{Ligand}} \left(\frac{R_{ij}}{r_{ij}^{12}} - \frac{A_{ij}}{r_{ij}^{6}} + \frac{q_i q_j}{\epsilon r_{ij}} \right), \tag{8}$$

which contains the empirical Pauli repulsion, the van-der-Waals attraction and the

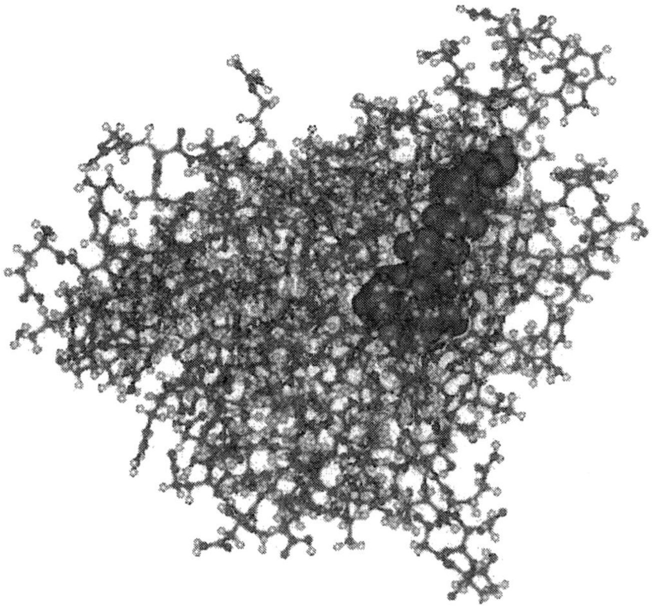

Fig. 8: The retinol docking protein with its ligand

electrostatic Coulomb potential. Neither entropic solvation effects nor dielectric screening were used in the simulations because such terms alter the specifics of the affinity of a given ligand to the receptor, but not the nature of the optimization problem. The ligands are simulated as rigid bodies, there are five degrees of freedom in the simulations. In cases where rotatable bonds exist, the x-ray crystallographic structures of the docked ligands were taken from the PDB database. The force field parameters R_{ij} and A_{ij} are taken from the OPLSAA force field [21] and the scoring function is pre-calculated on grids. The atomic affinity grids are interpolated using a logarithmic interpolation technique [34].

To localize the ligand in the vicinity of the receptor, we introduce a drift term $\mathbf{F}$

$$\mathbf{r}(t+\delta t) = \mathbf{r}(t) + \mathbf{F}(t)\,\delta t/f_t + \mathbf{X}, \tag{9}$$

where $\mathbf{X}$ is the random displacement sampled from a Gaussian with zero mean and width $\langle X^2\rangle = 2k_BT\,\delta t/f_t$. For the drift term we introduce a point $\mathbf{p}$ somewhere inside the cavity of the receptor. The drift force $\delta\mathbf{f}_d \sim -k_dT(\mathbf{r}-\mathbf{p})$ defines an additional, systematic contribution to the dynamics. The strength of the drift is proportional to the distance $|\mathbf{r}-\mathbf{p}|$ as it were in a harmonic oscillator field. If there were no further external forces, this drift would lead to a Gaussian localization of the center of mass coordinates of the ligand [35]. The advantage of this approach over the introduction of a penalty function is that the structure of the potential surface remains unchanged. No more than a bias in the sampling procedure is added

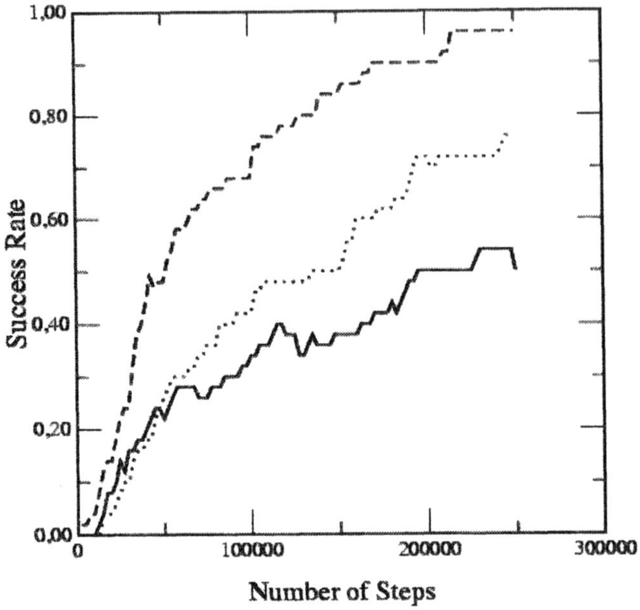

Fig. 9: Success Rate of SA (full line), PT (dotted line) and STUN (dashed line) in docking methotrexate versus the number of steps

to the random displacements. The ligand is localized in a way that its probability distribution fills the cavity of the receptor and is significantly reduced far outside the cavity. The localization volume does not depend on the step size δt and the temperature T.

In all simulations ligands were placed in a random position *outside the cavity* and we averaged the results of 50 runs of pre-described step number. A ligand was defined as 'docked' if the average RMS deviation of the atoms from the global minimum was less than 0.1 nm. The potential values were pre-calculated on cubic grids with a grid constant of 0.04 nm and a dimension of $3 \times 3 \times 3$ nm^3.

Methotrexate is a prolate shaped ligand with a axial ratio of $a/b = 2.5$ in the ellipsoidal approximation, i.e., a fat cigar. This system presents several problems: The ligand has a strong dipole moment and tends to dock wherever there are residues with partial electric charges. This leads to a rugged potential surface with a large number of local minima. To make things worse, there exists a very deep local minimum just at the entrance of the cavity. The global minimum, the energy of which is only a few percent lower, is separated from the metastable state by a barrier of hundreds kJ/mol. The ligand has to tunnel through the barrier, which requires a high temperature, and then to localize the minimum to an accuracy of a few percent in order to distinguish it from the local minimum in front of the barrier.

Figure 9 shows the success distribution, where STUN reached a reliability of 0.5 after 50,000 steps, PT required 150,000 while SA required 200,000 steps. In previous work [36] SA was reported to fail completely for this system in the absence of a drift term. During the simulations we observed frequently that the ligand reached the inner region of the docking site in early stages of the simulation, when the temperature was still too high to probe the potential minimum so that shortly afterwards a better score was found again in the low energy region in front of the cavity. This effect was less pronounced for STUN (tunnel parameter $\gamma = 0.05$), where the temperature is regulated by an automatic mechanism. The energy difference between the minima is resolved much earlier in the simulation.

Retinol (see Fig. 8) is a prolate shaped ligand with an axial ratio of $a/b = 4.5$, i.e., a slim cigar. There is no significant dipole moment in retinol. In the rigid receptor approximation the binding site is almost completely enclosed and molecule has to tunnel through a barrier of several thousands kJ/mol to reach the global minimum. On the other hand, this system presents no low-energy secondary minima. Since the cavity is quite elongated (length 1.65 nm) only a weak drift term was applied. In agreement with previous studies [34], SA completely failed to pass the ligand through the barrier, the same held true for PT. Among 50 runs there was no successful docking. The success distribution for STUN ($\gamma = 0.002$) demonstrates that this technique is capable for a fast and reliable docking of retinol, reaching a a success rate of 0.5 after about 40,000 energy evaluations.

5. Summary and conclusions

We have presented our motivation to use the stochastic optimization methods as a technique to predict the structure of complicated biomolecules. To implement this approach, a forcefield that parameterizes the free energy of the underlying model must be developed, such a forcefield must contain an implicit parameterization of the interactions of the biomolecule with the solvent. We have argued that there is a rational, decoy-based strategy to develop a biomolecular forcefield that can be used to predict the structure of short peptide fragments using stochastic optimization techniques such as the stochastic tunneling method. We have illustrated the success of this approach in the folding of short peptide fragments and presented an analysis of the difficulties encountered in the folding of the 36 head residues of 1VII. Stochastic optimization methods nevertheless permit an analysis of this problem and a systematic strategy for the improvement of the forcefield several orders of magnitude faster than competing simulation techniques. Finally we have illustrated the applicability of the stochastic tunneling method to a related problem of great practical interest in rational drug design.

Acknowledgments:

This work was funded by the Deutsche Forschungsgemeinschaft (We 1863/11-1), the BMBF and the Bode foundation.

References

[1] D. Baker and A. Sali, Science, **294**, 93 (2001).

[2] C. Branden and J. Tooze, *Introduction to Protein Folding* (Garland, 1999) 2nd edition.

[3] C. Walsh, Nature **409**, 226 (2001).

[4] The rcbs protein data bank: http://www.rcsb.org/pdb, 2001.

[5] K. Gubernator (Ed.), *Structure Based Ligand Design.* Wiley, 1998.

[6] B. Honig, J. Molec. Biol. **293**, 283 (1999).

[7] C. L. Brooks, J. N. Onuchic, and D. J. Wales, Science **293**, 612 (2001).

[8] A. R Dinner, A. Sali, L. J. Smith, C. M. Dobson, and M. Karplus, in *Trends in Structural Biology* **25**, 331 (2001).

[9] M. Daune, *Molecular Biophysics: Structures in Motion* (Oxford Scientific, 1999).

[10] B. Park and M. Levitt, J. Molec. Biol. **258**, 367 (1996).

[11] T. Lazaridis and M. Karplus, J. Molec. Biol. **288**, 447 (1998).

[12] IBM Blue Gene Team, IBM Systems Journal **40**, 310 (2001).

[13] Y. Duan and P. A. Kollman, Science **23**, 740 (1998).

[14] J. Pillardy, C. Czaplewski, A. Liwo, J. Lee, D. R. Ripoll, R. Kamierkiewicz, Stanislaw Oldziej, W. J. Wedemeyer, K. D. Gibson, Y. A. Arnautova, J. Saunders, Y.-J. Ye, and Harold A. Scheraga, Proc. Nat. Acad. Science (USA) **98**, 2329 (2001).

[15] D. Eisenberg and A.D. McLachlan, Nature **319**, 199 (1986).

[16] B. A. Berg and T. Neuhaus, Phys. Lett. B **267**, 249 (1991).

[17] K. Binder and A.P. Young, Rev. Mod. Phys. **58**, 801 (1986).

[18] H. J. Böhm and G. Schneider, (Eds.), *Virtual screening for bioactive molecules* (Wiley, 2001).

[19] W.F. van Gunsteren and H.J.C. Berendsen, *The groningen molecular simulation manual (gromos)* Technical report, Groningen University, 1987.

[20] MacKerell Jr. et al., J. Phys. Chem. B **102**, 3586 (1998).

[21] W. L. Jorgensen and N. A. McDonald, J. Mol. Struct. **424**, 145 (1998).

[22] Y. Duan, L. Wang, and P.A. Kollman, Proc. Nat. Acad. Science (USA) **95**, 9897 (1998).

[23] W. L. Jorgensen and J. Tirado-Rives, J. Amer. Chem. Soc. **110**, 1657 (1988).

[24] F. Avbelj and J. Moult, Biochemistry **34**, 755 (1995).

[25] F. Avbelj, Biochemistry **31**, 6290 (1992).

[26] D.J. Wales and H.A. Scheraga, Science **285**, 1368 (1999).

[27] S. Kirkpatrick, C.D. Gelatt, and M.P. Vecchi, Science **220**, 671 (1983).

[28] Nicholas Metropolis, Arianna W. Rosenbluth, Marshall N. Rosenbluth, Augusta H. Teller, and Edward Teller, J. Chem. Phys. **21**, 1087 (1953).

[29] A. P. Lyubartsev, A.A. Martinovski, S. V. Shevkunov, and P.N. Vorontsov-Velyaminov, J. Chem. Phys. **96**, 1776 (1992).

[30] E. Marinari and G. Parisi, Europhys. Lett. **451**, 19 (1992).

[31] Ulrich H.E. Hansmann and Yuko Okamoto, J. Comput. Chem **18**, 920 (1997).

[32] Kamal K. Bhattacharya and James P. Sethna, Phys. Rev. E **57**, 2553 (1998).

[33] W. Wenzel and K. Hamacher, Phys. Rev. Lett. **82**, 3003 (1999).

[34] David D. Diller and Christophe L.M.J. Verlinde, J. Comp. Chem. **20**, 1740 (1999).

[35] S. Chandrasekhar, Rev. Mod. Phys. **15**, 1 (1943).

[36] G. M. Morris, J. Comp. Chem. **19**, 1639 (1998).

Chapter 11

Electrical transport through a molecular nanojunction

Matthias H. Hettler[a*], Herbert Schoeller[b], and Wolfgang Wenzel[a]

[a]*Forschungszentrum Karlsruhe, Institut für Nanotechnologie, Postfach 3640, D-76021 Karlsruhe, Germany*
**E-mail: hettler@int.fzk.de*

[b]*RWTH Aachen, Theoretische Physik A, D-52056 Aachen, Germany*

Abstract

We consider electrical transport through a system of a molecule coupled to metallic electrodes. We give an overview of some of the issues involved in the problem. We discuss the two extreme regimes of transport. In the strong molecule-electrode coupling limit interaction effects beyond the Hartree-Fock level can be ignored. The transport can be described by a single particle scattering or Landauer approach. In the weak-coupling limit interaction effects dominate and the molecule must be treated as a many-body system. The transport is best described by incoherent sequential tunneling of single electrons. We discuss in general terms the relevance of spatial electronic structure, field effects and relaxation on the molecule. As an example, we consider a simple model for a molecule in the weak coupling limit. The model includes charging effects as well as aspects of the electronic structure of the molecule. The interplay of strong interactions and an asymmetry of the metal–molecule coupling can lead to various effects in non–linear electrical transport. In particular, strong negative differential conductance is observed under rather generic conditions.

---

1. Introduction

Transistors based on single molecules offer exciting perspectives for further miniturization of electronic devices with a potentially large impact in applications. To date several experiments have shown the possibility to attach individual molecules to leads and to measure the electrical transport. Two terminal transport through a single molecule [1–4] has been achieved by deposition of the object between two fixed electrodes or a conducting-tip STM above an object attached to a conducting substrate [5,6]. Among the most exciting effect observed in molecules so far is the negative differential conductance (NDC) observed in the experiment by Chen et al. [7]. Although, strictly speaking, an experiment on a molecule film, it is most likely that qualitatively similar effects should be displayed by a single molecule, too. Explanations of NDC so far have evoked mostly a conformational change of the molecule. One of the results of our work is that there are mechanisms of purely electronic origin which could lead to NDC in a fairly generic class of molecules.

In Sect. 2, we discuss the general issues of single molecule transport. We introduce the important energy scales and their relations. This will suggest the distinction between two pictures of transport, the "coherent" transport picture and the "tunneling" transport picture. We will discuss the limitations of each picture in some detail. In fairly general terms we then consider the relevance of spatial electronic structure of the molecule, effects due to the applied electric field and relaxation processes on the molecule. In Sect. 3, we concentrate on the limit of tunneling transport. We introduce a simple but generic molecular model and show how the current can be calculated by means of perturbation theory. In Sect. 4, we study a specific model that displays NDC in rather generic circumstances.

2. Transport through a molecule: General properties

One of the major theoretical problem in electrical transport through molecules comes from the fact that we deal with a "hybrid" system of materials with possibly very different electronic properties. In experiments today transport is measured mostly in setups where organic molecules are attached via thiol (S) groups to gold (Au) electrodes. The reasons for this choice have been the chemical feasibility and stability considerations. As all single molecule measurements so far have been performed at room temperature (at least for the break junction setup), a strong chemical bond like Au-S was helpful to provide the stability to make reproducible measurements. The diversity of the components in the transport experiment is huge. On one hand, we have gold electrodes of still relatively large size (20-50 nm cross section), a very good metal with well known electronic structure. On the other hand, there is an organic

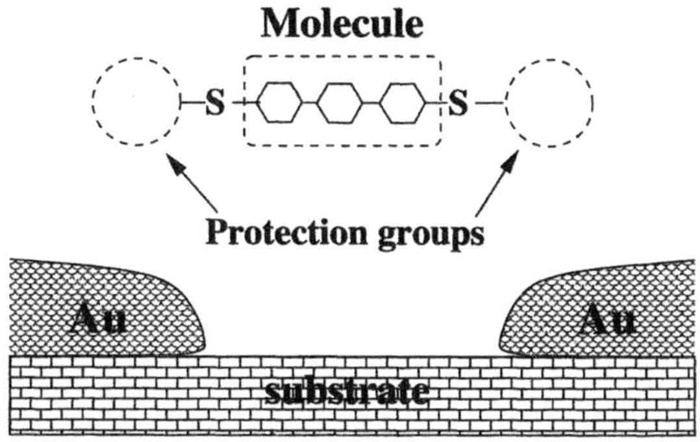

Fig. 1: Sketch of the hybrid system when molecule and electrodes are far apart. The protection groups are removed when the molecule comes close to the gold surface.

molecule of nanoscale size, with electronic structure that can be calculated by means of quantum chemistry, but often less studied experimentally and theoretically.

Because of this qualitative difference in size and structure the contact or interface properties of the components are poorly understood. Even less known is the relevance of the contact properties for transport. For example, the Au-S bond might be well studied experimentally and theoretically, but the relevance of these studies for transport is less clear. One must realize that the geometry of the transport experiment i.e., the topography of the electrode surface and the relative orientation of the molecule is not known. However, theoretical studies [8,9] show that the conductance can be different by orders of magnitude for different orientations.

The underlying fundamental problem (and also the main theoretical interest in this field) is that, in general, transport through a metal-molecule hybrid system is more than just the sum of transport the components. The different components interact with each other in many complex ways. Field effects, screening, dielectric effects, vibrations, electro-mechanical effects and relaxation via electromagnetic radiation can play a role. Which of these effects takes the dominant role in a given experiment can only be established a posteriori, if at all. In the following we try to put a perspective on when and how these aspects become important.

2.1 Strong vs weak electrode-molecule coupling

The basic issue can be posed as follows: Initially, as sketched in Fig. 1, the components of the hybrid system are far apart. In this case, the electronic structure of each component is understood, as sketched in Fig. 2. The electrodes can be considered as Fermi liquids, described by a density of states ρ_e and a chemical potential, or Fermi energy. As the experiments are done at room temperature, which for metals of interest is much smaller than the Fermi energy, the electronic states of the electrodes are filled up to the Fermi energy and empty above. The molecule can be described by a set of molecular orbitals (MOs) that are filled up to the HOMO (Highest Occupied

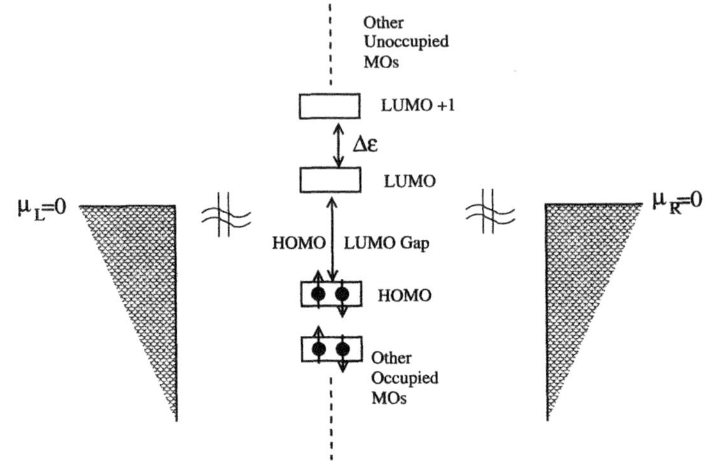

Fig. 2: Sketch of the electronic structure corresponding to Fig. 1. The electrodes are Fermi seas of electrons, the molecular orbitals are sharp quantum states.

MO). For a neutral organic molecule, each orbital up to the HOMO is (usually) filled by both a spin up and a spin down electron, and the ground state is usually a singlet. The energies and spatial distribution of the MOs can be computed by means of quantum chemistry for most molecules of interest.

Now, when the molecule comes in contact with one or both electrodes, the true quantum states are combinations of states on the electrodes and the molecule. However, it is useful to consider the problem from the molecule point of view as a "perturbation" by the electrodes of the MOs. Several effects are possible: (1) There can be energetic shifts, overall shifts as well as MO dependent shifts that can lift the degeneracies (many of the molecules of interest have a high symmetry, at least in parts of their structure). (2) Degenerate or nearly degenerate MOs might also mix to form new effective MOs in the presence of the electrodes. (3) Most importantly, because of the interaction the formerly sharp quantum states acquire a finite width in energy, and therefore a finite lifetime. This corresponds to the fact that the true eigenstates are also partly located at the electrode. If an electron of such an eigenstate "hops" from the molecule to the electrode, it appears like a decay of the electron from the molecule point of view. This is similar to the quasi-particle concept in solid state physics of metals.

To distinguish the different transport regimes, we introduce four energy scales that have proven useful in the related problem of transport through small quantum dots. First, $\Delta\epsilon$ is a measure of the typical energy difference of the MOs (ignoring degenerate or nearly degenerate MOs). Second, the contact between MOs and electrode might be described by a coupling strength Γ, though the actual coupling to a particular MO can be very much dependent on the MO in question. Third, the inverse of the the time of residence on the molecule τ_r of an electron participating in

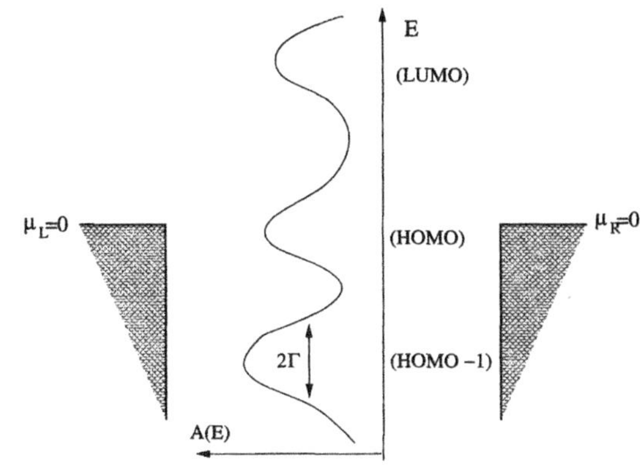

Fig. 3: Sketch of the electronic levels for strong coupling, the "coherent transport picture".

transport sets an energy scale $E_r = \hbar/\tau_r$. One could argue that this is inversely proportional to Γ, but, in principle, it is an independent quantity. Fourth, the energy to charge (or uncharge) the molecule by an additional electron we call E_C, the charging energy. The charging energy is related to the electron affinity or ionization energy of isolated molecules, but it is most likely much less than these energies because of the electrostatic effects of the electrodes, water and other dielectrics in the vicinity of the transport molecule.

It is mostly the relation of the coupling Γ to the other energies that governs the underlying physical picture of transport. Aside of possible intermediate regimes there are two basic scenarios: (i) Γ is larger than or comparable to $\Delta\epsilon, E_r, E_C$. In this case the formerly sharp sequence of quantum states on the molecule is smeared out to a continuous density of states (with maybe a few gaps remaining), the electrons spend more time in the electrodes than on the molecule, and charging effects are unimportant (except for an overall energy shift). Transport in this scenario happens via scattering states that are coherent quantum states over the entire system. The effect of the molecule is similar to a scatterer in a metallic constriction. The Landauer approach to conductance and its generalizations to nonlinear conductance seem appropriate. We call this scenario the "coherent transport picture".

(ii) Γ is much smaller $\Delta\epsilon, E_r, E_C$. This means that the molecular orbitals remain well defined and discrete states. Electrons spend enough time on the molecule for charging effects to take hold. The transport is best described as a sequence of incoherent hops of single electrons on and off the molecule. The Landauer approach breaks down as interaction effects on the molecule become dominant. We term this scenario the "tunneling transport picture".

In short, transport in the coherent transport picture is dominated by the contact, whereas in the tunneling transport picture it is dominated by the interactions on the molecule. Consequently, the theoretical approaches to the two regimes are nearly orthogonal.

It is to be noted that so far all theoretical work [8–13] has concentrated on the coherent transport regime, in the sense that the authors treated interactions at best in

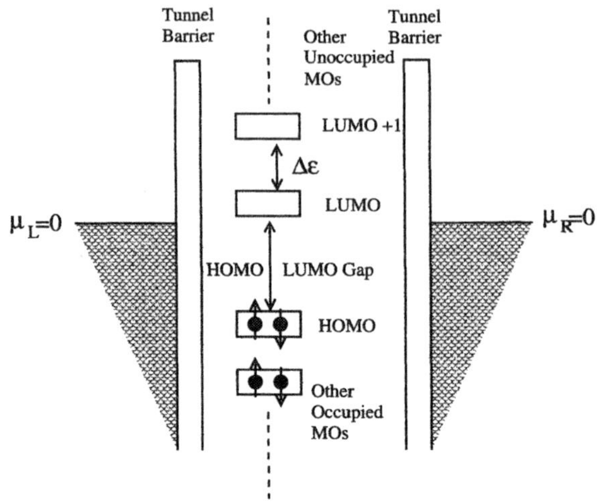

Fig. 4: Sketch of the electronic levels for weak coupling, the "tunneling transport picture".

a Hartree-Fock or mean field approach (to a certain point, this is also a reasonable qualification for the density functional approaches). One reason for this was that the sulfur-gold bond of the experiments is believed to be 'good' contact. Although basically all theoretical work overestimates the conductance by one order of magnitude, this is believed to be an issue of the geometry of the contact (cf., Ref. [8,9]). This is probably correct, although it is questionable whether a quantitative description of electronic transport with conductances of a few percent of the quantum of conductance $G_o = e^2/h$ is possible without better inclusion of interaction effects.

2.2 Failure of mean field theory

However, in the case of a tunneling contact the use of mean field type approaches is inadequate, even qualitatively. To demonstrate this, consider a model of just one molecular level of energy ϵ and interaction U (e.g., assuming that all other orbitals stay either occupied or unoccupied in the following Gedanken experiment). The Hamiltonian reads

$$H = \sum_{\sigma} \epsilon n_\sigma + U n_\downarrow n_\uparrow, \tag{1}$$

where n_σ is the number operator of electrons with spin σ. Let us consider the level occupation $\langle n \rangle = \sum_\sigma \langle n_\sigma \rangle$ as a function of the level energy ϵ. At temperature $T = 0$, it is clear that the exact solution shows a double step [see the right panel of Fig. 5 (solid line)]. In particular, for energies $-U < \epsilon < 0$ there is exactly one electron on the level. Now instead of "solving" the problem exactly let us perform a mean field approach. Here, we replace the interaction term in the Hamiltonian by $1/2U\langle n \rangle n$ leading to the non-interacting mean field Hamiltonian

$$H_{\mathrm{MFT}} = \left(\epsilon + \tfrac{1}{2}U\langle n \rangle\right) n. \tag{2}$$

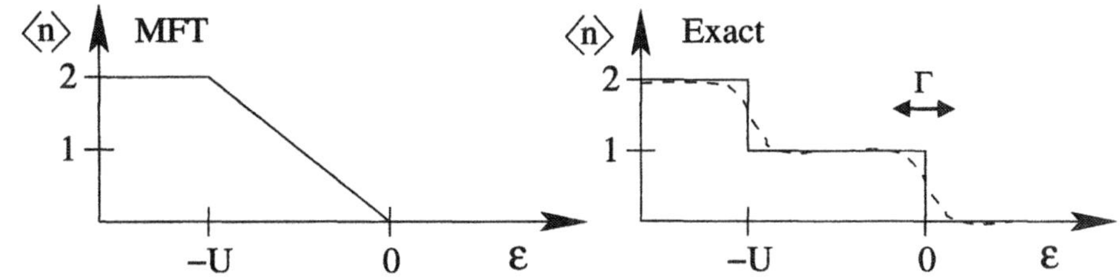

Fig. 5: Sketch of the density as a function of molecular orbital energy for a single interacting level.

This Hamiltonian must be supplemented by the self-consistency condition which reads

$$\langle n\rangle = 2f(\epsilon + \tfrac{1}{2}U\langle n\rangle) \tag{3}$$

where f is the Fermi function, $f(\epsilon) = 1/(\exp(\epsilon/k_BT) + 1)$. It is straightforward to see that Eq. 3 has a solution of non-integer $\langle n\rangle$ for any ϵ in the range $-U < \epsilon < 0$. Thus, the level occupancy for the mean field model looks like the left panel of Fig. 5, i.e., it is a *continuous* function of the level energy. This means that the mean field solution allows for continuous fluctuations of charge on the molecule, whereas the exact solutions allows only for integer changes of the molecule charge. The mean field approximation is therefore clearly inadequate.

These were the considerations for a toy model of a molecule in isolation. However, if we include the molecule-electrode coupling, the effective broadening Γ of the energy level produces a continuous level occupancy also for the "exact" solution, see the dashed line of the right panel in Fig. 5. Nevertheless, as long as Γ is sufficiently smaller than the charging energy (given by U in the toy model) there remains a plateau in the molecule occupancy which a mean field treatment would fail to predict even qualitatively. This defines the "Coulomb blockade" regime on which we will concentrate in the later parts of this work.

2.3 The trouble of having two contacts

The above discussion makes it clear that one needs to know at least the order of magnitude of the molecule-electrode coupling Γ to make qualitative theoretical predictions of electronic transport through molecules. Unfortunately, this is not easily determined experimentally, not even a posteriori for a given experiment. Clearly, the geometry of the molecule-electrode contact is not known, so theoretical estimates that depend strongly on geometry have serious limitations. Naively, one might hope to deduce the molecule-electrode coupling by the overall amount of measured current I. The simplest estimate would use the fact that Γ is basically the rate of transmitted electrons, so $I \sim e\Gamma/h$. This ignores the fact that the transmission coefficient of a transmission channel (in the Landauer sense) can be far less than unity, maybe just a few percent. Bearing this in mind, we look for the maximum of current per molecule that has been observed so far. It is of the order of 1 μA at about 1 Volt bias [2,4]. This leads to about 25 meV as a first estimate for Γ. If the transmission coefficient is indeed only a few percent, a Γ of the order of 1 eV is possible. This would be large enough to be considered on the strong coupling side, even if the charging energy could be of the same magnitude.

On the other hand, many measurements, in particular experiments on molecule films, show currents per molecule of the order of pA and less. The same reasoning as above would lead to a Γ of at most 1 μeV. This is clearly smaller than any imaginable charging energy. Can we conclude that in these experiments the molecule-electrode coupling is weak? Not necessarily! The point we have ignored so far, is that in fact we have two electrodes, and therefore two molecule couplings to the left and to the right electrode, Γ_L and Γ_R. As the circuit is basically a series of resistors, a better estimate can be achieved by using

$$I \sim \frac{e}{h} \frac{\Gamma_L \Gamma_R}{\Gamma_L + \Gamma_R}. \tag{4}$$

Because of the manufacturing process in the case of molecular films or geometry in the case of break junctions it is easy to imagine that the left and right couplings are very different in magnitude. In this case it is the smaller of the couplings (the larger resistor) that determines the magnitude of the current. But the broadening of energy levels is determined by the larger coupling! Therefore, even for experiments with very small currents the mean field approach might be appropriate. But because the overall current per molecule is so small in some experiments, it is quite possible that actually both couplings are weak compared to the charging energy. In that case the mean field theory would fail as discussed above. An overall strong asymmetry in the molecule-electrode coupling might be anticipated, especially in experiments in which only one side of the molecule has a thiol-gold bond, whereas the other side has a less definite contact, e.g., in the experiment of Ref. [7].

2.4 Impact of spatial electronic structure

So far, we have discussed "good" and "bad" contacts, weak and strong molecule-electrode coupling, without going into much detail of why some contacts are good and others are bad. Quantum mechanically, the quality of the contact is determined by the overlap of the wave functions on the electrode atoms with the wave function on the molecule atoms. Suppose for simplicity that the contact is simply made by a two atom bond, one atom from the electrode and one from the molecule. The wave functions at the corresponding atoms can be expanded on a basis set of atomic orbitals. Consequently, the overlap of the wave functions is a sum of overlaps of the atomic orbitals centered at the two atoms of the contact. As we do not have control about the geometry of the electrode surface and since the electrode is made of a simple metal, we assume that the orbital decomposition of the electrode atom can be estimated from e.g., the tight binding fits to the band structure in the bulk metal.

However, the molecule has a definite chemical structure and therefore well defined molecular orbitals that are superpositions of atomic orbitals of the molecule atoms. There are very different kind of orbitals, some consisting of very localized σ bonds, others exclusively consisting of delocalized π-bonds. Without elaborating too much

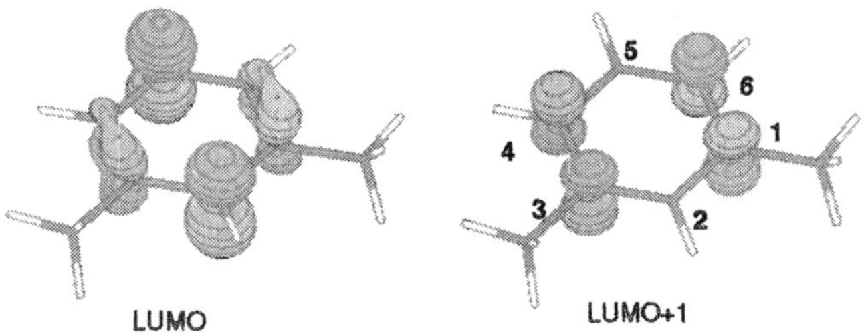

Fig. 6: The LUMO and LUMO+1 for a double methyl substituted benzene. By (anti)-symmetry, one of the MOs will have no coupling at the 2 position.

about all the possibilities, it is clear that molecular orbitals localized mainly inside the molecule will have very weak overlap to the electrode. Consequently, in zeroth approximation, they will not be able to contribute to the transport. The same holds for MOs localized at the contact atoms, but with negligible amplitude in the center of the molecule. In the language of the coherent transport picture, such MOs have a very small transmission coefficient. The basic dependence of transport on the spatial structure of the molecular orbitals has been nicely demonstrated in a recent work [14].

Because the spatial structure of the MOs generally depends on the chemical composition and specifically on the ligand groups attached to the "molecule backbone", this opens the door to the chemical design of electronic transport. One example is again Ref. [7] where NDC was observed in molecules that were equipped with nitro (NO_2) ligands at the central benzene ring. As was observed by the authors of Ref. [7] the nitro group induces an intrinsic dipole moment that was partially directed along the transport axis. This implies that some of the MOs must also be asymmetric along the transport axis. Following the reasoning above this might result in MOs that intrinsically couple with different strength to the left and the right electrode, independent of the actual contact geometry. We now give an explicit example of a molecule with such asymmetrically coupled MOs (though the reason here is even simpler than in the molecule of Ref. [7]).

Figure 6 shows the LUMO and LUMO+1 for a 1,3-dimethyl benzene (meta-xylene) that are closely spaced energetically and that couple very differently on various possible contact sites. There will be MOs which are antisymmetric with respect to the 2,5 - axis mirror symmetry with vanishing wave function amplitude at the 2 and the 5 position. In contrast, the symmetric MOs will in general have non-vanishing wave function at these positions. If one couples the molecule at the 2 and 6 positions to electrodes, the LUMO couples to both electrodes whereas the LUMO+1 would have no coupling to the electrode 'connected' to the 2 position [15]. Thus, the situation of a strongly MO and electrode dependent coupling seems to be generic for small aromatic molecules with ligand groups.

We will see below that such orbitals can have an enormous effect on transport, particularly in the case of weak molecule-electrode coupling [16]. We also point out that the definite and designable spatial structure of molecular orbitals is the major advantage to the otherwise similar (theoretical) problem of transport through semiconducting or metallic quantum dots [17,18].

2.5 Field and relaxation effects

To complete the discussion of general transport issues a few remarks on field and relaxation effects are in order. In an experiment, a bias of the order of volts is applied on a nanometre length scale. The resulting electric field is strong, and one should expect several effects due to it. Obviously, the energies of the MOs are shifted, similar to the Stark effect. More importantly, however, screening effects will take place due to polarization and movement of electrons. Because of the screening, the actual electric field along the molecule will be inhomogeneous. It has been suggested [6] that rather than being a ramp with fixed slope, the electrostatic potential follows a two-step profile, leading to a relatively weak electric field in the inside of the molecule. Although plausible, the actual field in a given experiment will depend on many parameters, and one should not rely too much on the two-step picture.

In general, however, the field will break the symmetry in direction of the transport axis (if there was any symmetry initially). Also, in an experiment the molecule is most likely *not* aligned perpendicular to two semi-infinite metal plates, as most sketches are drawn. This means that possible mirror symmetries along the transport or other axes are also broken by the field. The importance of such symmetry breaking is not clear and it strongly depends on the effectiveness of screening. Unfortunately, accounting for field effects within quantum chemistry is very time consuming, so this issue will remain unresolved for some time.

With the term "relaxation" we mean electronic transitions within the molecule, i.e., without changing the electron number on the molecule. This is possible by the coupling of the electrons to vibrational and electromagnetic degrees of freedom, i.e., phonons and photons. For the coherent transport picture, where the residence time τ_r of electrons on the molecule is short, relaxation is probably not important. One exception could be the excitation of a vibrational resonance. Then one might even expect the molecule to be destroyed, given the fact that up to $10^{12} - 10^{13}$ electrons are transmitted per second.

For the tunneling transport picture however, the residence time τ_r might be long enough for relaxation effects to take hold. Their greatest importance lies in the fact, that they allow transitions between molecular states that can not be achieved by a simple tunneling Hamiltonian. This is because the symmetry of the operators involved (e.g., of the dipole operator in case of photons) is different from tunneling. Therefore, molecular states that are inaccessible by electron tunneling (e.g., due to vanishing coupling of the relevant MOs) can be occupied by a relaxation process. On the other hand, the opposite situation is also possible, namely that a relaxation process occupies a state form which the molecule can not escape anymore by means of tunneling alone.

Below we discuss a model for which photon relaxation turns out to be ineffective. Relaxation by phonons, i.e., coupling to vibrations is not considered because it

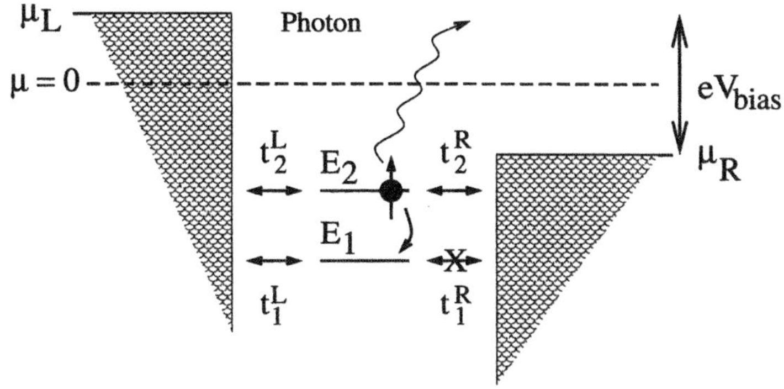

Fig. 7: Sketch of the couplings of the metal-molecule-metal system. The tunnel couplings can depend on both molecular orbital and electrode. The photons allow for on-molecule relaxation of excited states.

depends strongly on the molecule in question. One general feature, however, is the difference in energy scale between the two means of relaxation. Whereas photon relaxation involves energies in the infrared or optical range, phonon relaxation takes place at energies comparable to room temperature. This means that photons will in general be emitted, but not absorbed (since the number of available photons of optical energies are few at room temperature). On the other hand, phonons can be absorbed and emitted, unless the experiment is done at low temperature ($< 10K$) at which also phonons begin to freeze out. Temperature dependence in the transport is therefore most likely due to coupling of electrons to vibrations.

2.6 Preliminary summary

Though not entirely comprehensive, we have given a fairly broad overview of the basic issues in electronic transport through junctions consisting of single molecules. Several groups work intensively on the discussed problems, but very few works so far have considered transport in the weak coupling limit. On the other hand, from the above discussion, it should be clear that it is exactly for weakly coupled molecules for which the most interesting features might emerge, precisely because in the weak coupling limit the structure of the molecule (energetically as well as spatially) will emerge as the factor determining the transport. If molecules are supposed to become functional elements in electronic circuits rather than a small version of a semiconductor, then we have to study how transport happens for weakly coupled molecules. To deal with electronic structure and charging effects on equal footing is difficult problems, but are also exceedingly interesting.

3. The model and method of computation

We now turn our attention to the weak coupling limit. In this case we argued that charging effects are important and have to be taken into account, if possible non-

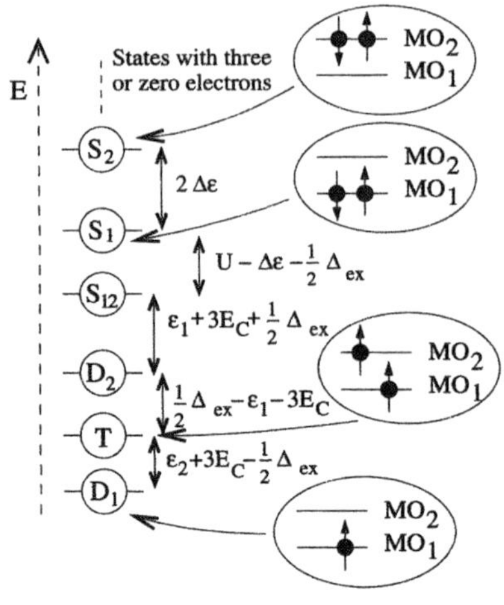

Fig. 8: Equilibrium molecule energies for model Hamiltonian (6). We choose energies such that the triplet states (T) lies below the singlets (S_{12} S_1 S_2) and between the two possible doublets.

perturbatively. Obviously, it is impossible to strictly do this even for a fairly small molecule like benzene. Instead of trying mean field type of approaches (which we argued is to fail qualitatively in the considered regime) we try to reduce the complexity by introducing a model that is easy enough to deal with and complex enough to be non-trivial. The description of the model and the method of computing the current at finite bias is given in the remainder of this section.

3.1 The model

For weak molecule-electrode coupling only a few of the molecular levels will contribute to transport in the low-bias regime. For the simplest non-trivial model, we assume that there are only two participating molecular levels that are both unoccupied at zero voltage, which we designate as LUMO and LUMO+1. (Other choices could be made, depending on where the chemical potential of the molecule is situated with respect to the electrode Fermi energy.) In the π-electron systems of aromatic molecules it is relatively easy to realize a situation, where two closely spaced MOs are separated far from both from the HOMO and the other LUMOs. We will compute transport in perturbation theory in the weak molecule-electrode coupling. Ignoring co-tunneling effects, which are of higher order in the perturbation theory we can assume that all other MOs stay inert (always occupied or always empty).

The MO Hamiltonian of the 'reduced' system can be written as

$$\mathcal{H} = \sum_{i\sigma} \epsilon_{i\sigma} c^\dagger_{i\sigma} c_{i\sigma} + \sum_{ijkl\sigma\sigma'} V_{ijkl} c^\dagger_{i\sigma} c^\dagger_{j\sigma'} c_{k\sigma'} c_{l\sigma}, \tag{5}$$

where the operators $c_{i\sigma}(c^\dagger_{i\sigma})$ destroy (create) electrons with spin σ in MO i. This

Hamiltonian contains a large number of parameters even for the two MO system, which can in principle be determined for a given molecule by quantum chemistry. Rather than specifying a particular molecule we seek to work out qualitative effects, presumably generic to whole classes of molecules. In order to make contact with the language of quantum dots it is useful to introduce an effective model for the molecular Hamiltonian

$$\begin{aligned} H_{\rm mol} = \epsilon_1 N_1 + \epsilon_2 N_2 + E_C\,(N_1+N_2)^2 \\ + \frac{U}{2}\sum_l N_l(N_l-1) + \frac{\Delta_{\rm ex}}{2}\sum_{\sigma,\sigma'} c^\dagger_{1\sigma}c^\dagger_{2\sigma'}c_{1\sigma'}c_{2\sigma}, \end{aligned} \tag{6}$$

where N_l is the occupation of the MO l. As unit of energy we take $\Delta\epsilon = \epsilon_2 - \epsilon_1$, the "bare" MO splitting. The diagonal electronic repulsion terms of Eq. (5) for the two MOs together with capacitive interactions with the leads can be rewritten as $E_C\,(N_1+N_2)^2 + \frac{U}{2}\sum_l N_l(N_l-1)$ (for simplicity, we assumed an orbital independent Hubbard-like repulsion U for double occupancy of a MO). $\Delta_{\rm ex}$ is a Hund's rule triplet-singlet splitting for the two electronic levels. In comparison to Eq. (5) we have neglected single electron hopping mediated by two-electron screening effects.

The bias is applied symmetrically over the molecule, so no capacitive shifts of energies appear (this is easily included). A likely energetic 'term scheme' for the single- and two-particle states described by the model Hamiltonian Eq. (6) is depicted in Fig. 8. The electrodes are considered as non–interacting electron reservoirs. The reservoirs are assumed to be occupied according to an equilibrium Fermi distribution function $f_\alpha(\omega) = f(\omega - \mu_\alpha)$, where μ_α denotes the electrochemical potential of electrode $\alpha = L, R$. The MOs couple to the electrodes via tunneling contacts with coupling strength t_l^α

$$H_{\rm mol-leads} = (\frac{\Gamma}{2\pi\rho_e})^{\frac{1}{2}} \sum_{\mathbf{k}\sigma\alpha l} \left(t_l^\alpha c^\dagger_{l\sigma} a_{\mathbf{k}\sigma\alpha} + h.c. \right) , \tag{7}$$

where ρ_e is the density of states (assumed constant) of the non-interacting electrons in the leads, described by operators $a_{\mathbf{k}\sigma\alpha}$. As before, Γ denotes the scale of the broadening of the MOs due to the coupling to the leads. As discussed in Sect. 2.4 the dimensionless couplings t_l^α can depend both on the molecular orbital considered as well as on the electrode (left or right).

We include a coupling of the molecule to a (broad band) boson field which simulates the relaxation of excited states in a real molecule by coupling of the electrons to an electromagnetic field (photons) and (though only crudely) vibrations (phonons).

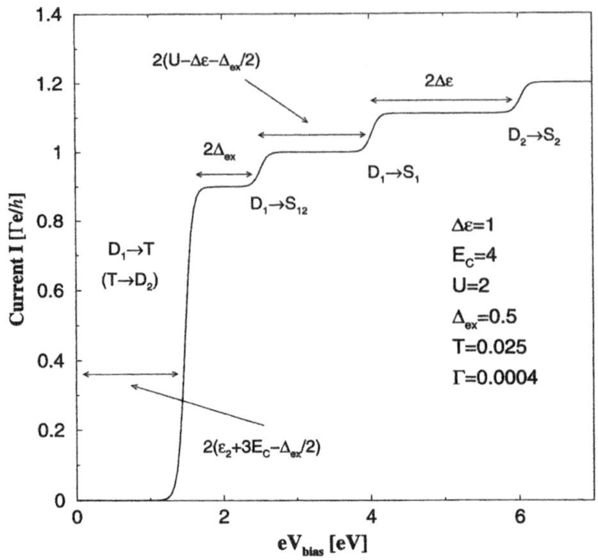

Fig. 9: Current-voltage characteristics if all molecule-electrode couplings are chosen equal. The parameters of the Hamiltonian (6) are $\epsilon_1 = -12$, $\epsilon_2 = -11$, $(\Delta\epsilon = 1)$, $U = 2$, $\Delta_{\rm ex} = 0.5$, $E_C = 4$, $\Gamma = 0.004$, $T = 0.025$. All energies are in units of eV.

3.2 Computational approach

We use a Master equation approach for the occupation probabilities P_s of the molecular many–body states [18]. The transition rate $\Sigma_{ss'}$ from state s' to s is computed up to linear order in Γ using golden rule (second order perturbation theory) in both the electrode-molecule t_l^α and the bosonic coupling. For the transition rates we have $\Sigma_{ss'} = (\sum_{\alpha,p=\pm} \Sigma_{ss'}^{\alpha p}) + \Sigma_{ss'}^b$ where $\Sigma_{ss'}^{\alpha p}$ is the tunneling rate to/from electrode α for creation $(p = +)$ or destruction $(p = -)$ of an electron on the molecule. We have

$$\Sigma_{ss'}^{\alpha +} = \Gamma f_\alpha(E_s - E_{s'}) \sum_\sigma | \sum_l t_l^\alpha \langle s|c_{l\sigma}^\dagger|s'\rangle |^2 \,, \tag{8}$$

and a corresponding equation for $\Sigma_{s's}^{\alpha -}$ by replacing $f_\alpha \to 1 - f_\alpha$. The boson-mediated rates $\Sigma_{ss'}^b$ describe absorption and emission of bosons. For photons we have

$$\Sigma_{ss'}^b = g_{ph} \frac{4e^2}{3\hbar^3 c^3} (E_s - E_{s'})^3 N_b(E_s - E_{s'}) \, |\langle s|\mathbf{d}|s'\rangle|^2 \,, \tag{9}$$

where $\mathbf{d}$ is the dipole operator and $N_b(E)$ denotes the equilibrium Bose function. g_{ph} is a parameter that allows us to modify the strength of the coupling to simulate increased dipole moment. $g_{ph} = 1$, unless specified otherwise. This value corresponds to a dipole of charge e and length 1Å. Increasing g_{ph} could also simulate relaxation by vibrations, but since the transition operators of a dipole and vibrations are different, this statement should be taken with a grain of salt.

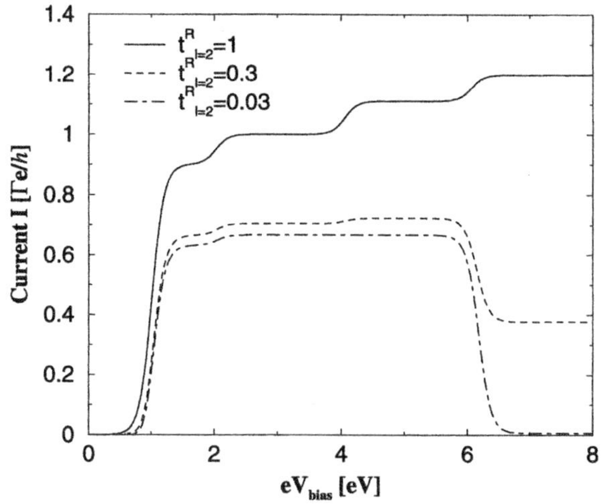

Fig. 10: I–V characteristics for various coupling t_2^R. A pronounced NDC effect is observed for reduced t_2^R.

We determine the P_s by solution of the stationarity condition

$$\dot{P}_s = 0 = \sum_{s'} (\Sigma_{ss'} P_{s'} - \Sigma_{s's} P_s). \tag{10}$$

The current in the left and right electrode can then be calculated via

$$I_\alpha = e \sum_{ss'} (\Sigma^{\alpha+}_{ss'} P_{s'} - \Sigma^{\alpha-}_{s's} P_s). \tag{11}$$

The bosonic transition rates do not contribute directly to the current, since they do not change the particle number on the molecule. However, they influence the state probabilities P_s, which also enters the current expression.

4. Results

The effective Hamiltonian Eq. (6) affords several generic scenarios for NDC. The NDC is generic in the sense that NDC will occur at some bias for an initially charged molecule [case (1)] as well as an initially neutral molecule [case (2)].

Case (1): Fig. 9 shows the I–V characteristics for equal tunneling couplings $t_l^\alpha = 1$ with a symmetric bias, $\mu_L = -\mu_R$. There are four characteristic steps which are related to the onset of the triplet and the three singlet states. From the plateau widths all characteristic energy scales can be deduced.

Strong NDC behavior is observed if one MO couples much more weakly to the right side than the other MO, e.g., $t_2^R = 0.03$; $t_1^L = t_1^R = t_2^L = 1$ (see Fig. 10). We see that the current decreases beyond a certain bias (negative differential conductance). In that region the current is suppressed by a factor $(t_2^R)^2$. The reason for this current decrease is the occupation of a molecule state (S_2 in this case) from which the

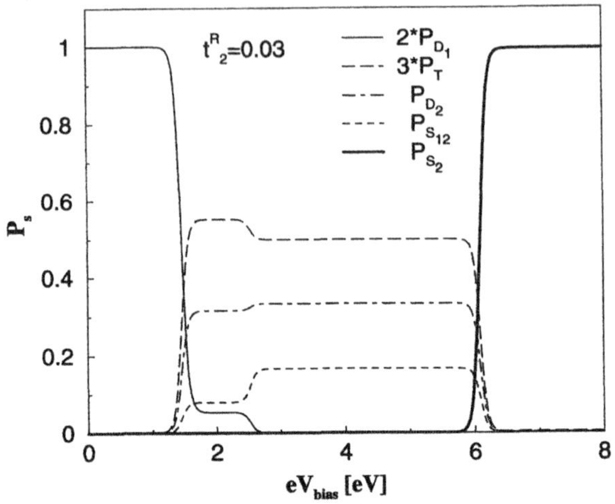

Fig. 11: Occupation probability P_s of the relevant molecule states for $t_2^R = 0.03$. The fat solid line indicating P_{S_2} reaches nearly unity in the blocking regime at bias $V_{\mathrm{bias}} > 6$. We multiply the probabilities with the corresponding degeneracy, so the P_s sum up to unity.

molecule has a hard time to escape due to a combination of blocking Fermi sea, Coulomb blockade and the small coupling of an MO to the electrode.

Initially, the molecule is singly occupied in state D_1. The current starts at a bias when the first two–electron state (triplet) becomes occupied (the "empty" state has higher energy for the given parameters). The current can flow via sequential hops through MO_1. The electron on MO_2 is essentially stuck since its tunneling time to the right reservoir is suppressed by a factor $(t_2^R)^2$. Tunneling to the left is suppressed for any electron because of the blocking Fermi sea (Pauli exclusion).

But at larger bias the electrons tunneling onto the molecule from the left can also form the state S_2, with both electrons in MO_2 as depicted in Fig. 8. No other electron can enter the molecule at this bias because of the charging energy. Since the relaxation due to the boson coupling is very slow, the only relevant decay of this state is via the small coupling to the right electrode. Consequently, the molecule is stuck for a long time in state S_2. Figure 11 shows that the average probability P_{S_2} is nearly unity. A relative suppression of t_2^R by 0.3 is sufficient to achieve a pronounced NDC effect. Increasing the temperature will broaden the plateau steps and shift the current maximum slightly to larger bias (not shown). At much larger bias (not shown), states with an additional electron become occupied, and the current rises again.

We also note that the NDC occurs at a bias corresponding to the $D_2 \rightarrow S_2$ transition, not the energy difference of the S_2 to the "ground state" D_1. This is because as soon as the triplets become occupied (onset of current) so does the state D_2 if the energy of D_2 is 'within bias range' of the energy of the triplets. Then, at the bias corresponding to $E_{S_2} - E_{D_2}$, S_2 gets occupied and the current decreases. Such *cascades* of transitions generally occur when the energies of states

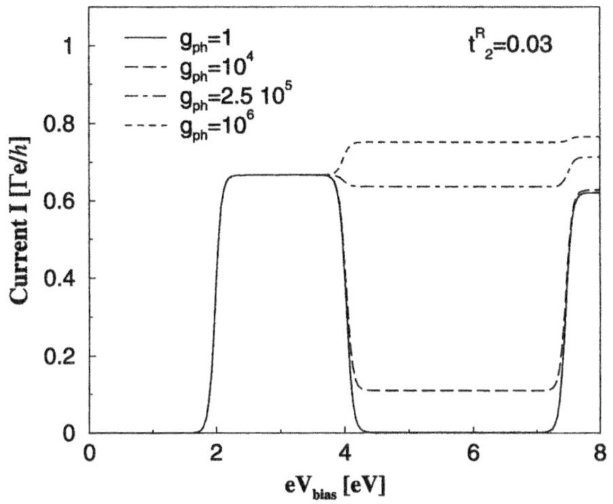

Fig. 12: I–V characteristics of the initially neutral molecule. $\epsilon_1 = -0.5$, $\epsilon_2 = 0.5$, $U = 1.5$, $\Delta_{ex} = 0.5$ and $E_C = 1.5$. NDC is observed for $t_2^R = 0.03$ involving D_2 as the blocking state. Relaxation by photon emission g_{ph} destroys NDC, but only if the coupling g_{ph} is increased by several orders of magnitude over the dipole approximation.

with different particle numbers mesh in some energy range [19]. For many MOs such cascades result in a quasi-ohmic I–V characteristics at high bias, even for the considered single tunneling picture of transport. In contrast, the first current step is always well defined by the energy differencc between the "ground state" and the first excitation with one additional electron on the molecule (or one electron less, depending on which has lower energy). For the same set of energy parameters NDC is observed also if t_1^R is suppressed instead of t_2^R. In this case the blocking state is S_1 [16].

Case (2): NDC is also observed if we start from an initially uncharged molecule (see the solid line in Fig. 12). The energy parameters for this set of curves are $\epsilon_1 = -0.5$, $\epsilon_2 = 0.5$, $U = 1.5$, $\Delta_{ex} = 0.5$ and $E_C = 1.5$. The blocking state in this case the doublet D_2, occupied when the symmetric bias reaches 4V. Also shown in Fig. 12 is the influence of internal molecule relaxation by increasing the boson coupling g_{ph} (case (1) shows similar behavior). An increase in the bosonic relaxation rate by *six orders of magnitude* over the one obtained in dipole approximation is necessary to completely eliminate NDC behavior. It is debatable whether e.g., coupling to vibrations of the molecule can provide such an enhancement of the on-molecule relaxation rate. However, even in a situation where the coupling g_{ph} is nominally large there can be selection rules that prevent decay of certain states. An example would be the inhibition of (direct) transitions between states of different total spin, i.e., singlet–triplet transitions [16].

5. Conclusions

We have given an overview of the physics involved in non–linear charge transport through a metal–molecule–metal system. Depending on the strength of molecule–metal coupling, two different descriptions of transport were identified: the coherent transport and the tunneling transport picture. We discuss the influence of electronic structure, field effects and relaxation on the transport for the different transport pictures. Concentrating on tunneling transport, we developed a model that includes charging effects as well as aspects of the spatially non–trivial electronic structure of the molecule, interplay of which can lead to current peaks and strong negative differential conductance. For a coupling to photons in dipole approximation the relaxation rate induced by the photons was found to be several orders of magnitude too small in comparison to typical tunneling rates to have an effect. We believe that the model is sufficiently generic to be realized in certain classes of aromatic molecules with tunnel contacts to the electrodes.

Acknowledgements

The authors gratefully acknowledge discussions with R. Ahlrichs, D. Beckmann, T. Koch, M. Mayor, J. Reichert, G. Schön, H. Weber and F. Weigend, the financial support from the Deutsche Forschungsgemeinschaft (DFG, WE 1863/8-1) and the von Neumann Center for Scientific Computing.

References

[1] M. A. Reed, C. Zhou, C. J. Muller, T. P. Burgin, and J. M. Tour, Science **278**, 252 (1997).

[2] C. Kergueris, J.-P. Bourgoin, S. Palacin, D. Esteve, C. Urbina, M. Magoga, and C. Joachim, Phys. Rev. B **59**, 12505 (1999).

[3] D. Porath, A. Bezryadin, S. de Vries, C. Dekker, Nature **403**, 635 (2000).

[4] J. Reichert, R. Ochs, D. Beckmann, H. B. Weber, M. Mayor, H. v. Loehneysen, cond-mat/0106219.

[5] L. A. Bumm, J. J. Arnold, M. T. Cygan, T. D. Dunbar, T. P. Burgin, L. Jones II, D. L. Allara, J. M. Tour, and P. S. Weiss, Science **271**, 1705 (1996).

[6] S. Datta, W. Tian, S. Hong, R. Reifenberger, J. I. Henderson, and C. P. Kubiak, Phys. Rev. Lett. **79**, 2530 (1997).

[7] J. Chen, M. A. Reed, A. M. Rawlett, and J. M. Tour, Science **286**, 1550 (1999).

[8] S. Yaliraki, A. E. Roitberg, C. Gonzalez, V. Mujica, and M. A. Ratner, J. Chem. Phys. **111**, 6997 (1999).

[9] M. Di Ventra, S. T. Pantelides and N. D. Lang, Phys. Rev. Lett. **84**, 979 (2000).

[10] V. Mujica, M. Kemp, A. Roitberg, and M. A. Ratner, J. Chem. Phys. **104**, 7296 (1996).

[11] M. P. Samanta, W. Tian, S. Datta, J. I. Henderson, and C. P. Kubiak, Phys. Rev. B **53**, R7626 (1996); [6]; S. Datta, and W. Tian, Phys. Rev. B **55**, R1914 (1997).

[12] M. Magoga and C. Joachim, Phys. Rev. B **56**, 4722 (1997).

[13] E. G. Emberly and G. Kirczenow, Phys. Rev. B **58**, 10911 (1998); Phys. Rev. Lett. **81**, 5205 (1998).

[14] J. Heurich, J. C. Cuevas, W. Wenzel, G. Schön, cond-mat/0110147.

[15] More realistically, we can say that the coupling of the antisymmetric MO will be much suppressed as compared to the symmetric MO. That is all that is necessary for the effect we present.

[16] M. H. Hettler, H. Schoeller and W. Wenzel, to appear in Europhys. Lett. (2001).

[17] H. Grabert and M.H. Devoret, *Single Charge Tunneling*, NATO ASI Series, Vol.294 (New York, Plenum Press 1992).

[18] L.L. Sohn *et al.* (Eds.), *Mesoscopic Electron Transport*, (Kluwer 1997).

[19] M. H. Hettler, H. Schoeller and W. Wenzel, in preparation.

Chapter 12

Single metalloproteins at work: Towards a single-protein transistor

Paolo Facci

Istituto Nazionale per la Fisica della Materia, Dipartimento di Fisica, Università di Modena e Reggio Emilia, 41100 Modena, Italy
E-mail: p.facci@unimo.it

Abstract

In this article we shall present recent results and research trends in the investigation of the functional properties of surface-immobilized metalloproteins towards their exploitation for assembling hybrid biomolecular electronic nanodevices. Particularly, scanning probe microscopy studies performed at the level of single molecule in an electrochemical cell address the functional behaviour of blue-copper proteins as biomolecular switches, while a series of spectroscopic and electrochemical experiments show relevant results on functional and structural properties of these molecules arranged in monolayers. Their potential use as channels of nanometer-size hybrid bio-FET all the way down to the single molecule level is also discussed. Results on monolayer formation on substrates of different nature, their characterization, and applications are reported together with possible short- and medium-term scenarios in biomolecular electronics research.

1. Introduction

The research activity in the field of molecular electronics [1] is becoming more and more exciting because in nanotechnology we are rapidly approaching the capability to manipulate single molecules to build nanometer size devices, and in this endevor, increasingly smarter molecules are being used [2]. Several types of organic molecules have been employed in recent works that are aimed at the demonstration of their potentials in implementing or even complementing the conventional materials used in solid state electronics [2]. These researches have led to fascinating results dealing with the possibility of using both organic synthetic molecules, supramolecular edifices and clusters for their intrinsic properties (e.g., wiring or switching capabilities, rectifying behavior) or in some special configurations achievable by state-of-the-art nanotechnology such as single electron transistors [3].

In this respect, biopolymers, and proteins in particular, bear different important features which could make them ideal candidates to be used in future hybrid nanoelectronic devices. In fact, proteins, being selected by evolution, turns out to be highly optimized functional units for performing special tasks in a very wide range of situations [4]. In particular, metalloproteins [5] are molecules which contain one or more metal ions inside their scaffold and are often devoted to or involved in processes connected with transferring electrons via intra- or intermolecular redox reactions through different metabolic pathways.

Protein engineering, with the help of site-directed mutagenesis, is another important resource which makes proteins very attracting for the development of biomolecular electronics. Indeed, these molecules can be modified and improved by properly engineering some structural or functional aspects such as, the mutation of protein surface residues for achieving molecular immobilization by chemical binding, etc. Of course, the molecules used in biomolecular electronics applications have also to be stable enough for operating successfully in a non-physiological environment and under rather artificial conditions. This point is indeed important and generally restricts the choice of the possible candidates beyond their functional characteristics.

We have used different metalloproteins, the blue-copper protein azurin [6], to face the problem of evaluating its potentialities as functional element in hybrid nanodevices such as single protein transistors, and a heme-based one (myoglobin [7]), to complement the data and to develop and generalize the chemical surface immobilization approach.

Azurin [6] (see Fig. 1), is an electron transfer metalloprotein (molecular mass 14600) involved in respiratory phosphorylation of the bacterium *Pseudomonas aeruginosa.* Its redox active site contains a copper ion liganded to 5 aminoacid atoms according to a peculiar ligand-field symmetry which endows the center with unusual spectroscopic and electrochemical properties. These include an intense electron absorption band at 628 nm (due to the S(Cys-σ)$\rightarrow$Cu charge transfer transition), a small hyperfine splitting in the electron paramagnetic spectrum [8] and an unusually large equilibrium potential [+116 mV vs saturated calomel electrode (SCE)] [9] in comparison to the Cu(II/I) aqua couple (-89 mV vs SCE) [10]. Moreover, Azurin

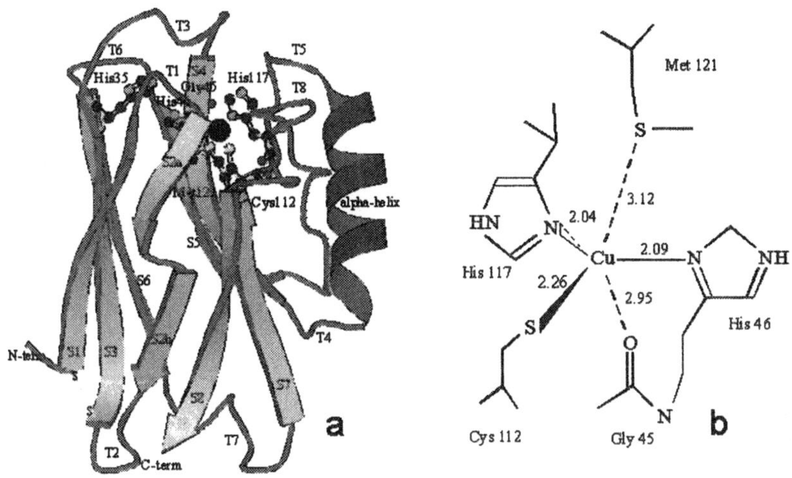

Fig. 1: The schematic representation (Molscript) of the structure of Azurin (coordinates from [20]) (a) and of its active site (b).

shows a smart self-assembling capability onto gold via a surface disulfide bridge formed between the two residues Cys3-Cys26 [11].

Myoglobin [7], one of the most studied metalloproteins, is a monomeric heme-containing protein found mainly in muscle tissues where it serves as an intracellular storage site for oxygen. Its molecular mass is 17000. Its optical absorption spectrum is characterized by a very intense Soret band (ε $\approx$200,000 M^{-1} cm^{-1}) located at 409 nm. Its redox equilibrium potential [couple Fe(III/II)] is typically (−110 mV vs SCE).

In this article, we describe our studies of the redox-behavior of azurin immobilized on Au (111) substrates by means of electrochemical STM/AFM [12] measurements which elucidate the underlying mechanism of electron tunneling through it and also evaluate its potentials for electronics applications. Towards that goal, possible scenarios involving single proteins and protein monolayers to be used as channels of innovative FET-like devices are discussed. Promising approaches allowing protein immobilization onto substrates of various nature (metal, insulating, semiconductor) are presented along with possible strategies for implementing in real devices. Morphological and spectroscopic results on both azurin and myoglobin samples help in assessing their structural and functional quality.

2. Materials and methods

Chemicals: Azurin from *Pseudomonas aeruginosa* was acquired from Sigma and used without further purification after having checked that the ratio OD_{628}/OD_{280} (OD_λ = optical density measured at λ nm) was in accordance with available values in the literature (0.53–0.58) [13]. Working solution was 10^{-4} M azurin in 50 mM NH_4Ac (Sigma) buffer (pH 4.6). The buffer was degassed with N_2 flow prior to use. Milli-Q grade water (resistivity 18.2 MΩcm) was used throughout all the experiments.

Myoglobin from skeletal horse muscle was acquired from Sigma, dissolved in 50 mM NH_4Ac (pH 4.6) and centrifuged (5 min at 14924 g) prior to use. The supernatant was collected and the resulting protein concentration of 3.2×10^{-5} M was used for the experiments. 3-aminopropyltriethoxysilane (3-APTS), 2-mercaptoethylamine (2-MEA), and glutaric dialdehyde (GD) were acquired from Sigma and diluted immediately prior to use to 6.6 % (V/V) in $CHCl_3$, or to final concentrations of 1.7×10^{-2} M or 4×10^{-4} M in H_2O (Milli-Q grade).

Substrates: Different substrates were used in order to obtain relevant results from the different methods applied: freshly cleaved mica sheets for SFM, glass slides for optical absorption spectroscopy, Si/SiO_2 for planar device implementation, and gold on mica for CV and ECSTM. Because of its lower roughness which facilitates molecular resolution, mica was preferred to silicon with native oxide in the case of SFM. Silicon was cleaned in acetone before use and glass was cleaned with 30 % H_2O_2 – 70 % H_2SO_4 solution. Au (111) substrates for CV and ECSTM measurements were prepared by evaporating 150-nm Au films onto freshly cleaved mica sheets. The sheets were first baked for 2 h at 450°C and 10^{-7} mbar. Gold was evaporated at a deposition rate of 0.3 nm/sec. After evaporation the films were annealed for 6 hours at 450°C and 10^{-7} mbar. After cooling down to room temperature, a moderate flame annealing was necessary to get large re-crystallized Au (111) terraces.

SFM-Probes: Rectangular Si_3N_4-sharpened Microlevers (Thermomicroscopes Co.) with an elastic constant of 0.02 N/m and an apex curvature radius of less than 20 nm were used for contact mode measurements and for producing and measuring steps in the sample. Magnetic Alternating Mode (MAC) cantilevers (Molecular Imaging Co.) with an elastic constant of 1.7 N/m and a resonant frequency of 155 kHz were used for intermittent contact measurements.

STM-Probes: STM tips were made from PtIr (80:20) by electrochemical etching of a 0.25 mm wire in a melt of $NaNO_3$ and NaOH [14]. Tips were then insulated with molten Apiezon W wax. Only tips displaying leakage levels below 10 pA were used for imaging.

Sample preparation: Sample for STM were prepared by incubating freshly evaporated Au (111) substrates with 10^{-4} M azurin for 20-40 min and then rinsed in abundant NH_4Ac buffer directly in the measuring cell. This always leaves an aqueous layer on the top of the substrate to prevent protein exposure to air-water surface tension. After several rinsing cycles the measuring cell was filled with the same buffer and immediately installed in the microscope for imaging.

Samples prepared via three-step chemical reaction (Fig. 2) were assembled as follows: (i) incubate the substrates (silicon, glass, or mica) for 2 min in 3-APTS. Then rinse in abundant $CHCl_3$ in order to remove 3-APTS molecules not linked to the surface; (ii) expose the silylated sample for 10 min to GD, followed by a

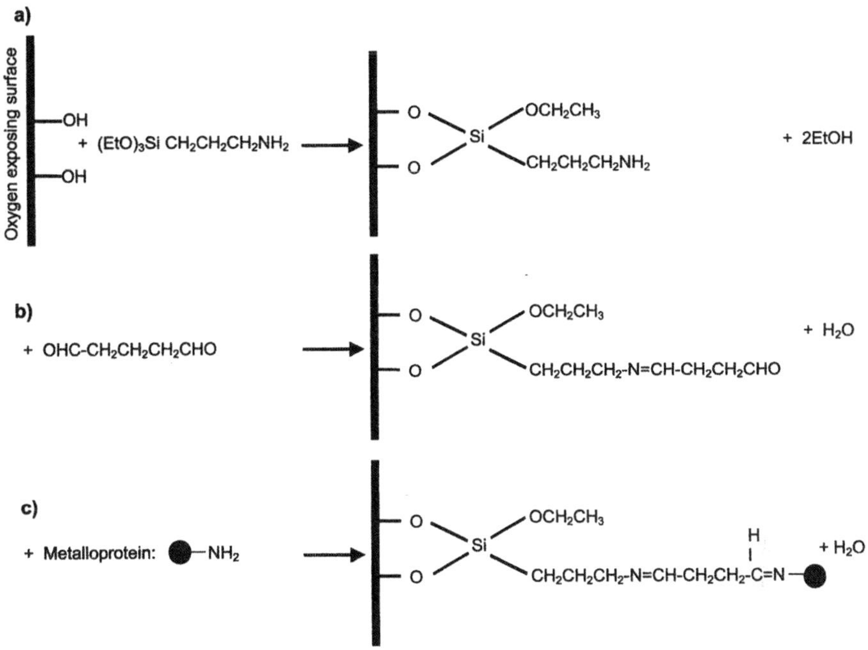

Fig. 2: Scheme of the three-step (a, b, c) chemical reaction used for immobilizing proteins.

thorough washing in ultra-pure H_2O; and (iii) expose the coated substrates for 20 min to azurin solution followed by rinsing with NH_4Ac solution in order to get rid of physically adsorbed molecules.

For the SFM measurements, the samples were installed in the measuring chamber and covered by a film of NH_4Ac solution. Samples on gold for CV were prepared via a three-step reaction that was similar to that used in the case of silicon, mica, and glass, except for the fact that in the first step a gold substrate was incubated for 2 min. with 2-mercaptoethylamine (which links to the gold surface by a SH group) and that the rinsing was performed in H_2O.

In situ STM: This kind of technique allows control of both substrate and tip Fermi levels at a given bias voltage, by sweeping the energy scale where molecular levels are fixed (Fig. 3). A Picoscan system (Molecular Imaging Co.) equipped with a Picostat (Molecular Imaging Co.) bipotentiostat was used to perform *in situ* STM investigation. The measuring cell consisted of a Teflon™ ring pressed over the Au (111) substrate operating as working electrode. A 0.5 mm Pt wire was used as counter electrode and a 0.5 mm Ag wire as quasi-reference electrode (AgQref). The AgQref potential vs SCE was measured before and after each experiment. In what follows, potentials will be always referred to as SCE. In order to minimize buffer evaporation, the cell was mounted into a sealed Pirex™ chamber. Images were acquired at room temperature under electrochemical control in the potential range −225 – +75 mV at steady state current conditions. A 10 μm scanner with a final

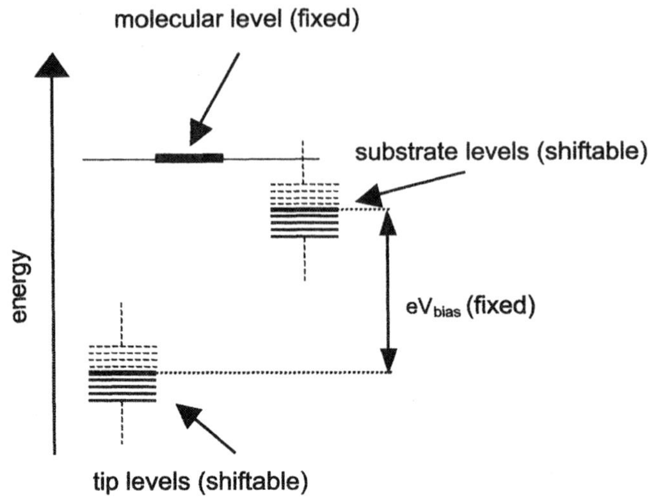

Fig. 3: The energy diagram showing the working principle of electrochemical STM.

preamplifier sensitivity of 1 nA/V was used for STM measurements. STM images were acquired in constant current mode with a typical tunneling current of 2 nA, a bias voltage of 400 mV (tip positive), and scan rate of 4 Hz.

In situ SFM: A Picoscan system (Molecular Imaging Co.) was used to perform SFM. The measuring cell consisted of a Teflon^{TM} ring pressed over the substrate. The cell was mounted into a sealed Pyrex^{TM} chamber in order to minimize buffer evaporation. A 6-μm scanner was used. The typical tip-sample interaction force was 1 nN at scan rates of 2-4 Hz.

Cyclic Voltammetry: CV has been performed with scan rates of 0.01 to 0.1 V/sec in 50 mM NH_4Ac (pH 4.6) using a Pt net as the counter-electrode, a Ag wire as (quasi) reference, and a 0.28 cm^2 Au (111) electrode (evaporated on mica, see "substrates" section for details) as working electrode. The Ag wire potential was measured against SCE before and after the experiment.

Optical absorption spectroscopy: Optical absorption spectra have been measured with a JASCO J514 dual-beam spectrometer in the wavelength range of 350 to 600 nm with 5 nm bandwidth in order to enhance the signal-to-noise ratio.

3. Results and discussion

In situ STM measurements of Azurin adsorbed on Au (111) yield images similar to those reported in Fig. 4 on areas of different sizes. A series of bright spots, 4-5 nm in diameter, appear on the underlying gold terraces (a). These are ascribed [11] to the presence of single Azurin molecules chemisorbed onto gold via the surface disulfide bridge. Figure 4 (b) shows a higher resolution image of a similar sample. As it has been demonstrated recently [15], the contrast is achieved by tuning the substrate potential (hence the Fermi level of the substrate) to the protein equilibrium potential

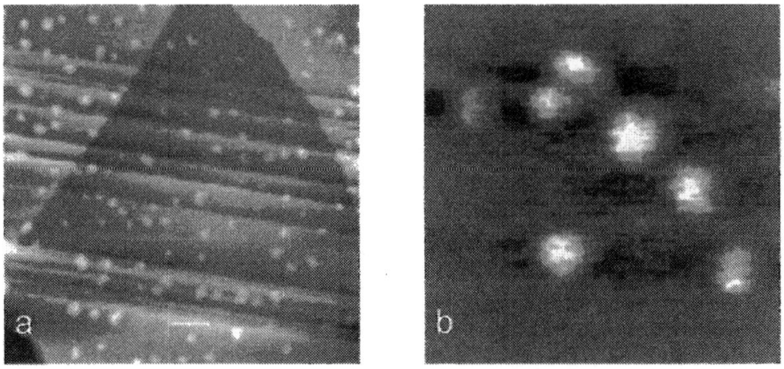

Fig. 4: Electrochemical STM images of Azurin, self-chemisorbed onto Au (111). (a) Image size 290×290 nm^2, substrate potential +75 mV, bias voltage +400 mV (tip positive), tunneling current 2 nA, Δz = 1.8 nm, scanning frequency 3.9 Hz; (b) image size 47×47 nm^2, substrate potential +75 mV, bias voltage +400 mV (tip positive), tunneling current 2 nA, Δz = 1 nm, scanning frequency 3.9 Hz.

(broadened by ~300 mV due to the presence of the aqueous solvent [12, 15]). The contrast in ECSTM of redox species on a metal surface is known to depend upon the value of the substrate potential [15]. It is worth noting that a similar electrochemical experiment performed with SFM do not show any dependence of the visible features upon the substrate potential (vide infra), indicating the purely electronic origin of the bright spots observed by electrochemical STM [15].

Although, in general, the solution equilibrium potential of a metalloprotein could differ from that of the immobilized species, all the available data for azurin do not show any appreciable difference [15]. This fact indicates the weak effect that the immobilization procedure (and hence the non-physiological environment) play on the geometry of the active site, which is known to determine in a very sensitive way the equilibrium potential of the molecule [16]. These considerations and the robustness of the molecule shown under repeated tip scans suggest that Azurin could be a good candidate for biomolecular electronics purposes.

In order to go into more details, we have focussed our attention on a more restricted area and we have imaged azurin as a function of the substrate potential. This approach yields information which are spectroscopic in nature and plays, in the liquid environment, the role that is usually played by V/I measurements in UHV STM experiments. Figure 5 reports three different images acquired on the same area at −25 mV (a and b) (close enough to azurin equilibrium potential), and at −125 mV (c) (well off this range). The bumps, clearly evident in the first image (a), disappear in the second (b) and appear again in the third (c), once the potential is re-established. Thus, the effect of tuning the substrate potential to the equilibrium potential of azurin is that of eliciting tunneling through the molecule, in a fashion similar to that of resonant tunneling devices. Analyzing these images in more details, it is also remarkable that in the image at −125 mV (b) some weak depressions

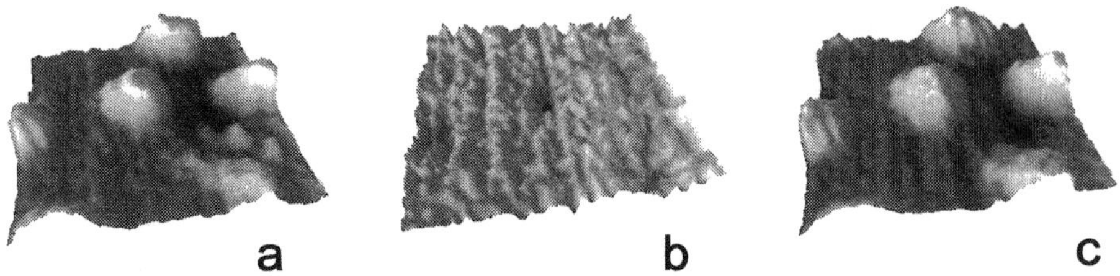

Fig. 5: Electrochemical STM images of the same physical area of a sample of Azurin, self-chemisorbed onto Au (111) as a function of the substrate potential. (a) and (c) substrate potential −25 mV; (b) substrate potential −125 mV. Image size 22×22 nm^2, bias voltage +400 mV (tip positive), tunneling current 2 nA, $\Delta z = 1$ nm (a) and (c), 0.2 nm (b), scanning frequency 3.9 Hz.

appear in correspondence to the bumps in the other two images. These effective depressions can be interpreted as a consequence of the STM feedback response to the local variation in the sample conductivity, suggesting once more that if the substrate potential is not tuned to the azurin equilibrium potential, the protein cannot let a current flow through it; it behaves as an insulating barrier. This evidence is somehow confirmed also by the occurrence of a sort of blurring in the features of Fig. 5 (c) which could be consistently due to the interaction of the tip apex with the protein globule when scanning the surface in de-tuned conditions. This effect has been quantified in terms of full width at half maximum of the features visible in Fig. 5 (a) and (c) providing a value of 4.3±0.1 nm and 4.8±0.1 nm, respectively. However, several imaging cycles can be performed without a drastic loss of resolution.

These evidences prove that it is possible to make azurin "conducting" by properly aligning the substrate Fermi level to the molecular levels. To be more specific, the situation in Fig. 5 (b) corresponds to that of a resonant diode in which, by the action of the bias voltage, the Fermi level of the source exceeds the resonant level, giving rise to a decrease of the resonant tunneling current.

This switching behavior shown by azurin molecules is of course very interesting in itself, since it represents the first demonstration of resonant tunneling via redox levels of a single biomolecule. Furthermore, it opens up the way to interesting potentialities for exploitation of this molecule in hybrid solid-state nanoelectronics. In fact, since the described mechanism remains substantially unaltered if one keeps the substrate levels fixed and shifts the molecular ones, the reported situation allows us to predict that, by electrostatically coupling the molecule with a gate electrode it will also be possible to tune the current flow through it. This allows us to exploit protein redox properties to yield a hybrid biomolecular switch. Such a device could be even based on the operation of a single metalloprotein in between two electrodes (source and drain) provided that state-of-the-art nanolithography can produce nanogates with gaps below 10 nm. In Fig. 6, a possible scheme for such a device is reported, along with its operating principles. The action of the electrostatic coupling provided by applying a potential to the gate electrode modulates the electrochemical potential

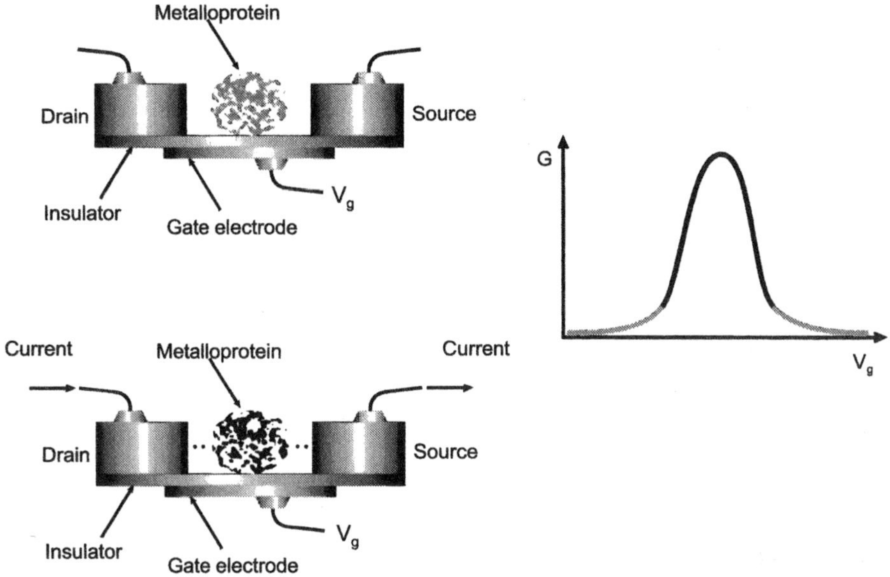

Fig. 6: Scheme and operating principle of a single-metalloprotein nanotransistor.

in the molecule with respect to that in the metal leads, allowing (or blocking) the electron flow through it.

The realization of this kind of devices requires that a number of conditions are fulfilled on many different levels. They range from the assessment of the stability of azurin in a completely non-physiological, dry environment, to the capabilities of producing nanogates with gaps not larger than 10 nm, to the development of suitable approaches for immobilizing metalloproteins onto substrates of different nature (e.g., gold, SiO_2). Some of these problems have been already faced and solved (protein immobilization, azurin stability in UHV). Some others (ultimate lithographic resolution) are subject of continuous effort and progress. However, so far electron beam lithography approach to the fabrication of nanogates has provided us with reproducible nanogates having at best 30 nm gap width [17]. Such a resolution does not allow us to build devices operating with a single molecule, but a single monolayer is required. To decrease nanogate gaps below 30 nm, a controlled electrolitical growth of metal on the electron beam lithography (EBL)-fabricated leads is likely to be nedeed.

Our approach so far has been that of devising Au leads on Si/SiO_2 substrates by means of EBL [17]. This will help in adding a gate electrode by exploiting a back side contact on the doped silicon. So far, we have faced the problem of immobilizing an azurin carpet in between the two Au leads. We want, moreover, to deal with one-molecule-thick layer in order to reduce the gap size to achieve single-molecule operation. Therefore, we have developed a general strategy for protein immobilization on oxygen exposing surfaces (Fig. 2), which works in principle with any protein since it exploits the outer amino-groups that are present in any protein but in differ-

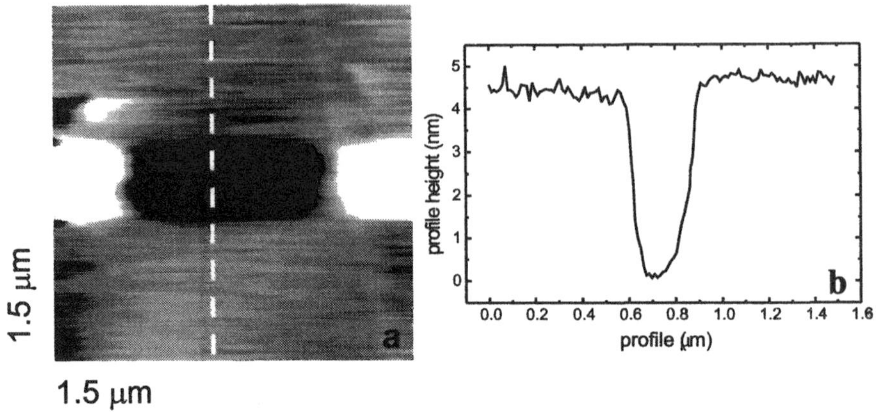

Fig. 7: SFM image (a) and the corresponding profile (b) of a $1.5 \times 1.5\ \mu m^2$ surface area element on which a $750 \times 350\ nm^2$ rectangle had been engraved by the tip action at high load.

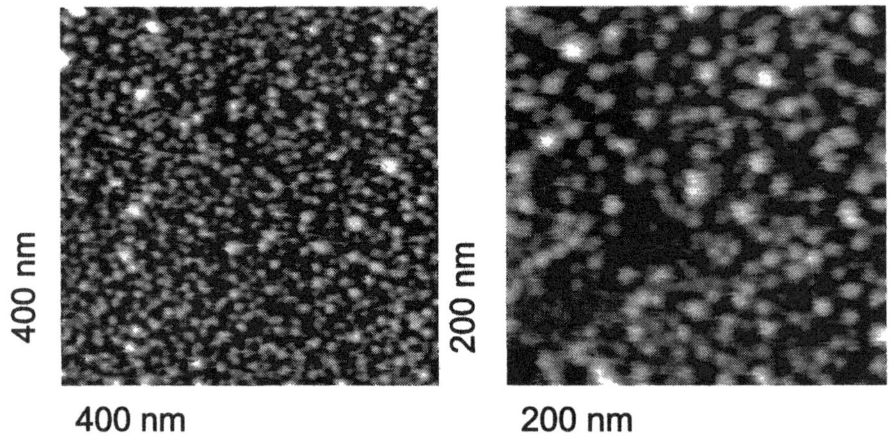

Fig. 8: Topographic SFM images of an azurin monolayer on mica. Contact mode under NH_4Ac buffer. (a) $400 \times 400\ nm^2$, $\Delta z = 3.2$ nm; (b) $200 \times 200\ nm^2$, $\Delta z = 3.9$ nm; scanning frequency 2.9 Hz.

ent amount. In a recent work [18], we have characterized the multiple step chemical synthesis by means of spectroscopic ellipsometry and X-ray photo-electron spectroscopy (XPS). We have confirmed the expected film growth as far as its thickness and the involvement of the correct chemical elements are concerned. These results also suggest that azurin redox site is stable in UHV conditions, since XPS signal from copper confirmed its presence in the site.

For more detailed morphological information on the protein layers, we have performed an extended SFM characterization. We first studied a sample at low resolution under ambient conditions. The effect of a high-load scan across a rectangular surface area element on the sample is reported in Fig. 7. The adsorbed layers had been removed from the scanned area and piled up on the sides [Fig. 7 (a)]. The profile, measured along the dashed white line in Fig. 7 (a), revealed a film thickness of 4.5 nm [Fig. 7 (b)], which is consistent with the theoretical thickness of a structure involving one monolayer of azurin on top of the two preparatory (anchoring) layers. Therefore, these structures appear to be genuine monolayers involving only molecules that are covalently linked, with no protein aggregates.

In order to gain more information on layer morphology and organization at the molecular level, SPM imaging was also performed in a liquid cell containing physiological buffer solution. These conditions reduce tip-sample interaction [19] and preserve protein integrity while yielding high-resolution images. Figures 8 (a) and (b) show the results of scans performed in the contact mode on area elements of different size. A film consisting of bumpy structures is clearly visible on the surface. These features have a lateral extent of about $10 - 15$ nm and can readily be attributed to a typical tip-sample convolution around the azurin molecules. The adsorbate size is very uniform, suggesting that the native protein structure is retained. Interestingly, the image corrugation is about 3.5 nm, and the brighter spots do not protrude above the average height by more than 1 nm. This observation is actually a confirmation of the adsorption mechanism. In fact, the chemical approach used here does not impart any particular orientation on the protein molecules, since they become linked to the preformed layer by their surface-exposed amino groups. According to X-ray crystallography [20], twelve of these groups (the terminal NH_2 group and surface exposed lysines) are situated at different points around the azurin molecule. Therefore, different molecular orientations are reflected in different apparent heights of the bumps seen by SFM.

Intermittent contact measurements have also been performed in order to increase resolution and minimize tip-sample interaction. We worked in the MAC (Molecular Imaging Co.) operating mode using a magnetic cantilever oscillating in an electromagnetic AC field. This operating mode, which minimizes spurious resonances coming from the experimental set-up [21], is appropriate when looking for high-resolution results from soft adsorbates in liquid cells. MAC mode topographic results acquired on different sample regions and on area elements of different size are shown in Figs. 9 (a)-(c). None of these images reveals any negative effects due to tip-sample interaction. As a result, the average bump size has now been reduced to 10 nm while the relative distribution of heights is retained and the dynamical range of the image is now 5 nm.

In summary, the SPM measurements document the presence of a compact protein layer on the substrate. The measured step heights and the absence of piles of physically adsorbed molecules deduced from the arguments involving the distribution of relative heights of the adsorbates confirm that the layer prepared by our method constitutes a genuine chemisorbed protein monolayer. In all the different investigations performed so far, the data have been obtained for the structure and chemical composition of the protein ensemble, but evidence for the retention of functional activity of these molecules has not been obtained thus far. This is actually a very

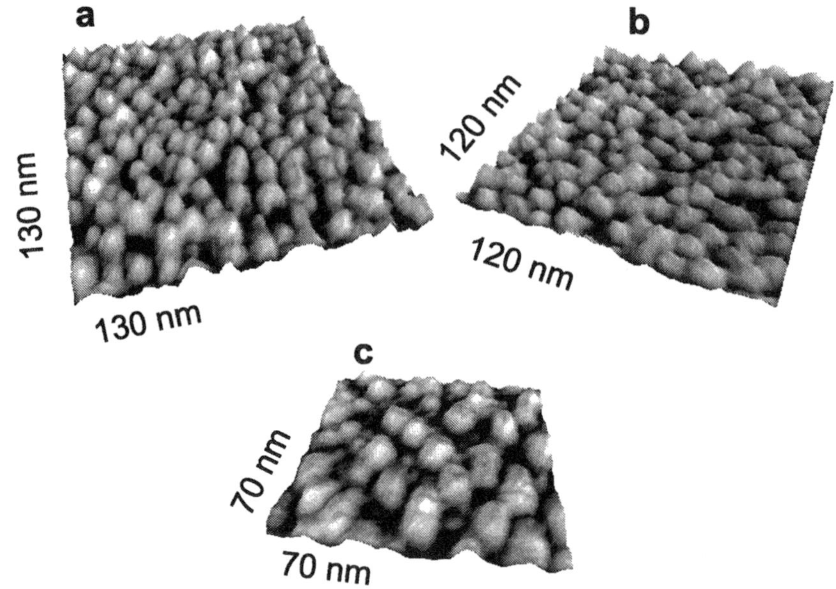

Fig. 9: Topographic SFM images of an azurin monolayer on mica. Intermittent contact mode (MAC mode, see Results and Discussion section) under NH_4Ac buffer. a) 130×130 nm^2, $\Delta z = 5$ nm; b) 120×120 nm^2, $\Delta z = 5$ nm; c) 70×70 nm^2, $\Delta z = 5$ nm; scanning frequency 1.02 Hz.

important point in view of possible biomolecular applications of these monolayers. Unfortunately, it is rather difficult to resolve this issue, since the number of molecules involved is very small and their specific redox activity cannot be studied on an insulating surface.

For a test of the effects of the immobilization procedure on the redox activity of these molecules, we modified the immobilization process by substituting the first linker (3-aminopropyltriethoxysilane) linking to oxygen with another molecule (2-mercaptoethylamine) binding to gold while bearing the same terminal group ($-NH_2$) at the other end of the molecule. This molecule will therefore mediate surface immobilization on gold while leaving all the other chemistry unaltered and providing the same molecular architecture (linker + GD + metalloprotein).

These samples were used to perform cyclic voltammetry (CV). Figure 10 reports results obtained with a scan rate of 0.05 V/sec for both cases of azurin and myoglobin. In Fig. 10 (a), an anodic wave corresponding to azurin oxidation is clearly visible while the corresponding cathodic wave is less pronounced and more diffuse. The nature of slow electron transfer in this process can be inferred from the peak separation (180 mV). The nonconducting linkers present between the azurin and the electrode are likely to further reduce the electron transfer rate. The redox midpoint (+130 mV vs SCE) matches fairly well with the values available in the literature,

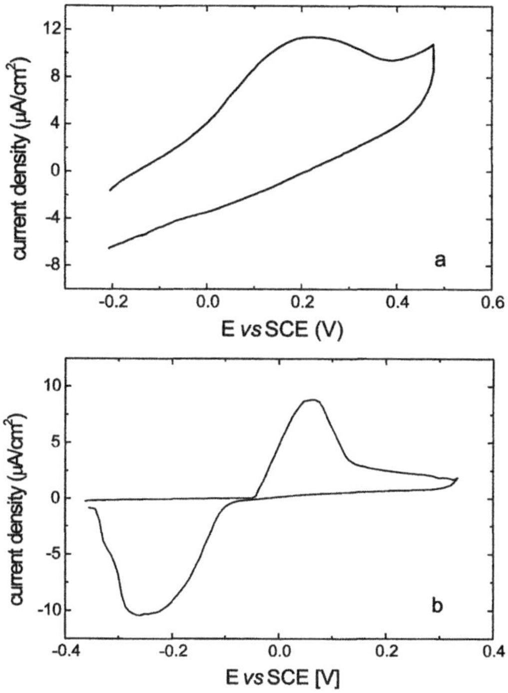

Fig. 10: Cyclic voltammetry of (a) a monolayer of azurin immobilized on gold, redox midpoint + 130 mV vs SCE, peak separation 180 mV, scan rate 0.05 V/sec; (b) a monolayer of myoglobin on gold, redox midpoint −110 mV, scan rate 0.05 V/sec.

for both the gold-immobilized [15] and the dissolved [9] counterparts of azurin. This confirms the native-like redox properties of the immobilized molecules. Figure 10 (b) shows the corresponding results for myoglobin. In this case, CV curves are also very similar to those appeared in the literature for the solution counterpart [22]. It has thus been demonstrated that our immobilization procedure yields functional protein monolayers on flat surfaces.

A further confirmation came from UV-Vis absorption spectroscopy on myoglobin layers. Myoglobin, bearing a heme group characterized by a very intense Soret band ($\varepsilon_{409} \approx$ 200,000 M^{-1} cm^{-1}) should be detectable even in a single monolayer by absorption spectroscopy (differently from azurin). Figure 11 reports the absorption spectrum of such a sample on glass. The Soret band centered at 409 nm is clearly visible. From the measured intensity it is possible to estimate a surface molecular density of 3.03×10^{12} molecule/cm^2, which corresponds to a submonolayer covering about 75% of the surface. Shifts due to solid-state effects cannot be detected in these Soret bands [23], which suggests that protein aggregates are absent. These data are in excellent agreement with the hypothesis that we are dealing with at most one protein monolayer. They also suggest that our immobilization approach is generally valid and applicable to all kinds of proteins.

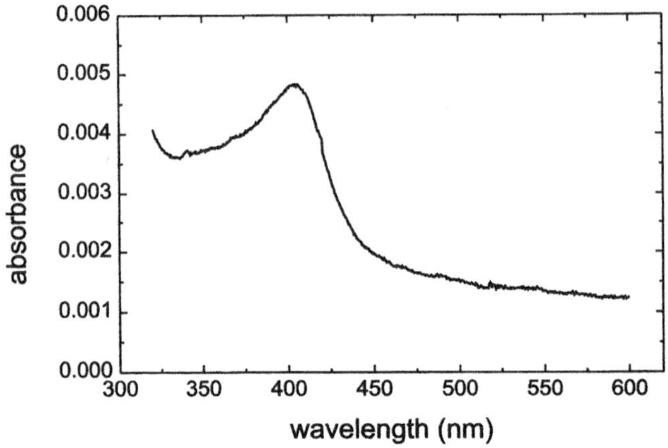

Fig. 11: Absorption spectrum of two monolayers of myoglobin immobilized on a 0.05 mm thick glass slide (1 monolayer on each side).

Transport measurements on two- and three-terminal planar devices with gaps of 30 – 50 nm have been performed recently and preliminary results seem to confirm that azurin monolayers immobilized by this technique display a highly rectifying behaviour with remarkable currents (few microamperes at 20 V) and good reproducibility and stability [17].

4. Conclusions

Investigations of azurin, self-chemisorbed onto gold and investigated by STM under full potentiostatic control discloses very interesting features of this electron transfer molecule. In particular, the possibility of switching on and off repeatedly the current flow through it and the robustness of the molecule open up the possibility for exploiting this metalloprotein as a molecular switch.

Redox protein monolayers self-chemisorbed on oxygen-exposing surfaces by three-step chemical reaction have been successfully built up. The results are compatible with the presence of a (sub)monolayer of proteins on a layer consisting of 3-APTS and GD. The nature and morphology of the protein layer has also been assessed by SFM at various resolution levels under ambient conditions (air) as well as in liquid buffer solution. A modified linking strategy suitable for immobilization on gold which retains the various chemical steps has been implemented for the purposes of CV measurements on the films. This assay has confirmed the presence of redox-active molecules on the surface, and thus verified our immobilization method as a powerful approach to protein monolayer formation on substrates. This approach is suitable for all kinds of proteins such as other metalloproteins, enzymes, antibodies etc. which is useful for both basic and applied research. Preliminary results on the implementation of 30 – 50 nm devices indicates that azurin is a suitable candidate for biomolecular electronics purposes.

Acknowledgement

This work has been supported by INFM through the Advanced Research Project "SINPROT" and by the EC project "SAMBA".

References

[1] A. Aviram, M. Ratner, (eds.), *Molecular Electronics: Science and Technology* (Annals of the New York Academy of Sciences, New York, 1998).

[2] C. Joachim, J. K. Gimzewsky, and A. Aviram, Nature **408**, 541 (2000).

[3] D. L. Klein, et al., Nature **389**, 699 (1997).

[4] J. Brash, and L. Horbertt, (eds.), *Proteins at interfaces II*; ACS Symposium Series 602; American Chemical Society: Washington, DC, 1995.

[5] M. Alper, H. Bayley, D. Kaplan, and M. Navia (eds.), *Biomolecular Materials by Design*, Symposium held during November 29–December 3, 1993, Boston, Massachusetts, U.S.A. (Materials Research Society Symposium, V); Material Research Society, 1994.

[6] E. T. Adman, in *Topics in Molecular and Structural Biology: Metalloporteins*, edited by P. M. Harrison, (Chemie Verlag, Weinheim, 1985).

[7] M. Brunori, *Hemoglobin and Myoglobin* (North-Holland, Amsterdam, 1971).

[8] A. S. Brill. *Transition metals in Biochemistry* (Springer Verlag, Berlin 1977).

[9] Q. Chi, J. Zhang, E. P. Friis, J. E. T. Andersen, and J. Ulstrup, Electrochemistry Communications **1**, 91 (1999).

[10] CRC Handbook of Physics and Chemistry - 74th Edition ; D. R. Lide, Editor.; CRC Press: Boca Raton, 1993.

[11] E.P. Friis, et al., Proc. Natl. Acad. Sci. (USA) **96**, 1379 (1999).

[12] N. J. Tao, Phys. Rev. Lett. **76**, 4066 (1996); H. Siegenthaler, in *Scanning Tunneling Micrscopy II*, edited by R.Wiesendanger and H.-J. Güntherodt (Springer-Verlag, Berlin, Heidelberg, 1995).

[13] B. G. Karlsson, et al., FEBS Lett. **246**, 211 (1989).

[14] D. Alliata, Investigation of nanoscale intercalation into graphite and carbon materials by in situ scanning probe microscopy. Ph.D. Dissertation, University of Bern, 2000.

[15] P. Facci, D. Alliata, and S. Cannistraro, Ultramicroscopy, in press, (2001).

[16] M. A. Webb, C. M. Kwong, and G. R. Loppnow, J. Phys. Chem. B **101**, 5062 (1997).

[17] R. Rinaldi, et al., manuscript in preparation, (2001).

[18] P. Facci, D. Alliata, L. Andolfi, B. Schnyder, and R. Koetz, Surf. Sci., submitted, (2001).

[19] O. Marti, and M. Amrein (eds.), *STM and SFM in Biology* (Academic Press, San Diego, CA, 1993).

[20] H. Nar, A. Messerschmitd, R. Huber, M. van de Kamp, and G. W. Canters, J. Mol. Biol. **221**, 765 (1991).

[21] W. Han, S. M. Lindsay, and T. Jing, Appl. Phys. Lett. **69**, 4111 (1996).

[22] G. Li, L. Chen, J. Zhu, D. Zhu, and D. F. Utereker, Electroanalysis, **11**, 139 (1999).

[23] P. Facci, M. P. Fontana, E. Dal Canale, M. Costa, and T. Sacchelli, Langmuir **16**, 7726 (2000).

Chapter 13

Towards synthetic evolution of nanostructures

Hod Lipson

Mechanical & Aerospace Engineering and Computing & Information Science
Cornell University, Ithaca NY 14853, USA
E-mail: Hod.Lipson@cornell.edu

Abstract

This article begins to consider key ingredients necessary to apply non-biological evolutionary processes at the nano scale. In such processes, large numbers of building blocks spontaneously self-organize into new irregular forms that were not directly predetermined by a designer, but rather form solutions to a given functional requirements. Such structures will be able to adapt their configuration and behavior in response to changing requirements and changing conditions, ultimately leading to a form of evolutionary materials. The motivation for looking into self-organizing phenomena like evolution at the nano-scale is both for finding new ways do design structures, and to synthetically recreate and examine the evolutionary processes that gave rise to primordial life. This article describes some lessons learnt from applying evolutionary processes at macro-level robotic structures, and postulates as to key ingredients that will be necessary to complete an evolutionary cycle at the nano scale.

1. Introduction

The understanding of nano-structure processes is taking place at several levels simultaneously. Some research is focused on understanding the physics of single building-block units, such as the electrical, mechanical and chemical properties of carbon nanotubes and DNA strands. At the next level there is interest in the interface physics that allows single building blocks to connect to each other in a controllable way to make junctions, structures and machines. At a still higher level, we are interested in self-assembly processes that allow very large numbers of building blocks to form semi-regular structures according to a predetermined set of rules. Finally, we may now start considering techniques that will allow large numbers of building blocks to adaptively self-organize into irregular forms that were not directly predetermined by a designer, but rather form solutions to some given functional requirements. Such structures will be able to adapt their internal configuration and behavior in response to changing requirements and changing conditions, ultimately leading to a form of *evolutionary materials.*

The motivation for looking into self-organizing phenomena like evolution at the nano-scale is twofold. First, the difficulty in manipulation and design of small-scale structures calls for new techniques of design, construction and testing. Evolutionary mechanisms offer an alternative paradigm for automating this cycle while avoiding low-level discrete manipulation. In particular, with proper setup, some self-organizing phenomena can allow for closing the design-test loop by harnessing the massively parallel nature of the nanoscale substrate to search the design space. The second reason for looking into self-organizing processes at the nanoscale is the potential to synthetically recreate and examine physical evolutionary processes similar to those that gave rise to primordial life. Such experimentation will allow testing of some basic hypotheses on the emergence of biological life, and possibly even recreating very simple artificial life forms in a synthetic substrate.

This brief paper does not present any experimental results achieved at the nano-structure scale. Instead, I will describe evolutionary robotics experiments carried out in scale-less simulation and verified at the physical macro level. Although many aspects of macro-scale physics to not apply to nanoscale systems, some of the phenomena, especially those concerned with the emergence of complexity, are broadly applicable to self organizing systems in general. Lessons learnt from evolution of structures in simulation may thus shed some light on possible paths towards application of evolutionary processes at the nano scale, and specifically the roles of modularity and hierarchy in achieving structures with complex functionality. Finally, I will conclude with some general ideas about possible ingredients necessary for implementation of evolutionary processes in the physical nanoscale substrate.

2. Evolution of structures

Genetic algorithms (GAs) – a subset of evolutionary computation involving mutation and crossover in a population of fixed length bit strings - have been applied for several decades in many engineering problems as an optimization technique for a fixed set of parameters. But open-ended systems, in which the process is allowed to add more and more building blocks and parameters, seem particularly adequate to describe evolutionary processes that might occur at the nano scale. Open-ended evolutionary design systems have been demonstrated for a variety of simple design problems, including structures, mechanisms, software, optics, robotics, control, and many others (for overviews see, for example, Refs. [1,2]). Yet these accomplishments remain simple compared to what teams of human engineers can design and what nature has produced. The evolutionary design approach is often criticized as scaling badly when challenged with design requirements of higher complexity.

In a set of experiments, we investigated the possibilities of evolving locomotive structures out of bars, linear actuators and neurons. We then used commercial rapid prototyping techniques to transfer the evolved machines into reality to test their physical viability. Details of these experiments and their results can be found in earlier reports [3]. Here I will briefly overview the findings and then discuss their scaling properties.

In these experiments, the physics model considered combinations of bars, actuators and neurons. The bars connect to each other through free joints, neurons can connect to other neurons through synaptic connections, and neurons can connect to bars. In the latter case, the length of the bar becomes governed by the output of the neuron, essentially making it a linear actuator. This model was chosen because it allows for a large variety of structure: Trusses can form arbitrary rigid, flexible and articulated structures as well as multiple detached structures. Bars connected with free joints can form configuration acting as revolute, linear and planar joints at various levels of hierarchy. Similarly, sigmoidal neurons can connect to create arbitrary control architectures such as feed-forward and recurrent nets, state machines and multiple independent brains. A schematic illustration of a possible architecture is shown in Fig. 1.

The goal was to evolve machines (structure and control) that could locomote over an horizontal plane. Starting with a population of 200 blank machines that were initially comprised of zero bars, zero actuators and zero neurons, we conducted evolution in simulation. The fitness of a machine was determined by its locomotion ability: the net distance its center of mass moved on an infinite plane in a fixed duration. The process iteratively selected fitter machines, created offspring by randomly adding, modifying and removing building blocks, and replaced them into the population. The process was typically carried out for over 600 generations. Some results are shown in Fig. 2.

3. Scaling

While there are still many poorly understood factors that determine the success of evolutionary design - such as starting conditions, variation operators, primitive building blocks and fidelity of simulation - one problem is that the design space is exponentially large, because there is an exponentially increasing number of ways a linearly increasing set of components can be assembled. Consequently, evolutionary

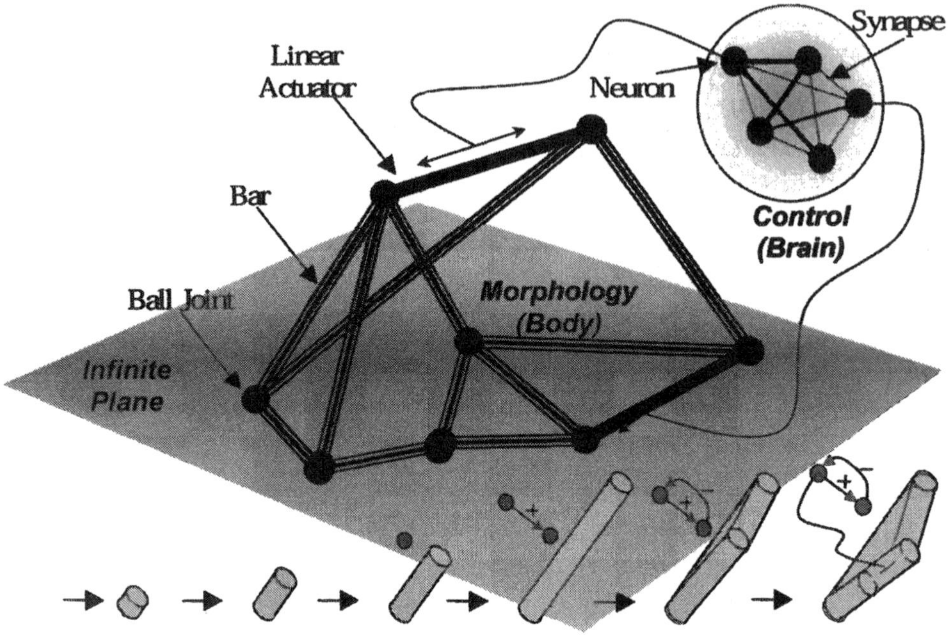

Fig. 1: Schematic illustration of an evolvable robot, containing only linear bar/actuators and control neurons. Some bar architectures form rigid substructures.

approaches that operate on direct encodings quickly become intractable because of combinatorial complexity.

This is evident in our own work: Fig. 3 shows a typical progress of fitness of evolved machines as function of generations. The abscissa represents evolutionary time (generations), the ordinate measures fitness (net movement on a horizontal plane) and each point in the scatter plot represents one candidate robot. In general, after an initial period of drift, with zero fitness, we observe rapid growth followed by a logarithmic slowdown in progress, characterized by longer and longer durations between successive step-improvements in the fitness. This real time lingering is amplified by the fact that evaluation time or duration of a generation (in simulation or in physical reality) also increases as solutions become more complex. However, note that because of the stochastic nature of the process, it is hard to determine definitely whether progress has actually halted, and improvements may still occur after long periods of apparent stagnation (Fig. 3c, 3d).

Looking at the kind of structures that result after leaving the system running for extended periods of time (weeks, in practical terms), we see that the kind of structures that evolve exhibit high internal coupling (Fig. 4). While these machines do have a slightly higher fitness than their predecessors, they are complex to the point where it is difficult for them to continue evolving efficiently. In other words, their *evolvability* is impaired.

From engineering design principles and from evidence in nature, we know that

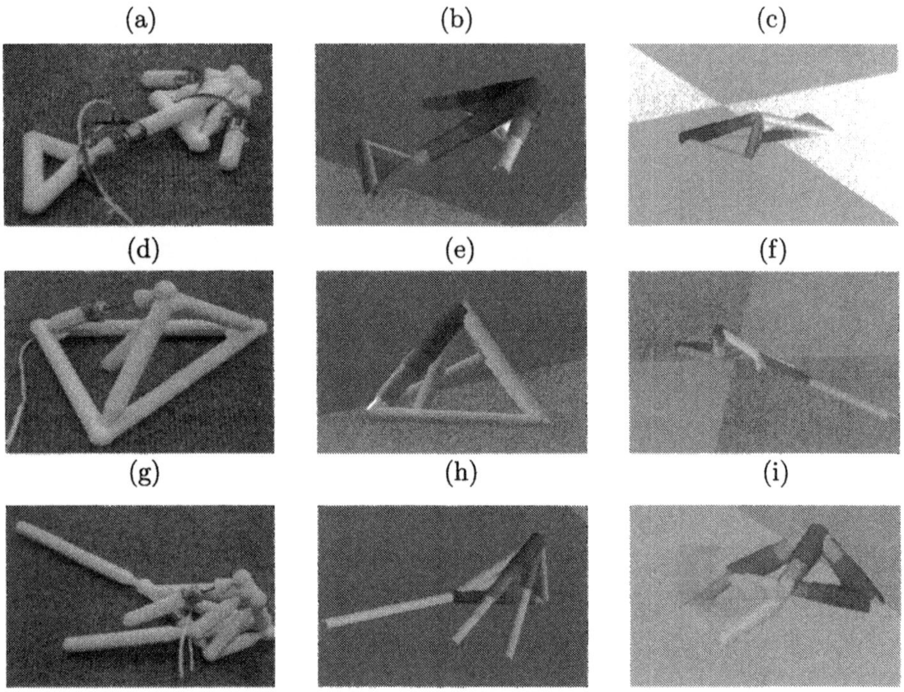

Fig. 2: Some results of the evolutionary process (from [3]): (a,b) This surprisingly symmetric machine uses a 7-neuron network to drive the center actuator in perfect anti-phase with the two synchronized side limb actuators. While the upper two limbs push, the central body is retracted, and vice versa. (d,e) A tetrahedral mechanism that produces hinge-like motion and advances by pushing the central bar against the floor. (g,h) This mechanism has an elevated body, from which it pushes an actuator down directly onto the floor to create ratcheting motion. It has a few redundant bars dragged on the floor, which might be contributing to its stability. (c,f,i) other converged machines, all produces with the same parameter settings. These machines perform in reality is the same way they perform in simulation. Motion videos of these robots and others can be viewed at http://www.demo.cs.brandeis.edu/golem

one of the keys to maintaining evolvability is the use of architectural principles of regularity, modularity and hierarchy. Below I will briefly discuss some of these new approaches and their implementation in evolutionary computation. However, these methods require a genotype-phenotype separation that is not necessarily available in non-biological physical nano-scale evolutionary processes. And so I will propose an alternative weak approach that may induce modularity without requiring a direct genotype-phenotype representation.

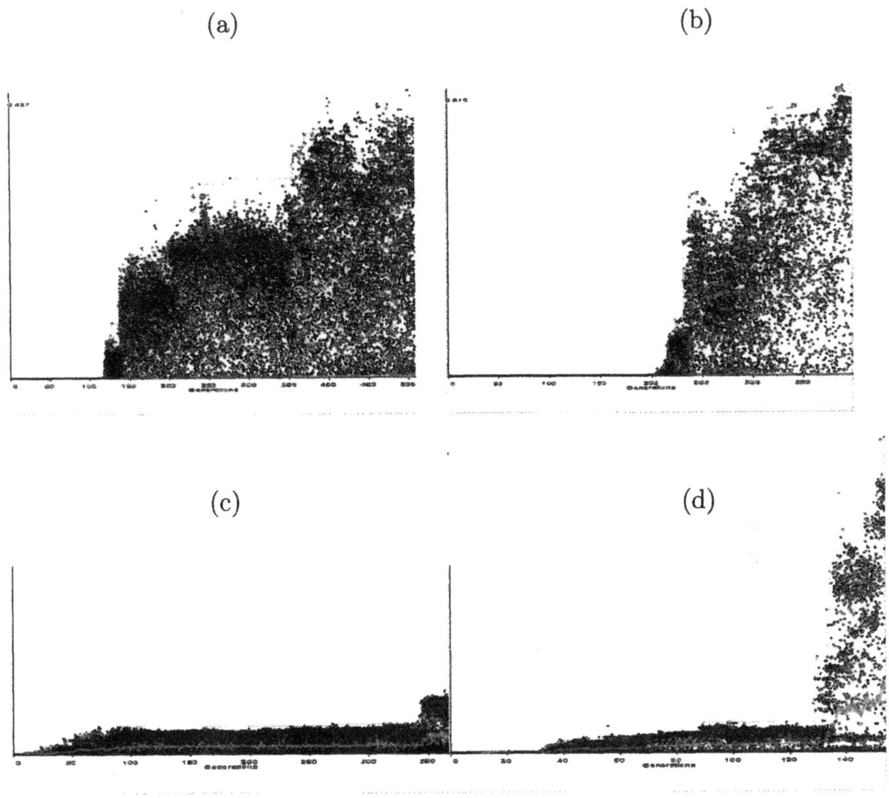

Fig. 3: Progress of a typical evolutionary design run. The abscissa represents evolutionary time (generations) and the ordinate measures fitness. Each point in the scatter plot represents one candidate design. In general, a logarithmic slowdown in progress can be observed, characterized by longer and longer durations between successive step-jumps in the fitness (a,b). Occasionally, however, progress is made after long periods of stagnation (c,d).

4. Modularity

It has long been recognized that architectures that exhibit functional separation into modules are more robust and amenable to design and adaptation [4,5]. Modularity creates a separation that reduces the amount of coupling between internal and external changes, allowing evolution to rearrange inputs to modules without changing their intrinsic behaviors, and so to reuse modules as high-level building blocks. In nature this idea is supported by theoretical arguments such as that proteins are difficult to evolve once they are participating in many different interactions, and by observations of phenomena such as tight coordination of the expression of groups of genes functioning in a common process, as well as evidence of modules appearing as a robust segmentation mechanism for handling developmental noise. Conversely,

(a) (b)

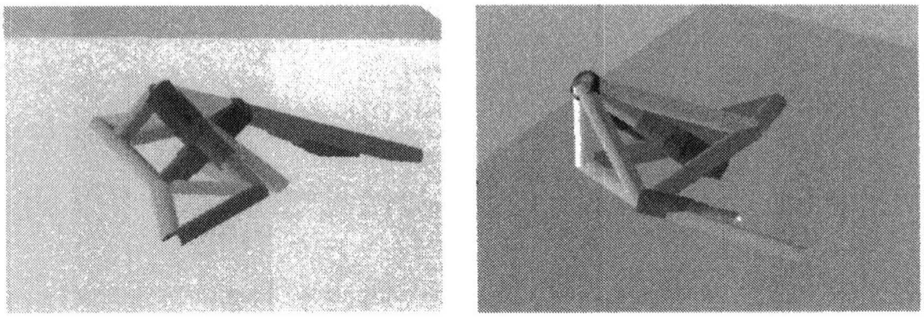

Fig. 4: Machines evolved after very large number of generations of mutations may have a slightly higher fitness, but generally highly coupled internally.

there is evidence that proteins which interact with many other proteins, such as histones, actin and tubulin, have changed very little during evolution. In artificial systems modularity is critical too: Herbert Simon [6] noted, in his famous "Tempus and Hora" fable, that the evolution of complex forms from simple elements depends critically on the numbers and distribution of potential stable intermediate forms. Modularity has also been recognized as a primary facilitating characteristic of system engineering (e.g., Ref. [7]), economics [8], and named as one of the principles of design.

Perhaps one of the first arguments to the importance of building blocks was put forward by Holland [9] through the building block hypothesis. The building block hypothesis suggests that genetic algorithms are able to identify low-order schemata (low-level modules) with above-average fitness and combine them, through crossover operators, to produce high-order schemata (high-level modules), and continue doing so recursively. However, it has been shown that this powerful property relies on proper diversity of the population and, more importantly, on genetic linkage - the proximity along the genotype of genes that encode for related function. In absence of these assumptions genetic algorithms will not scale well [10]. It is clear that when we address synthesis problems where the initial building blocks are scattered around, as in physical nanoscale setup, there is no way to guarantee genetic linkage. Moreover, when there is no genotype at all, the term is undefined.

One approach to avoiding reliance on genetic linkage is based on Genetic Programming [11] principles, which deal with partial specifications directly and manipulate them in coherent tree structures. This concept is enhanced through a mechanism for automatically defining functions (ADF) that maintains partial solutions (modules) in separate populations. However, direct manipulation of modules may be difficult at the physical nanoscale.

Another alternative that allows production of modularity without assuming genetic linkage is the generative approach. Here a structure is not specified directly, but is specified through an evolving coding that in turn generates the structure. Like

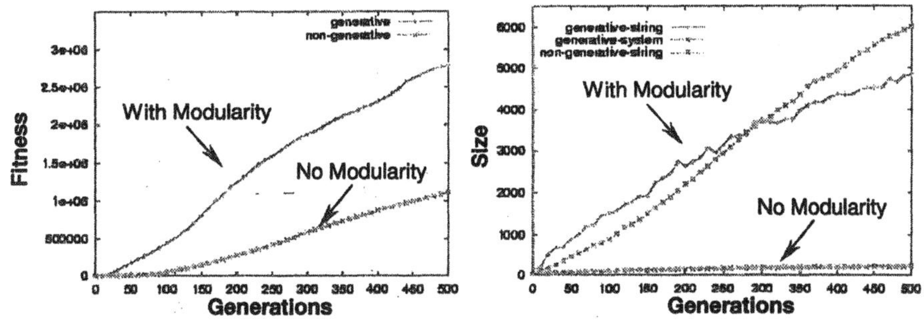

Fig. 5: Comparison of fitness (left) and complexity (right) as function of generation, in a generative versus non-generative substrate, for evolution of locomoting structures [12].

a structured computer program, a generative specification can allow the definition of re-usable sub-procedures allowing the design system to use loops and recursion to produce large regular structures by reiterating a small set of commands. The DNA, combined with a developmental process, is an example of this kind of process. Generative systems can promote both modularity (separation of function) and regularity (reuse), through this indirect process. In a preliminary set of experiments [12], we used Lindenmayer systems (L-systems) as the generative encoding to be evolved: Evolution produces L-system programs, those programs in turn generate construction sequences that in turn construct robots. Figure 5 shows a comparison of fitness and complexity (measured by description length) as function of generation, in a generative versus non-generative substrate, for evolution of static structures. Although this does not prove scalable behavior, it provides support for the notion that modularity and regularity can significantly accelerate progress.

A more recent approach to enhancing modularity and hierarchy is the Symbiogenic Evolutionary Adaptation Model [13]. This model works with a population of partially specified solutions and tests them in context of each other to combine them into higher-level partial solutions. This combinative process relies on co-existence of the subcomponents for evaluation and Pareto-dominance criterion to decide on transition between hierarchy levels. It has been shown to solve efficiently complex test problems that do not assume genetic linkage yet have an exponential number of local minima, and are therefore more closely applicable to physical nanoscale evolution.

5. An indirect method for inducing modularity

The physical nanoscale evolution substrate is characterized by three properties that do not allow direct application of standard modularity enhancing methods. First, there is no linkage, i.e., the building blocks are scattered around randomly and we cannot assume that building blocks that happen to lie in proximity are functionally related. Thus, methods that rely on linkage (like standard Genetic Algorithms) cannot be applied. Second, there is no genotype (only a phenotype), so methods that require a genome (like generative approaches) cannot be directly applied, unless they use an intermediate representation like a living cell. Third, it is difficult to directly

manipulate modules, so that methods that rely on module manipulation, like Genetic Programming, would also be difficult to implement. We seek an indirect method for enhancing modularity that does not rely on a particular representation.

In a recent study [14], we have chosen to examine the spontaneous emergence of modularity using a simple and abstract model of an adaptive system as a transformation of a set of resources into a set of arbitrary functional requirements for survival. We model a natural evolving system as a transfer matrix **A** that is required to transform a vector of given environmental resources **E** into some vector **F** representing a set of arbitrary functional requirements: $\mathrm{F} = \mathrm{A} \times \mathrm{E}$ (see Fig. 6 a). The figure of merit $f(\mathbf{A})$ of the design candidate **A** is how well it satisfies the above vector equation, given a vector of requirements **F** and a vector of resources **E**. This figure of merit (or *fitness*, in evolutionary terms) can be quantified as the magnitude of deviation $|\mathbf{F} - \mathbf{AE}|$. Based on the definitions above, we quantify modularity as inversely proportional to the amount of coupling $C(A)$ in the system. We simulated a simple evolutionary process where a population of **A**'s was evolved for a given pair of **F** and **E**. We observed the dependency of the average coupling $C(A)$ on the rate of change $d\mathbf{E}/dt$ of the environment resources.

The main results are summarized in a plot of internal coupling versus environment change rate, shown in Fig. 6 b. Each point in the plot is averaged for 100 experiments, each of 20,000 generations of a population of 200 matrices. This plot shows a clear linear-log correlation that can be quantified as an empirical modularity law that relates the amount of coupling in a system C to the rate of change of the problem:

$$C = -k \log \frac{dE}{dt} + C_0$$

where k and C_0 are constants that are dependent on the mutation bias, the amount of interaction between the system and the environment, and other specifics of the substrate. This experiment suggests that modularity arises spontaneously in evolutionary systems in response to a changing environment, and that the amount of modular separation is logarithmically proportional to the rate of change of the environment. This quantitative model can shed light on the evolution of modularity in nature, and predicts that modular architectures would appear in correlation with high environmental change rates. In the nanoscale evolution context, these results also suggest that modularity can be induced within physical systems by evolving in a changing environment. This approach does not assume linkage, is not based on a particular representation, and does not require direct manipulation of modules.

6. Towards physical implementation

Evolutionary processes require a set of founding blocks that can be varied and replicated in a fitness-dependent way. The existence of founding building blocks whose interfaces can be preprogrammed has already been established and has been demonstrated to assemble into large regular structures like DNA crystals [15]. The bonding energy of these structures is at the same approximate of level of ambient

$$\begin{bmatrix} -1 \\ +1 \\ -1 \\ -1 \\ +1 \\ +1 \\ +1 \\ -1 \end{bmatrix} \stackrel{?}{=} \begin{bmatrix} +1 & 0 & -1 & -1 & 0 & 0 & -1 & 0 \\ 0 & 0 & 0 & 0 & 0 & -11 & +1 & 0 \\ -1 & 0 & +1 & 0 & 0 & 0 & +1 & 0 \\ 0 & -1 & 0 & 0 & +1 & 0 & 0 & 0 \\ -1 & 0 & 0 & +1 & 0 & 0 & 0 & -1 \\ 0 & 0 & 0 & 0 & 0 & 0 & -1 & 0 \\ 0 & 0 & +1 & -1 & 0 & 0 & -1 & 0 \\ +1 & 0 & 0 & 0 & 0 & 0 & -1 & 0 \end{bmatrix} \times \begin{bmatrix} +1 \\ -1 \\ +1 \\ +1 \\ -1 \\ +1 \\ +1 \\ -1 \end{bmatrix}$$

F — An arbitrary set of functional requirements or survival requirements

A — A design candidate or individual trying to transform building blocks to meet requirements

E — An arbitrary set of building blocks or resources in environment

(a)

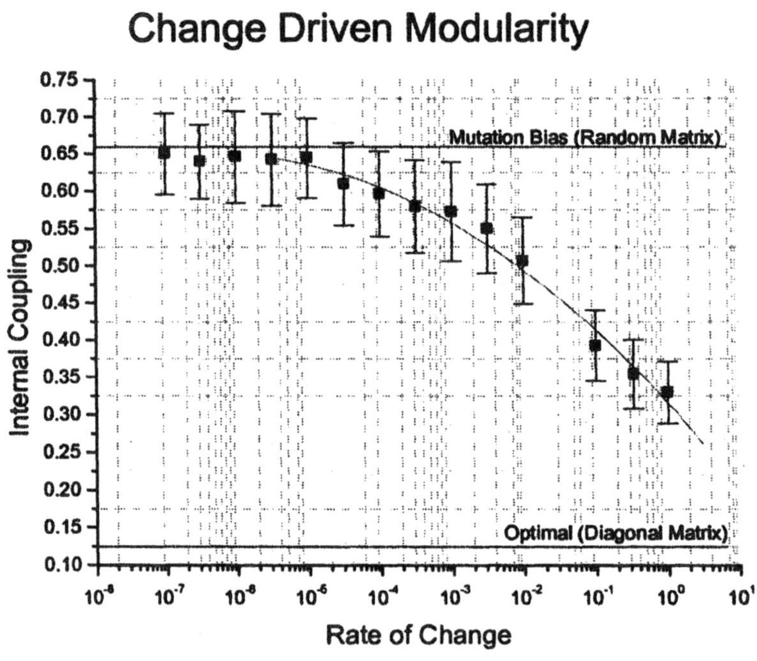

(b)

Fig. 6: Environment driven modularity: (top) An abstract general design problem, where a matrix **A** is required to transform a set of building blocks E to meet a set of requirements **F**. Many solutions exist, and, (bottom) Internal coupling versus environment change rate shows a linear-log correlation.

energy of the system (kT), and so 'design variations' are relatively easy to induce. While in most cases this source of variation and error is considered a problem, for evolution it is in fact a source of innovation. The trick is inversely linking the rate of variation with the design goal, or the 'fitness' of the structure, so that good designs will be stable, while poor design will be subject to more variation. This fitness-dependent-variation link needs to be established in a non-centralized way in order to exploit the massively parallel nature of the nanoscale substrate. For example, if we are interested in evolving machines that can move, say, we could expose the substrate to a moving energy field that would transfer more energy into static structures than to moving structure. This energy would thus induce more variation in lower fitness structures. Another example might be evolving structure to sustain a load. Here, we need to design markers that attach to DNA sequences and absorb more of a radiating field when the structure is under high stress.

7. Conclusions

There are still many key aspects of evolutionary process that require a physical implementation before an entire evolutionary cycle can be completed in nanoscale reality. While there is much effort and significant results in establishing good building blocks whose interface can be controlled, we still need to address ways of producing fitness dependent variation and replication. Even when we are able to do this, experiments in other domains indicate that the level of complexity we would be able to achieve would still be limited. Early results in DNA computing that did not consider scaling were deemed impractical when applied to more complex problems. Similarly, in search of practical implementation of evolutionary processes we must consider ways to promote modularity and hierarchy if we expect these methods to scale to higher complexities. One way to achieve modularity is through the use of biological agents like cells, who have already 'solved' much of the representation and replication issues. However, to induce a fully synthetic evolutionary process we need to do so without relying on intermediate representations and on direct manipulation, and some ideas towards these goals are presented here.

References

[1] P. J. Bentley, (Ed.), *Evolutionary Design by Computers* (Morgan Kaufman 1999).

[2] P. Husbands and J. A. Meyer, *Evolutionary Robotics* (Springer Verlag 1998).

[3] H. Lipson and J. B. Pollack, Nature **406**, 974 (2000).

[4] L. H. Hartwell, J. H. Hopfield, S. Leibler, and A. W. Murray, Nature **402**, C47 (1999).

[5] G. P. Wagner, American Zoologist, **36**, 36 (1996).

[6] H. A. Simon, *The Sciences of the Artificial* (MIT Press 1996), 3rd edition, (The fable of Tempus and Hora).

[7] C. C. Huang and A. Kusiak, IEEE Transactions on Systems, Man, and Cybernetics, Part A, **28**, 66 (1998).

[8] R. N. Langlois, Journal of Economic Behavior and Organization, in press (2001).

[9] J. Holland, *Adaptation in natural and artificial systems* (University of Michigan Press 1975).

[10] R. A. Watson and J. B. Pollack, GECCO-99 Late Breaking Papers, 292 (1999).

[11] J. Koza, *Genetic Programming* (MIT Press 1992).

[12] G. S. Hornby, H. Lipson, J. B. Pollack, IEEE International Conference on Robotics and Automation (2001).

[13] R. A. Watson andJ. B. Pollack, Parallel problem solving from nature 2000, PPSN VI (2000).

[14] H. Lipson, J. B. Pollack, and N. P. Suh, Proceedings of DETC'01 2001 ASME Design Engineering Technical Conferences, September 9-12, 2001, Pittsburgh, Pennsylvania, USA (2001).

[15] E. Winfree, F. Liu, L. A. Wenzler, N. C. Seeman, Nature **394**, 539 (1998).

Subject Index

Kolibri 360 GmbH
Ferdinand-Jühlke-Straße 7
99095 Erfurt, Deutschland
produktsicherheit@kolibri360.de